Spectral Discretization-Based Eigen-Analysis of Time-Delay Systems

Hua Ye · Yutian Liu · Xiaofan Jia

Spectral Discretization-Based Eigen-Analysis of Time-Delay Systems

Numerical Methods and Scalable Applications

Science Press
Beijing

Springer

Hua Ye
School of Electrical Engineering
Shandong University
Jinan, China

Yutian Liu
School of Electrical Engineering
Shandong University
Jinan, China

Xiaofan Jia
School of Electrical Engineering
Shandong University
Jinan, China

ISBN 978-981-96-0737-2 ISBN 978-981-96-0738-9 (eBook)
https://doi.org/10.1007/978-981-96-0738-9

Jointly published with Science Press
The print edition is not for sale in China mainland. Customers from China mainland please order the print book from: Science Press.

This Springer imprint is published by the registered company Springer Nature Singapore Pte Ltd.
The registered company address is: 152 Beach Road, #21-01/04 Gateway East, Singapore 189721, Singapore

To our families

Preface

Background

Time delays are intrinsic of a variety of biological, chemical, economic, and electrical systems. Since the introduction of past-dependence in the evolution of a system, time delays may change system dynamics drastically. For example, wide-area damping controllers (WADCs) commonly applied in the interconnected power systems are fed by remote synchrophasor measurements and thus inevitably introduce time delays into the control loop of the power system. These delays ranging from tens to several hundreds of milliseconds can deteriorate the performance of the controllers and further jeopardize the power system stability. Hence, the modeling, analysis, optimization, and control of time-delay systems have received much attention in the literature for many years.

To a deeper understanding of system dynamics, time-delay systems can be described as a set of delayed differential equations (DDEs). Opposite to ordinary differential equations (ODEs), DDEs are infinite-dimensional and the study of their stability properties requires much more effort. The time domain Lyapunov-Krasovskii functional-based methods are expected to obtain the conservative stability boundary for the magnitude of time delays via relatively complicated procedures. The straight-forward frequency domain approaches, namely, the Padé approximation and Rekasius transformation, employ low-order rational polynomials to substitute time delays, while their accuracy decreases fast with the increase in the delays' magnitudes. In recent years, there are increasing interests in the spectral discretization-based eigenvalue analysis methods regarding two infinite-dimensional operators on the Banach space, i.e., the solution operator and the infinitesimal generator. The methods are expected to obtain the accurate stability boundary of time-delay systems with affordable computational burden.

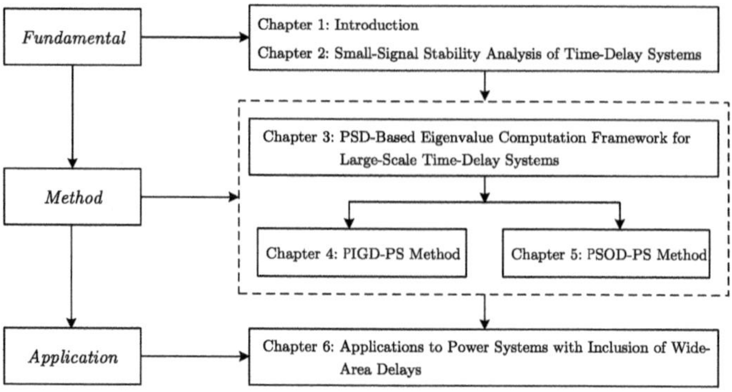

Fig. 1 Structure schematic of the book

Book Outline

The monograph provides a corpus of theoretical and practical results relative to the partial spectral discretization (PSD)-based eigenvalue computation methods for large-scale time-delay systems.

The book is organized in three parts, as shown in Fig. 1. The *Fundamental* part presents the modeling of time-delay systems for small-signal stability analysis. In the *Method* part, Chap. 3 establishes the eigenvalue computation framework for time-delay systems based on the PSD idea. The subsequent Chaps. 4 and 5 provide two eigenvalue computation methods based on the partial infinitesimal generator discretization (PIGD) and the partial solution operator discretization (PSOD), respectively, where the pseudospectral (PS) method is applied. The *Application* part elaborates applications of these methods to compute rightmost or least-damped eigenvalues of power systems with inclusion of wide-area communication and control delays.

Jinan, China
August 2024

Hua Ye
Yutian Liu
Xiaofan Jia

Acknowledgements The authors greatly acknowledge the financial support of the National Natural Science Foundation of China under Grants 51107073, 51677107, 52077126, 52207120, and the Taishan Scholars Program helped us to continue and to finish the monograph.

We would like to thank all who have helped us by useful suggestions and discussions and by encouraging us to pursue the realization of this book.

Contents

Acronyms

CPPS	Cyber-Physical Power System
DAE	Differential-Algebraic Equation
DDAE	Delayed Differential-Algebraic Equation
DDE	Delayed Differential Equation
EIGD	Explicit Infinitesimal Generator Discretization
FACTS	Flexible Alternative Current Transmission System
GPS	Global Positioning System
HVDC	High Voltage Direct Current
IGD	Infinitesimal Generator Discretization
IRA	Implicitly Restarted Arnoldi
LQR	Linear Quadratic Regulator
MIVP	Matrix Inversion-Vector Product
MVP	Matrix-Vector Product
ODE	Ordinary Differential Equation
PDC	Phasor Data Concentrator
PDE	Partial Differential Equation
PIGD	Partial Infinitesimal Generator Discretization
PMU	Phasor Measurement Unit
PS	Pseudospectral
PSOD	Partial Solution Operator Discretization
PSS	Power System Stabilizer
RFDE	Retarded Functional Differential Equation
SCADA	Supervisory Control and Data Acquisition
SOD	Solution Operator Discretization
SVC	Static Var Compensator
TCSC	Thyristor Controlled Series Compensator
WADC	Wide-Area Damping Controller
WAMS	Wide-Area Measurement System

Notation

$\Delta \boldsymbol{x}$, $\Delta \boldsymbol{y}$	System state and algebraic variables
$\Delta \hat{\boldsymbol{x}}$	Augmented system state variables
$\Delta \boldsymbol{x}^{(1)}$, $\Delta \boldsymbol{x}^{(2)}$	System delay-free and retarded state variables
$\Delta \boldsymbol{y}^{(1)}$, $\Delta \boldsymbol{y}^{(2)}$	System delay-free and retarded algebraic variables
$\Delta \hat{\boldsymbol{x}}^{(1)}$, $\Delta \hat{\boldsymbol{x}}^{(2)}$	Augmented system delay-free and retarded state variables
$\boldsymbol{\varphi}$, $\boldsymbol{\psi}$	System state and the corresponding derivative
α, θ	Magnification factor and coordinate rotation angle in the rotation-and-multiplication preconditioning
h	Transfer step-length of the solution operator
m	Number of time delays
n, l	Dimensions of $\Delta \boldsymbol{x}$ and $\Delta \boldsymbol{y}$
d	Dimension of $\Delta \hat{\boldsymbol{x}}$
n_1, n_2	Dimensions of $\Delta \boldsymbol{x}^{(1)}$ and $\Delta \boldsymbol{x}^{(2)}$
l_1, l_2	Dimensions of $\Delta \boldsymbol{y}^{(1)}$ and $\Delta \boldsymbol{y}^{(2)}$
d_1, d_2	Dimensions of $\Delta \hat{\boldsymbol{x}}^{(1)}$ and $\Delta \hat{\boldsymbol{x}}^{(2)}$
λ, $\boldsymbol{u}$, $\boldsymbol{v}$	Eigenvalue, left eigenvector and right eigenvector of the time-delay system or the infinitesimal generator
μ	Characteristic multiplier, namely eigenvalue of the solution operator
$\hat{\lambda}$, $\hat{\mu}$	Eigenvalues of the spectral discretization matrices

$\hat{\boldsymbol{v}}$	Right eigenvector of the spectral discretization matrix/augmented right eigenvector of the time-delay system
λ_s	Shift point
$\tau_i\ (i = 1, 2, \cdots, m)$	Time delays, satisfying $\tau_1 < \tau_2 < \cdots < \tau_m \triangleq \tau_{\max}$
$\boldsymbol{J}_i = \begin{bmatrix} \boldsymbol{A}_i & \boldsymbol{B}_i \\ \boldsymbol{C}_0 & \boldsymbol{D}_0 \end{bmatrix} (i = 0, 1, \cdots, m)$	Augmented state matrices of the time-delay system
$\tilde{\boldsymbol{A}}_i = \boldsymbol{A}_i - \boldsymbol{B}_i \boldsymbol{D}_0^{-1} \boldsymbol{C}_0\ (i = 0, 1, \cdots, m)$	State matrices of the time-delay system
$\boldsymbol{J}_i^{(1)}, \boldsymbol{J}_i^{(2)}\ (i = 0, 1, \cdots, m)$	Delay-free and retarded submatrices of $\boldsymbol{J}_i\ (i = 0, 1, \cdots, m)$
$\tilde{\boldsymbol{A}}_i^{(1)}, \tilde{\boldsymbol{A}}_i^{(2)}\ (i = 0, 1, \cdots, m)$	Delay-free and retarded submatrices of $\tilde{\boldsymbol{A}}_i\ (i = 0, 1, \cdots, m)$
$\mathcal{A}$	Infinitesimal generator
$\mathcal{A}_N$	Partial discretization matrix of the infinitesimal generator
$\mathcal{T}(h)$	Solution operator
$\boldsymbol{T}_{M,N}$	Partial discretization matrix of the solution operator

Chapter 1
Introduction

This chapter briefly addresses the stability analysis of time-delay systems.

1.1 Differential-Algebraic Equations with Delays

The dynamic model of a conventional system without delays can be expressed as a set of ordinary differential-algebraic equations (DAEs)

$$\begin{cases} \dot{\boldsymbol{x}} = \boldsymbol{f}(\boldsymbol{x},\ \boldsymbol{y}) \\ \boldsymbol{0} = \boldsymbol{g}(\boldsymbol{x},\ \boldsymbol{y}) \end{cases} \tag{1.1}$$

where $\boldsymbol{x} \in \mathbb{R}^{n\times 1}$ and $\boldsymbol{y} \in \mathbb{R}^{l\times 1}$ are the vectors of system state and algebraic variables, respectively, $\boldsymbol{f}$ represents the set of differential equations describing the dynamics of the system, and $\boldsymbol{g}$ denotes the set of algebraic equations describing the statics of the system.

The delayed DAEs (DDAEs) formulation is further obtained by introducing time delays in (1.1) [1]

$$\begin{cases} \dot{\boldsymbol{x}} = \boldsymbol{f}(\boldsymbol{x},\ \boldsymbol{y},\ \boldsymbol{x}_{\mathrm{d}1},\ \boldsymbol{y}_{\mathrm{d}1},\ \ldots,\ \boldsymbol{x}_{\mathrm{d}m},\ \boldsymbol{y}_{\mathrm{d}m}) \\ \boldsymbol{0} = \boldsymbol{g}(\boldsymbol{x},\ \boldsymbol{y},\ \boldsymbol{x}_{\mathrm{d}1},\ \boldsymbol{y}_{\mathrm{d}1},\ \ldots,\ \boldsymbol{x}_{\mathrm{d}m},\ \boldsymbol{y}_{\mathrm{d}m}) \end{cases} \tag{1.2}$$

where

$$\boldsymbol{x}_{\mathrm{d}i} = \boldsymbol{x}(t - \tau_i),\ \ \boldsymbol{y}_{\mathrm{d}i} = \boldsymbol{y}(t - \tau_i),\ \ i = 1,\ 2,\ \ldots,\ m$$

are delayed state and algebraic variables, with time delays $\tau_i > 0$.

H. Ye et al., *Spectral Discretization-Based Eigen-Analysis of Time-Delay Systems*,
https://doi.org/10.1007/978-981-96-0738-9_1

1.2 Stability Analysis Methods of Time-Delay Systems

1.2.1 Time-Domain Method

In the time domain, Lyapunov-Krasovskii functional-based delay-dependent stability criteria have been proposed for analyzing the asymptotic stability of time-delay systems. Especially, the maximum delay that the system can tolerate and remain stable, i.e., delay margin, can be obtained by solving a series of generalized eigenvalue minimization problems with the help of the popular linear matrix inequality techniques [2, 3].

The criteria, however, are merely sufficient conditions for the system stability and inherently conservative due to their dependence on the selection of cost function and their coefficients. Furthermore, the criteria are usually used in conjunction with the model reduction technique for faster convergence and acceptable computational burden when analyzing large-scale time-delay systems, which inevitably deteriorates the accuracy. In addition, only a few criteria, e.g., [4, 5], can adapt to multiple time delays.

1.2.2 Predictive Compensation Method

The predictive compensation method, such as Smith predictor and model prediction, estimates the dynamic characteristics of the controlled object and then compensates it with a predictive model. As a result, a regulated signal is fed to the controller and thus the whole closed-loop system can be modeled without time delays. Specifically, the Smith predictor method moves the delay unit out of the control loop [6, 7], and the model prediction method predicts the future trajectory of the system [8–10]. However, the performance of the two methods relies heavily on the accuracy of the predictive mathematical model, which may negatively affect the performance of controllers. In the meanwhile, the error introduced by the system model reduction technique used in the application of the two methods, can also result in the poor robustness of the whole closed-loop system.

In addition to the Smith predictor and model prediction methods, [11–13] propose the trajectory extrapolation and phase compensation methods to overcome the adverse effects of communication delays on the wide-area damping control.

1.2.3 Value Set Method

The value set method is used to obtain the small-signal stability region of time-delay systems against parametric and delay uncertainties [14]. Specifically, this method first maps the arbitrary stability region in the complex plane and high-dimensional parameter space to the value set of the system's characteristic polynomial. The zero exclusion principle is then applied to assess system stability.

The team led by Professor Deqiang Gan from Zhejiang University has conducted in-depth researches on the application of the value set method in low-frequency oscillation analysis and control of actual large-scale power systems with parametric uncertainties, such as [15–17]. Reference [18] adopts the value set method to heuristically determine the maximum step length of solution operator discretization (SOD)-based methods. In [19], the edge theorem and mapping theorem are first applied to calculate the external and internal approximations of the value set. Then, the branch and bound algorithm is utilized to determine whether the exact value set of the system characteristic quasi-polynomial contains zero, so that one can obtain the small-signal parametric and delay stability region of the system.

The advantages of the value set method mainly lie in two aspects: ① the method avoids repeatedly solving zeros of the characteristic (quasi) polynomial or eigenvalues, by transforming it into the problem whether the value set of the system characteristic (quasi) polynomial contains zero; ② based on the edge theorem and mapping theorem, the method avoids meshing and scanning the high-dimensional space composed of delays and parameters. However, the bottleneck that restricts the application of the value set method lies in how to efficiently calculate the characteristic (quasi) polynomial of large-scale systems.

1.2.4 Frequency-Domain Method

In the frequency domain, delays are naturally transformed into exponential terms. Correspondingly, the characteristic equation of the time-delay system becomes transcendental and has an infinite number of eigenvalues, which is basically unsolvable by the traditional eigenvalue method. For ease of stability analysis and control synthesis, lots of efforts have been made on reducing the infinite-dimensional eigenvalue problem into a finite-dimensional one. Subsequently, a subset of critical eigenvalues (such as the rightmost ones) of the system can be accurately computed. According to different ways of reducing the eigen-problem, the frequency-domain methods can be further classified as delay substitution/approximation-based methods and spectral discretization-based methods.

The core of the delay substitution/approximation-based methods is to estimate the exponential delay terms with suitable rational polynomials or Lambert function, such as the Rekasius substitution (also known as bilinear transformation) [20–25], the Lambert-W function [26–28] and the Padé approximation [29–33]. Reference [34] presents a sufficient stability condition for time-delay systems with multiple incommensurate delays based on the Rekasius substitution. The Padé approximation is popular in approximating exponential delay terms, so that efficient stability analysis [32, 35] and controller design [32, 36] can be achieved.

However, the delay substitution/approximation-based methods have the following shortcomings: ① the Lambert-W function method only suits for a particular class of systems with commensurate delays whose state matrices can be simultaneously triangularized [27]; ② the Rekasius substitution can only accurately compute eigenvalues located on the imaginary axis whereas it cannot catch eigenvalues elsewhere; ③ the Padé approximation introduces a non-minimum phase artifact ("wrong way" effect) in the initial transient response; ④ the accuracy of the Rekasius substitution and the Padé approximation decreases fast with the increase of the magnitudes of time delays [37, 38].

The spectral discretization-based methods have been emerged as a prospective way for efficient and accurate eigen-analysis of time-delay systems [39, 40]. The methods are characterized by the discretization of two spectral operators associated with the time-delay system, i.e., the solution operator and the infinitesimal generator. The spectral discretization-based methods have been extensively studied in the field of numerical analysis and computational mathematics. Several relevant MATLAB toolboxes are also available, including DDE-BIFTOOLS, TRACE-DDE, etc. Up to the recent few years, the methods have been applied to power engineering community. An infinitesimal generator discretization (IGD)-based method is used in [41] for the first time to compute eigenvalues of time-delay power systems, so that the impact of time delays on the system stability is evaluated. Since then, the spectral discretization-based methods have been endowed with the scalability of dealing with real-life time-delay systems, through applying the partial spectral discretization (PSD) idea, adopting the preconditioning techniques, exploiting the sparsity underlying the augmented state matrices of the system and so on [42–48].

References

1. Milano, F., & Dassios, I. (2016). Small-signal stability analysis for non-index 1 Hessenberg form systems of delay differential-algebraic equations. *IEEE Transactions on Circuits and Systems-I: Regular Papers, 63*(9), 1521–1530.
2. Yao, W., Jiang, L., Wu, Q. H., et al. (2011). Delay-dependent stability analysis of the power system with a wide-area damping controller embedded. *IEEE Transactions on Power Systems, 26*(1), 233–240.
3. Yao, W., Jiang, L., Wen, J., et al. (2014). Wide-area damping controller of FACTS devices for inter-area oscillations considering communication time delays. *IEEE Transactions on Power Systems, 29*(1), 318–329.
4. Li, J., Chen, Z., Cai, D., et al. (2016). Delay-dependent stability control for power system with multiple time-delays. *IEEE Transactions on Power Systems, 31*(3), 2316–2326.
5. He, Y., Wu, M., & She, J.-H. (2006). Delay-dependent stability criteria for linear system with multiple time delays. *IEE Proceedings-Control Theory and Applications, 153*(4), 447–452.
6. Zhong, Q., & Weiss, G. (2004). A unified Smith predictor based on the spectral decomposition of the plant. *International Journal of Control, 77*(15), 1362–1371.
7. Chaudhuri, B., Majumder, R., & Pal, B. C. (2004). Wide-area measurement-based stabilizing control of power system considering signal transmission delay. *IEEE Transactions on Power Systems, 19*(4), 1971–1979.
8. Chaudhuri, B., Majumder, R., & Pal, B. C. (2004). Application of multiple-model adaptive control strategy for robust damping of interarea oscillations in power system. *IEEE Transactions on Control Systems Technology, 12*(5), 727–736.

9. Majumder, R., Chaudhuri, B., & Pal, B. C. (2005). A probabilistic approach to model-based adaptive control for damping of interarea oscillations. *IEEE Transactions on Power Systems, 20*(1), 367–374.
10. Yao, W., Jiang, L., Wen, J., et al. (2015). Wide-area damping controller for power system interarea oscillations: A networked predictive control approach. *IEEE Transactions on Control Systems Technology, 23*(1), 27–36.
11. Chaudhuri, N. R., Ray, S., Majumder, R., et al. (2010). A new approach to continuous latency compensation with adaptive phasor power oscillation damping controller (POD). *IEEE Transactions on Power Systems, 25*(2), 939–946.
12. Wang, S., Gao, W., Wang, J., et al. (2012). Synchronized sampling technology-based compensation for network effects in WAMS communication. *IEEE Transactions on Smart Grid, 3*(2), 837–845.
13. Cheng, L., Chen, G., Gao, W., et al. (2014). Adaptive time delay compensator (ATDC) design for wide-area power system stabilizer. *IEEE Transactions on Smart Grid, 5*(6), 2957–2966.
14. Jacob, K. (1995). *Robust stability and convexcity: An introduction*. Heidelberg: Springer.
15. Zhou, J., Shi, P., Gan, D., et al. (2017). Large-scale power system robust stability analysis based on value set approach. *IEEE Transactions on Power Systems, 32*(5), 4012–4023.
16. Shi, P., Zhou, J., Gan, D., et al. (2018). A rational fractional representation method for wind power integrated power system parametric stability analysis. *IEEE Transactions on Power Systems, 33*(6), 7122–7131.
17. Zhou, J., Zheng, T., & Gan, D. (2019). Value-set-based power system robust stability analysis: Further results. *IEEE Transactions on Power Systems, 34*(2), 1383–1392.
18. Engelborghs, K., & Roose, D. (2003). On stability of LMS methods and characteristic roots of delay differential equations. *SIAM Journal on Numerical Analysis, 40*(2), 629–650.
19. Schauss, T., Peer, A., & Buss, M. (2018). Parameter-space stability analysis of LTI time-delay systems with parametric uncertainties. *IEEE Transactions on Automatic Control, 63*(11), 3927–3934.
20. Walton, K., & Marshall, J. E. (1987). Direct method for TDS stability analysis. *IEE Proceedings D—Control Theory and Applications, 134*(2), 101–107.
21. Olgac, N., & Sipahi, R. (2002). An exact method for the stability analysis of time-delayed linear time-invariant (LTI) systems. *IEEE Transactions on Automatic Control, 47*(5), 793–797.
22. Olgac, N., & Sipahi, R. (2005). The cluster treatment of characteristic roots and the neutral type time-delayed systems. *Journal of Dynamic Systems, Measurement and Control, 127*(1), 88–97.
23. Ebenbauer, C., & Allgower, F. (2006). Stability analysis for time-delay systems using Rekasius's substitution and sum of squares. In *Proceedings of 45th IEEE Conference on Decision and Control* (pp. 5376–5381), San Diego.
24. Liu, Z., Zhu, C., & Jiang, Q. (2008). Stability analysis of time delayed power system based on cluster treatment of characteristic roots method. In *Proceedings of IEEE Power and Energy Society General Meeting* (Vol. 1, pp. 1–6). Pittsburgh.
25. Sonmez, S., Ayasun, S., & Nwankpa, C. O. (2016). An exact method for computing delay margin for stability of load frequency control systems with constant communication delays. *IEEE Transactions on Power Systems, 31*(1), 370–377.
26. Cheng, Y. C., & Hwang, C. (2006). Use of the Lambert *W* function for time-domain analysis of feedback fractional delay systems. *IEE Proceedings—Control Theory and Applications, 153*(2), 167–174.
27. Jarlebring, E., & Damm, T. (2007). The Lambert *W* function and the spectrum of some multidimensional time-delay systems. *Automatica, 43*(12), 2124–2128.
28. Yi, S., Nelson, P. W., & Ulsoy, A. G. (2010). *Time-delay systems: Analysis and control using the Lambert W function*. Singapore: World Scientific Publishing Company.
29. Chow, J. H., Sanchez-Gasca, J. J., Ren, H. X., et al. (2000). Power system damping controller design-using multiple input signals. *IEEE Control Systems Magazine, 20*(4), 82–90.
30. Wu, H., Ni, H., & Heydt, G. T. (2002). The impact of time delay on control design in power systems. In *Proceedings of IEEE Power and Energy Society Winter Meeting* (Vol. 2, pp. 1511–1516). New York.

31. Partington, J. R. (2004). Some frequency-domain approaches to the model reduction of delay systems. *Annual Reviews in Control, 28*(1), 65–73.
32. Yuan, Y., Li, G., Cheng, L., et al. (2010). A phase compensator for SVC supplementary control to eliminate time delay by wide area signal input. *International Journal of Electrical Power and Energy Systems, 32*(3), 163–169.
33. Golub, G. H., & Loan, C. F. V. (2012). *Matrix Computations* (4th ed.). Baltimore: The Johns Hopkins University Press.
34. Munz, U., Ebenbauer, C., Haag, T., et al. (2009). Stability analysis of time-delay systems with incommensurate delays using positive polynomials. *IEEE Transactions on Automatic Control, 54*(5), 1019–1024.
35. Ke, D. P., Chung, C. Y., & Xue, Y. (2011). An eigenstructure-based performance index and its application to control design for damping inter-area oscillations in power systems. *IEEE Transactions on Power Systems, 26*(4), 2371–2380.
36. Zhang, S., & Vittal, V. (2013). Design of wide-area power system damping controllers resilient to communication failures. *IEEE Transactions on Power Systems, 28*(4), 4292–4300.
37. Ghosh, S., Folly, K. A., & Patel, A. (2018). Synchronized versus non-synchronized feedback for speed-based wide-area PSS: Effect of time-delay. *IEEE Transactions on Smart Grid, 9*(5), 3976–3985.
38. Lou, G., Gu, W., Lu, X., et al. (2020). Distributed secondary voltage control in islanded microgrids with consideration of communication network and time delays. *IEEE Transactions on Smart Grid, 11*(5), 3702–3715.
39. Breda, D. (2004). *Numerical computation of characteristic roots for delay differential equations*. Veneto, Italy: Universitá degli Studi di Padova.
40. Michiels, W., & Niculescu, S.-I. (2014). *Stability, control, and computation for time-delay systems: An eigenvalue-based approach* (2nd ed.). Philadelphia: SIAM.
41. Milano, F., & Anghel, M. (2012). Impact of time delays on power system stability. *IEEE Transactions on Circuit and Systems-I: Regular Papers, 59*(4), 889–900.
42. Breda, D. (2006). Solution operator approximations for characteristic roots of delay differential equations. *Applied Numerical Mathematics, 56*(3–4), 305–317.
43. Breda, D., Maset, S., & Vermiglio, R. (2012). Approximation of eigenvalues of evolution operators for linear retarded functional differential equations. *SIAM Journal on Numerical Analysis, 50*(3), 1456–1483.
44. Milano, F. (2016). Small-signal stability analysis of large power systems with inclusion of multiple delays. *IEEE Transactions on Power Systems, 31*(4), 3257–3266.
45. Ye, H., Mou, Q., & Liu, Y. (2017). Enabling highly efficient spectral discretization-based eigen-analysis methods by Kronecker product. *IEEE Transactions on Power Systems, 32*(5), 4148–4150.
46. Ye, H., Mou, Q., Wang, X., et al. (2018). Eigen-analysis of large delayed cyber-physical power system by time integration-based solution operator discretization methods. *IEEE Transactions on Power Systems, 33*(6), 5968–5978.
47. Ye, H., Liu, Y., & Zhang, P. (2016). Efficient eigen-analysis for large delayed cyber-physical power system using explicit infinitesimal generator discretization. *IEEE Transactions on Power Systems, 31*(3), 2361–2370.
48. Li, C., Wu, J., Duan, C., et al. (2019). Development of an effective model for computing rightmost eigenvalues of power systems with inclusion of time delays. *IEEE Transactions on Power Systems, 34*(6), 4216–4227.

Chapter 2
Small-Signal Stability Analysis of Time-Delay Systems

This chapter focuses on the small-signal stability of time-delay systems, encompassing the small-signal modeling and eigenvalue-based stability analysis.

2.1 Small-Signal Modeling

2.1.1 Overview

When the time delay is taken into account, the dynamic model of the system is described by a set of DDAEs. Subsequently, DDAEs are required to be transformed into delayed differential equations (DDEs), so as to analyze the small-signal stability of the system.

It was shown in [1] that the most general DDAEs of time-delay systems take the non-index-1 Hessenberg form, where algebraic equations depend on both the delayed state and algebraic variables [2, 3]. By eliminating the algebraic variables, the non-index-1 Hessenberg DDAEs can be transformed into DDEs with an infinite number of delayed terms. If the algebraic equations were assumed to be independent of the delayed algebraic variables, the non-index-1 Hessenberg DDAEs can be reduced to the index-1 Hessenberg ones [4, 5]. In [6] and [7], further assumption was made where the algebraic equations rely on neither the delayed state nor the delayed algebraic variables. Accordingly, the alternative index-1 Hessenberg DDAEs of time-delay systems were derived.

H. Ye et al., *Spectral Discretization-Based Eigen-Analysis of Time-Delay Systems*,
https://doi.org/10.1007/978-981-96-0738-9_2

2.1.2 *Transforming Non-index-1 Hessenberg DDAEs into DDEs*

This section theoretically addresses how to transform the non-index-1 Hessenberg DDAEs of time-delay systems into DDEs.

2.1.2.1 The Non-index-1 Hessenberg DDAEs

At time t, a dynamic system with m delays can be modeled by the following general DDAEs:

$$\begin{cases} \dot{\boldsymbol{x}} = \boldsymbol{f}(\boldsymbol{x},\ \boldsymbol{y},\ \boldsymbol{x}_{\mathrm{d}1},\ \boldsymbol{y}_{\mathrm{d}1},\ \ldots,\ \boldsymbol{x}_{\mathrm{d}m},\ \boldsymbol{y}_{\mathrm{d}m}) \\ \boldsymbol{0} = \boldsymbol{g}(\boldsymbol{x},\ \boldsymbol{y},\ \boldsymbol{x}_{\mathrm{d}1},\ \boldsymbol{y}_{\mathrm{d}1},\ \ldots,\ \boldsymbol{x}_{\mathrm{d}m},\ \boldsymbol{y}_{\mathrm{d}m}) \end{cases} \tag{2.1}$$

where $\boldsymbol{x} = \boldsymbol{x}(t) \in \mathbb{R}^{n\times 1}$ and $\boldsymbol{y} = \boldsymbol{y}(t) \in \mathbb{R}^{l\times 1}$ are system state and algebraic variables,

$$\boldsymbol{x}_{\mathrm{d}i} = \boldsymbol{x}(t-\tau_i),\quad \boldsymbol{y}_{\mathrm{d}i} = \boldsymbol{y}(t-\tau_i),\quad i = 1,\ 2,\ \ldots,\ m$$

are delayed state and algebraic variables, and $\tau_i > 0$ $(i = 1,\ 2,\ \ldots,\ m)$ are time delay constants. It is noteworthy that $\boldsymbol{g}$ in (2.1) depends on $\boldsymbol{y}_{\mathrm{d}i}$, $i = 1,\ 2,\ \ldots,\ m$. Consequently, $\boldsymbol{y}$ and $\boldsymbol{y}_{\mathrm{d}i}$ $(i = 1,\ 2,\ \ldots,\ m)$ in $\boldsymbol{f}$ cannot be eliminated by a finite number of differentiation and substitutions. Hence, (2.1) represents the non-index-1 Hessenberg DDAEs. In addition, though (2.1) seems of the retarded type, it is essentially of the neutral type for $\boldsymbol{y}$ (see the Appendix of [8]).

Linearizing (2.1) at the equilibrium point

$$(\bar{\boldsymbol{x}},\ \bar{\boldsymbol{y}}) = \left(\boldsymbol{x}^{(0)},\ \boldsymbol{y}^{(0)}, \boldsymbol{x}_{\mathrm{d}1}^{(0)},\ \boldsymbol{y}_{\mathrm{d}1}^{(0)},\ \ldots,\ \boldsymbol{x}_{\mathrm{d}m}^{(0)},\ \boldsymbol{y}_{\mathrm{d}m}^{(0)}\right)$$

yields

$$\Delta\dot{\boldsymbol{x}} = \boldsymbol{A}_0\Delta\boldsymbol{x} + \boldsymbol{B}_0\Delta\boldsymbol{y} + \boldsymbol{A}_1\Delta\boldsymbol{x}_{\mathrm{d}1} + \boldsymbol{B}_1\Delta\boldsymbol{y}_{\mathrm{d}1} + \cdots + \boldsymbol{A}_m\Delta\boldsymbol{x}_{\mathrm{d}m} + \boldsymbol{B}_m\Delta\boldsymbol{y}_{\mathrm{d}m} \tag{2.2}$$

and

$$\boldsymbol{0} = \boldsymbol{C}_0\Delta\boldsymbol{x} + \boldsymbol{D}_0\Delta\boldsymbol{y} + \boldsymbol{C}_1\Delta\boldsymbol{x}_{\mathrm{d}1} + \boldsymbol{D}_1\Delta\boldsymbol{y}_{\mathrm{d}1} + \cdots + \boldsymbol{C}_m\Delta\boldsymbol{x}_{\mathrm{d}m} + \boldsymbol{D}_m\Delta\boldsymbol{y}_{\mathrm{d}m} \tag{2.3}$$

where $\boldsymbol{A}_0 = \left.\dfrac{\partial \boldsymbol{f}}{\partial \boldsymbol{x}}\right|_{(\bar{\boldsymbol{x}},\bar{\boldsymbol{y}})}$, $\boldsymbol{B}_0 = \left.\dfrac{\partial \boldsymbol{f}}{\partial \boldsymbol{y}}\right|_{(\bar{\boldsymbol{x}},\bar{\boldsymbol{y}})}$, $\boldsymbol{C}_0 = \left.\dfrac{\partial \boldsymbol{g}}{\partial \boldsymbol{x}}\right|_{(\bar{\boldsymbol{x}},\bar{\boldsymbol{y}})}$, $\boldsymbol{D}_0 = \left.\dfrac{\partial \boldsymbol{g}}{\partial \boldsymbol{y}}\right|_{(\bar{\boldsymbol{x}},\bar{\boldsymbol{y}})}$, $\boldsymbol{A}_i = \left.\dfrac{\partial \boldsymbol{f}}{\partial \boldsymbol{x}_{\mathrm{d}i}}\right|_{(\bar{\boldsymbol{x}},\bar{\boldsymbol{y}})}$, $\boldsymbol{B}_i = \left.\dfrac{\partial \boldsymbol{f}}{\partial \boldsymbol{y}_{\mathrm{d}i}}\right|_{(\bar{\boldsymbol{x}},\bar{\boldsymbol{y}})}$, $\boldsymbol{C}_i = \left.\dfrac{\partial \boldsymbol{g}}{\partial \boldsymbol{x}_{\mathrm{d}i}}\right|_{(\bar{\boldsymbol{x}},\bar{\boldsymbol{y}})}$ and $\boldsymbol{D}_i = \left.\dfrac{\partial \boldsymbol{g}}{\partial \boldsymbol{y}_{\mathrm{d}i}}\right|_{(\bar{\boldsymbol{x}},\bar{\boldsymbol{y}})}$ $(i =$

1, 2, ..., m) are Jacobian matrices, also known as the augmented state matrices of the system.

2.1.2.2 DDEs with an Infinite Number of Delayed Terms

Theorem 2.1 *The DDAEs given by (2.2) and (2.3) can be transformed into the following DDEs with an infinite number of delayed terms:*

$$\Delta\dot{\boldsymbol{x}} = \tilde{\boldsymbol{A}}_0\Delta\boldsymbol{x} + \sum_{i=1}^{m}\tilde{\boldsymbol{A}}_i\Delta\boldsymbol{x}_{\mathrm{d}i} + \sum_{k=2}^{\infty}\sum_{d_1,d_2,\ldots,d_k=1}^{m}\tilde{\boldsymbol{A}}_{\mathrm{d}\Sigma k}\Delta\boldsymbol{x}_{\mathrm{d}\Sigma k} \tag{2.4}$$

where

$$\Delta\boldsymbol{x}_{\mathrm{d}\Sigma k} = \Delta\boldsymbol{x}(t-\tau_{d_1}-\tau_{d_2}-\cdots-\tau_{d_k}) = \Delta\boldsymbol{x}\left(t-\sum_{i=1}^{k}\tau_{d_i}\right)$$

$$\tilde{\boldsymbol{A}}_0 = \boldsymbol{A}_0 + \boldsymbol{B}_0\boldsymbol{K}_0 \tag{2.5}$$

$$\tilde{\boldsymbol{A}}_i = \boldsymbol{A}_i + \boldsymbol{B}_i\boldsymbol{K}_{0,\mathrm{d}i} + \boldsymbol{B}_0\boldsymbol{M}_{i,\mathrm{d}i} \tag{2.6}$$

$$\tilde{\boldsymbol{A}}_{\mathrm{d}\Sigma k} = \boldsymbol{N}_{d_1}\prod_{j=2}^{k-1}\boldsymbol{L}_{d_j,\mathrm{d}\Sigma(j-1)}\boldsymbol{M}_{d_k,\mathrm{d}\Sigma k} \tag{2.7}$$

with

$$\begin{cases} \boldsymbol{K}_0 = -\boldsymbol{D}_0^{-1}\boldsymbol{C}_0,\ \boldsymbol{K}_i = -\boldsymbol{D}_0^{-1}\boldsymbol{C}_i \\ \boldsymbol{L}_i = -\boldsymbol{D}_0^{-1}\boldsymbol{D}_i,\ \boldsymbol{N}_i = \boldsymbol{B}_i + \boldsymbol{B}_0\boldsymbol{L}_i \\ \boldsymbol{K}_{0,\mathrm{d}j} = -\boldsymbol{D}_{0,\mathrm{d}j}^{-1}\boldsymbol{C}_{0,\mathrm{d}j},\ \boldsymbol{K}_{0,\mathrm{d}\Sigma k} = -\boldsymbol{D}_{0,\mathrm{d}\Sigma k}^{-1}\boldsymbol{C}_{0,\mathrm{d}\Sigma k} \\ \boldsymbol{K}_{i,\mathrm{d}j} = -\boldsymbol{D}_{0,\mathrm{d}j}^{-1}\boldsymbol{C}_{i,\mathrm{d}j},\ \boldsymbol{K}_{i,\mathrm{d}\Sigma k} = -\boldsymbol{D}_{0,\mathrm{d}\Sigma k}^{-1}\boldsymbol{C}_{i,\mathrm{d}\Sigma k} \\ \boldsymbol{L}_{i,\mathrm{d}j} = -\boldsymbol{D}_{0,\mathrm{d}j}^{-1}\boldsymbol{D}_{i,\mathrm{d}j},\ \boldsymbol{L}_{i,\mathrm{d}\Sigma k} = -\boldsymbol{D}_{0,\mathrm{d}\Sigma k}^{-1}\boldsymbol{D}_{i,\mathrm{d}\Sigma k} \\ \boldsymbol{M}_{i,\mathrm{d}i} = \boldsymbol{K}_i + \boldsymbol{L}_i\boldsymbol{K}_{0,\mathrm{d}i},\ \boldsymbol{M}_{i,\mathrm{d}\Sigma k} = \boldsymbol{K}_{i,\mathrm{d}\Sigma(k-1)} + \boldsymbol{L}_{i,\mathrm{d}\Sigma(k-1)}\boldsymbol{K}_{0,\mathrm{d}\Sigma k} \end{cases} \tag{2.8}$$

where

$$\begin{cases} C_{0,\mathrm{d}j} = \left.\dfrac{\partial g}{\partial x}\right|_{(\bar{x}(t-\tau_j),\bar{y}(t-\tau_j))}, & D_{0,\mathrm{d}j} = \left.\dfrac{\partial g}{\partial y}\right|_{(\bar{x}(t-\tau_j),\bar{y}(t-\tau_j))} \\ C_{i,\mathrm{d}j} = \left.\dfrac{\partial g}{\partial x_{\mathrm{d}i}}\right|_{(\bar{x}(t-\tau_j),\bar{y}(t-\tau_j))}, & D_{i,\mathrm{d}j} = \left.\dfrac{\partial g}{\partial y_{\mathrm{d}i}}\right|_{(\bar{x}(t-\tau_j),\bar{y}(t-\tau_j))} \\ C_{0,\mathrm{d}\Sigma k} = \left.\dfrac{\partial g}{\partial x}\right|_{\left(\bar{x}\left(t-\sum_{j=1}^{k}\tau_{d_j}\right),\bar{y}\left(t-\sum_{j=1}^{k}\tau_{d_j}\right)\right)} \\ D_{0,\mathrm{d}\Sigma k} = \left.\dfrac{\partial g}{\partial y}\right|_{\left(\bar{x}\left(t-\sum_{j=1}^{k}\tau_{d_j}\right),\bar{y}\left(t-\sum_{j=1}^{k}\tau_{d_j}\right)\right)} \\ C_{i,\mathrm{d}\Sigma k} = \left.\dfrac{\partial g}{\partial x_{\mathrm{d}i}}\right|_{\left(\bar{x}\left(t-\sum_{j=1}^{k}\tau_{d_j}\right),\bar{y}\left(t-\sum_{j=1}^{k}\tau_{d_j}\right)\right)} \\ D_{i,\mathrm{d}\Sigma k} = \left.\dfrac{\partial g}{\partial y_{\mathrm{d}i}}\right|_{\left(\bar{x}\left(t-\sum_{j=1}^{k}\tau_{d_j}\right),\bar{y}\left(t-\sum_{j=1}^{k}\tau_{d_j}\right)\right)} \end{cases}.$$

2.1.2.3 Proof of (2.5) and (2.6)

Proof The non-singularity of D_0 allows one to solve (2.3) for Δy. Then by applying the notations defined in (2.8), one can obtain

$$\Delta y = K_0\Delta x + \sum_{i=1}^{m} K_i\Delta x_{\mathrm{d}i} + \sum_{i=1}^{m} L_i\Delta y_{\mathrm{d}i}. \tag{2.9}$$

Note that (2.9) essentially represents a recursion for Δy. In the same vein, at time $t-\tau_i$ and $t-\tau_i-\tau_j$ $(i,\ j=1,\ 2,\ \ldots,\ m)$, $\Delta y_{\mathrm{d}i}$ and $\Delta y_{\mathrm{d}\Sigma 2}$ are respectively solved as

$$\Delta y_{\mathrm{d}i} = K_{0,\mathrm{d}i}\Delta x_{\mathrm{d}i} + \sum_{j=1}^{m} K_{j,\mathrm{d}i}\Delta x_{\mathrm{d}\Sigma 2} + \sum_{j=1}^{m} L_{j,\mathrm{d}i}\Delta y_{\mathrm{d}\Sigma 2} \tag{2.10}$$

and

$$\Delta y_{\mathrm{d}\Sigma 2} = K_{0,\mathrm{d}\Sigma 2}\Delta x_{\mathrm{d}\Sigma 2} + \sum_{k=1}^{m} K_{k,\mathrm{d}\Sigma 2}\Delta x_{\mathrm{d}\Sigma 3} + \sum_{k=1}^{m} L_{k,\mathrm{d}\Sigma 2}\Delta y_{\mathrm{d}\Sigma 3} \tag{2.11}$$

where $\Delta x_{\mathrm{d}\Sigma 2} = \Delta x(t-\tau_i-\tau_j) = \Delta x(t-\tau_{d_1}-\tau_{d_2})$, $\Delta y_{\mathrm{d}\Sigma 2} = \Delta y(t-\tau_i-\tau_j) = \Delta y(t-\tau_{d_1}-\tau_{d_2})$, $\Delta x_{\mathrm{d}\Sigma 3} = \Delta x(t-\tau_i-\tau_j-\tau_k) = \Delta x(t-\tau_{d_1}-\tau_{d_2}-\tau_{d_3})$, $\Delta y_{\mathrm{d}\Sigma 3} = \Delta y(t-\tau_i-\tau_j-\tau_k) = \Delta y(t-\tau_{d_1}-\tau_{d_2}-\tau_{d_3})$, $i,\ j,\ k = 1,\ 2,\ \ldots,\ m$.

By substituting (2.9) and (2.10) into (2.2) to eliminate Δy and $\Delta y_{\mathrm{d}i}$, one has

$$\begin{aligned}\Delta\dot{\boldsymbol{x}} =&(\boldsymbol{A}_0+\boldsymbol{B}_0\boldsymbol{K}_0)\Delta\boldsymbol{x}+\sum_{i=1}^{m}((\boldsymbol{A}_i+\boldsymbol{B}_0\boldsymbol{K}_i)+(\boldsymbol{B}_i+\boldsymbol{B}_0\boldsymbol{L}_i)\boldsymbol{K}_{0,\mathrm{d}i})\Delta\boldsymbol{x}_{\mathrm{d}i}\\ &+\sum_{i=1}^{m}\sum_{j=1}^{m}(\boldsymbol{B}_i+\boldsymbol{B}_0\boldsymbol{L}_i)\boldsymbol{K}_{j,\mathrm{d}i}\Delta\boldsymbol{x}_{\mathrm{d}\Sigma2}+\sum_{i=1}^{m}\sum_{j=1}^{m}(\boldsymbol{B}_i+\boldsymbol{B}_0\boldsymbol{L}_i)\boldsymbol{L}_{j,\mathrm{d}i}\Delta\boldsymbol{y}_{\mathrm{d}\Sigma2}\\ =&(\boldsymbol{A}_0+\boldsymbol{B}_0\boldsymbol{K}_0)\Delta\boldsymbol{x}+\sum_{i=1}^{m}(\boldsymbol{A}_i+\boldsymbol{B}_i\boldsymbol{K}_{0,\mathrm{d}i}+\boldsymbol{B}_0\boldsymbol{M}_{i,\mathrm{d}i})\Delta\boldsymbol{x}_{\mathrm{d}i}\\ &+\sum_{i=1}^{m}\sum_{j=1}^{m}(\boldsymbol{B}_i+\boldsymbol{B}_0\boldsymbol{L}_i)\boldsymbol{K}_{j,\mathrm{d}i}\Delta\boldsymbol{x}_{\mathrm{d}\Sigma2}+\sum_{i=1}^{m}\sum_{j=1}^{m}(\boldsymbol{B}_i+\boldsymbol{B}_0\boldsymbol{L}_i)\boldsymbol{L}_{j,\mathrm{d}i}\Delta\boldsymbol{y}_{\mathrm{d}\Sigma2}.\end{aligned} \tag{2.12}$$

Similarly, substituting (2.11) into (2.12) to eliminate $\Delta\boldsymbol{y}_{\mathrm{d}\Sigma2}$, one can obtain

$$\begin{aligned}\Delta\dot{\boldsymbol{x}} =&(\boldsymbol{A}_0+\boldsymbol{B}_0\boldsymbol{K}_0)\Delta\boldsymbol{x}+\sum_{i=1}^{m}(\boldsymbol{A}_i+\boldsymbol{B}_i\boldsymbol{K}_{0,\mathrm{d}i}+\boldsymbol{B}_0\boldsymbol{M}_{i,\mathrm{d}i})\Delta\boldsymbol{x}_{\mathrm{d}i}\\ &+\sum_{d_1,d_2=1}^{m}\boldsymbol{N}_{d_1}\boldsymbol{M}_{d_2,\mathrm{d}\Sigma2}\Delta\boldsymbol{x}_{\mathrm{d}\Sigma2}+\sum_{d_1,d_2,d_3=1}^{m}\boldsymbol{N}_{d_1}\boldsymbol{L}_{d_2,\mathrm{d}d_1}\boldsymbol{K}_{d_3,\mathrm{d}\Sigma2}\Delta\boldsymbol{x}_{\mathrm{d}\Sigma3}\\ &+\sum_{d_1,d_2,d_3=1}^{m}\boldsymbol{N}_{d_1}\boldsymbol{L}_{d_2,\mathrm{d}d_1}\boldsymbol{L}_{d_3,\mathrm{d}\Sigma2}\Delta\boldsymbol{y}_{\mathrm{d}\Sigma3}.\end{aligned} \tag{2.13}$$

It can found by comparing the first two terms of (2.13) with those of (2.4) that, (2.5) and (2.6) hold. □

2.1.2.4 Proof of (2.7)

Lemma 2.1 *The sum of the index-k and index-(k + 1) delayed terms ($k \geq 2$) of $\Delta\dot{\boldsymbol{x}}$ can be expressed as*

$$\begin{aligned}&\sum_{d_1,d_2,\ldots,d_k=1}^{m}\boldsymbol{N}_{d_1}\prod_{j=2}^{k-1}\boldsymbol{L}_{d_j,\mathrm{d}\Sigma(j-1)}\boldsymbol{M}_{d_k,\mathrm{d}\Sigma k}\Delta\boldsymbol{x}_{\mathrm{d}\Sigma k}\\ &+\sum_{d_1,d_2,\ldots,d_{k+1}=1}^{m}\boldsymbol{N}_{d_1}\prod_{j=2}^{k}\boldsymbol{L}_{d_j,\mathrm{d}\Sigma(j-1)}\boldsymbol{K}_{d_{k+1},\mathrm{d}\Sigma k}\Delta\boldsymbol{x}_{\mathrm{d}\Sigma(k+1)}\\ &+\sum_{d_1,d_2,\ldots,d_{k+1}=1}^{m}\boldsymbol{N}_{d_1}\prod_{j=2}^{k+1}\boldsymbol{L}_{d_j,\mathrm{d}\Sigma(j-1)}\Delta\boldsymbol{y}_{\mathrm{d}\Sigma(k+1)}.\end{aligned} \tag{2.14}$$

Proof The well-known mathematical induction method is utilized here to prove the above lemma.

First, when $k = 2$, (2.14) becomes

$$\sum_{d_1,d_2=1}^{m} \boldsymbol{N}_{d_1}\boldsymbol{M}_{d_2,\mathrm{d}\Sigma 2}\Delta\boldsymbol{x}_{\mathrm{d}\Sigma 2} + \sum_{d_1,d_2,d_3=1}^{m} \boldsymbol{N}_{d_1}\boldsymbol{L}_{d_2,\mathrm{d}d_1}\boldsymbol{K}_{d_3,\mathrm{d}\Sigma 2}\Delta\boldsymbol{x}_{\mathrm{d}\Sigma 3} + \sum_{d_1,d_2,d_3=1}^{m} \boldsymbol{N}_{d_1}\boldsymbol{L}_{d_2,\mathrm{d}d_1}\boldsymbol{L}_{d_3,\mathrm{d}\Sigma 2}\Delta\boldsymbol{y}_{\mathrm{d}\Sigma 3}. \tag{2.15}$$

It is clear that (2.15) is identical to the last two terms of (2.13), i.e., the sum of the index-2 and index-3 delayed terms of $\Delta\dot{\boldsymbol{x}}$.

Next, based on (2.14), the expression of the sum of the index-$(k+1)$ and index-$(k+2)$ delayed terms of $\Delta\dot{\boldsymbol{x}}$ can be derived by eliminating the algebraic variables $\Delta\boldsymbol{y}_{\mathrm{d}\Sigma(k+1)}$. It can be obtained by propagating time back from t to $t-\sum_{i=1}^{k}\tau_{d_i}$ $(k=2,\ 3,\ \cdots)$ as was done when arriving at (2.9) $\sim$ (2.11):

$$\Delta\boldsymbol{y}_{\mathrm{d}\Sigma(k+1)} = \boldsymbol{K}_{0,\mathrm{d}\Sigma(k+1)}\Delta\boldsymbol{x}_{\mathrm{d}\Sigma(k+1)} + \sum_{d_{k+2}=1}^{m} \boldsymbol{K}_{d_{k+2},\mathrm{d}\Sigma(k+1)}\Delta\boldsymbol{x}_{\mathrm{d}\Sigma(k+2)} + \sum_{d_{k+2}=1}^{m} \boldsymbol{L}_{d_{k+2},\mathrm{d}\Sigma(k+1)}\Delta\boldsymbol{y}_{\mathrm{d}\Sigma(k+2)}. \tag{2.16}$$

Considering $i=1,\ 2,\ \ldots,\ k+1$ and $d_i=1,\ 2,\ \ldots,\ m$, $\Delta\boldsymbol{y}_{\mathrm{d}\Sigma(k+1)}$ generally represents m^{k+1} entries. By substituting them into (2.14), the last two terms of the equation become

$$\begin{aligned}
&\sum_{d_1,d_2,\ldots,d_{k+1}=1}^{m} \boldsymbol{N}_{d_1}\prod_{j=2}^{k}\boldsymbol{L}_{d_j,\mathrm{d}\Sigma(j-1)}\boldsymbol{K}_{d_{k+1},\mathrm{d}\Sigma k}\Delta\boldsymbol{x}_{\mathrm{d}\Sigma(k+1)} \\
&+ \sum_{d_1,d_2,\ldots,d_{k+1}=1}^{m} \boldsymbol{N}_{d_1}\prod_{j=2}^{k+1}\boldsymbol{L}_{d_j,\mathrm{d}\Sigma(j-1)}\Delta\boldsymbol{y}_{\mathrm{d}\Sigma(k+1)} \\
=& \sum_{d_1,d_2,\ldots,d_{k+1}=1}^{m} \boldsymbol{N}_{d_1}\prod_{j=2}^{k}\boldsymbol{L}_{d_j,\mathrm{d}\Sigma(j-1)}\boldsymbol{K}_{d_{k+1},\mathrm{d}\Sigma k}\Delta\boldsymbol{x}_{\mathrm{d}\Sigma(k+1)} \\
&+ \sum_{d_1,d_2,\ldots,d_{k+1}=1}^{m} \boldsymbol{N}_{d_1}\prod_{j=2}^{k+1}\boldsymbol{L}_{d_j,\mathrm{d}\Sigma(j-1)}\Bigg(\boldsymbol{K}_{0,\mathrm{d}\Sigma(k+1)}\Delta\boldsymbol{x}_{\mathrm{d}\Sigma(k+1)} \\
&+ \sum_{d_{k+2}=1}^{m} \boldsymbol{K}_{d_{k+2},\mathrm{d}\Sigma(k+1)}\Delta\boldsymbol{x}_{\mathrm{d}\Sigma(k+2)} + \sum_{d_{k+2}=1}^{m} \boldsymbol{L}_{d_{k+2},\mathrm{d}\Sigma(k+1)}\Delta\boldsymbol{y}_{\mathrm{d}\Sigma(k+2)}\Bigg) \\
=& \sum_{d_1,d_2,\ldots,d_{k+1}=1}^{m} \boldsymbol{N}_{d_1}\prod_{j=2}^{k}\boldsymbol{L}_{d_j,\mathrm{d}\Sigma(j-1)}(\boldsymbol{K}_{d_{k+1},\mathrm{d}\Sigma k} + \boldsymbol{L}_{d_{k+1},\mathrm{d}\Sigma k}\boldsymbol{K}_{0,\mathrm{d}\Sigma(k+1)})\Delta\boldsymbol{x}_{\mathrm{d}\Sigma(k+1)}
\end{aligned}$$

$$
\begin{aligned}
&+\sum_{d_1,d_2,\ldots,d_{k+1}=1}^{m} \boldsymbol{N}_{d_1}\prod_{j=2}^{k+1}\boldsymbol{L}_{d_j,\mathrm{d}\Sigma(j-1)}\sum_{d_{k+2}=1}^{m}\boldsymbol{K}_{d_{k+2},\mathrm{d}\Sigma(k+1)}\Delta\boldsymbol{x}_{\mathrm{d}\Sigma(k+2)}\\
&+\sum_{d_1,d_2,\ldots,d_{k+1}=1}^{m} \boldsymbol{N}_{d_1}\prod_{j=2}^{k+1}\boldsymbol{L}_{d_j,\mathrm{d}\Sigma(j-1)}\sum_{d_{k+2}=1}^{m}\boldsymbol{L}_{d_{k+2},\mathrm{d}\Sigma(k+1)}\Delta\boldsymbol{y}_{\mathrm{d}\Sigma(k+2)}\\
=&\sum_{d_1,d_2,\ldots,d_{k+1}=1}^{m} \boldsymbol{N}_{d_1}\prod_{j=2}^{k}\boldsymbol{L}_{d_j,\mathrm{d}\Sigma(j-1)}\boldsymbol{M}_{d_{k+1},\mathrm{d}\Sigma(k+1)}\Delta\boldsymbol{x}_{\mathrm{d}\Sigma(k+1)}\\
&+\sum_{d_1,d_2,\ldots,d_{k+2}=1}^{m} \boldsymbol{N}_{d_1}\prod_{j=2}^{k+1}\boldsymbol{L}_{d_j,\mathrm{d}\Sigma(j-1)}\boldsymbol{K}_{d_{k+2},\mathrm{d}\Sigma(k+1)}\Delta\boldsymbol{x}_{\mathrm{d}\Sigma(k+2)}\\
&+\sum_{d_1,d_2,\ldots,d_{k+2}=1}^{m} \boldsymbol{N}_{d_1}\prod_{j=2}^{k+2}\boldsymbol{L}_{d_j,\mathrm{d}\Sigma(j-1)}\Delta\boldsymbol{y}_{\mathrm{d}\Sigma(k+2)}.
\end{aligned}
\tag{2.17}
$$

By comparing (2.17) and (2.14), it can be seen that they have the same structure. The only difference is that k has been replaced by $k+1$. Therefore, (2.14) is proved and the proof of Lemma 2.1 is completed. □

From Lemma 2.1, it is known that the first term of (2.14) is the k-th order delayed term of $\Delta\dot{\boldsymbol{x}}$, i.e.,

$$
\sum_{d_1,d_2,\ldots,d_k=1}^{m}\tilde{\boldsymbol{A}}_{\mathrm{d}\Sigma k}\Delta\boldsymbol{x}_{\mathrm{d}\Sigma k} \triangleq \sum_{d_1,d_2,\ldots,d_k=1}^{m}\boldsymbol{N}_{d_1}\prod_{j=2}^{k-1}\boldsymbol{L}_{d_j,\mathrm{d}\Sigma(j-1)}\boldsymbol{M}_{d_k,\mathrm{d}\Sigma k}\Delta\boldsymbol{x}_{\mathrm{d}\Sigma k}
$$

where the coefficient matrix coincides with (2.7).

Since the delayed algebraic equations expressed in (2.10), (2.11) and (2.16) are derived through the time delay operation of the algebraic equation expressed in (2.9) without delay, one can obtain $\boldsymbol{C}_0=\boldsymbol{C}_{0,\mathrm{d}j}=\cdots=\boldsymbol{C}_{0,\mathrm{d}\Sigma k}$, $\boldsymbol{D}_0=\boldsymbol{D}_{0,\mathrm{d}j}=\cdots=\boldsymbol{D}_{0,\mathrm{d}\Sigma k}$, $\boldsymbol{C}_i=\boldsymbol{C}_{i,\mathrm{d}j}=\cdots=\boldsymbol{C}_{i,\mathrm{d}\Sigma k}$, $\boldsymbol{D}_i=\boldsymbol{D}_{i,\mathrm{d}j}=\cdots=\boldsymbol{D}_{i,\mathrm{d}\Sigma k}$, $i=1,\ 2,\ \ldots,\ m;\ k=2,\ 3,\ \cdots$. Therefore, $\tilde{\boldsymbol{A}}_i$ and $\tilde{\boldsymbol{A}}_{\mathrm{d}\Sigma k}$ expressed in (2.6) and (2.7) become

$$
\begin{cases}
\tilde{\boldsymbol{A}}_i=\boldsymbol{A}_i+\boldsymbol{B}_i\boldsymbol{D}_0^{-1}\boldsymbol{C}_0-\boldsymbol{B}_0\boldsymbol{D}_0^{-1}\boldsymbol{C}_i+\boldsymbol{B}_0\boldsymbol{D}_0^{-1}\boldsymbol{D}_i^{-1}\boldsymbol{D}_0\boldsymbol{C}_0\\
\tilde{\boldsymbol{A}}_{\mathrm{d}\Sigma k}=-(\boldsymbol{B}_{d_1}-\boldsymbol{B}_0\boldsymbol{D}_0^{-1}\boldsymbol{D}_{d_1})\prod_{j=2}^{k-1}(\boldsymbol{D}_0^{-1}\boldsymbol{D}_{d_j})(\boldsymbol{D}_0^{-1}\boldsymbol{C}_i-\boldsymbol{D}_0^{-1}\boldsymbol{D}_i^{-1}\boldsymbol{D}_0\boldsymbol{C}_0)
\end{cases}.
$$

2.1.2.5 Eliminating the Index-2 and Higher Index Delayed Terms

The following presents a sufficient condition under which the index-2 and higher index delayed terms in (2.4) can be readily eliminated.

Theorem 2.2 *If the condition*

$$\boldsymbol{C}_i = \boldsymbol{C}_0,\ \boldsymbol{D}_i = \boldsymbol{D}_0,\ \ i = 1,\ 2,\ \ldots,\ m \tag{2.18}$$

is satisfied, (2.4) can be transformed into the following DDEs consisting of a no-delay term and m index-1 delayed terms:

$$\Delta\dot{\boldsymbol{x}} = \tilde{\boldsymbol{A}}_0\Delta\boldsymbol{x} + \sum_{i=1}^{m}\tilde{\boldsymbol{A}}_i\Delta\boldsymbol{x}_{\mathrm{d}i} \tag{2.19}$$

where

$$\begin{cases} \tilde{\boldsymbol{A}}_0 = \boldsymbol{A}_0 + \boldsymbol{B}_0\boldsymbol{K}_0 \\ \tilde{\boldsymbol{A}}_i = \boldsymbol{A}_i + \boldsymbol{B}_i\boldsymbol{K}_0 \\ \boldsymbol{K}_0 = -\boldsymbol{D}_0^{-1}\boldsymbol{C}_0 \end{cases}.$$

Proof Taking (2.18) into account, (2.8) becomes

$$\begin{cases} \boldsymbol{K}_i = -\boldsymbol{D}_0^{-1}\boldsymbol{C}_0 = \boldsymbol{K}_0 \\ \boldsymbol{L}_i = -\boldsymbol{D}_0^{-1}\boldsymbol{D}_0 = -\boldsymbol{I}_l \\ \boldsymbol{N}_i = \boldsymbol{B}_0\boldsymbol{L}_i + \boldsymbol{B}_i = -\boldsymbol{B}_0 + \boldsymbol{B}_i \\ \boldsymbol{K}_{0,\mathrm{d}j} = \boldsymbol{K}_{0,\mathrm{d}\Sigma k} = \boldsymbol{K}_{i,\mathrm{d}j} = \boldsymbol{K}_{i,\mathrm{d}\Sigma k} = -\boldsymbol{D}_0^{-1}\boldsymbol{C}_0 = \boldsymbol{K}_0 \\ \boldsymbol{L}_{i,\mathrm{d}j} = \boldsymbol{L}_{i,\mathrm{d}\Sigma k} = -\boldsymbol{D}_0^{-1}\boldsymbol{D}_0 = -\boldsymbol{I}_l \\ \boldsymbol{M}_{i,\mathrm{d}i} = \boldsymbol{M}_{i,\mathrm{d}\Sigma k} = \boldsymbol{K}_0 + \boldsymbol{L}_i\boldsymbol{K}_0 = \boldsymbol{0} \end{cases}. \tag{2.20}$$

Substituting (2.20) into (2.6) and (2.7) yields

$$\tilde{\boldsymbol{A}}_i = \boldsymbol{A}_i + \boldsymbol{B}_i\boldsymbol{K}_{0,\mathrm{d}i} + \boldsymbol{B}_0\boldsymbol{M}_{i,\mathrm{d}i} = \boldsymbol{A}_i + \boldsymbol{B}_i\boldsymbol{K}_0$$

$$\tilde{\boldsymbol{A}}_{\mathrm{d}\Sigma k} = \boldsymbol{N}_{d_1}\prod_{j=2}^{k-1}\boldsymbol{L}_{d_j,\mathrm{d}\Sigma(j-1)}\boldsymbol{M}_{d_k,\mathrm{d}\Sigma k} = \boldsymbol{0}. \tag{2.21}$$

It can be seen from (2.21) that all delayed terms of index-2 and higher indices have been eliminated. The proof of (2.19) is therefore completed. □

The characteristic polynomial of (2.19) is given by

$$|\Delta(\lambda)| = 0$$

where

$$\Delta(\lambda) = \lambda\boldsymbol{I}_n - \tilde{\boldsymbol{A}}_0 - \sum_{i=1}^{m}\tilde{\boldsymbol{A}}_i\mathrm{e}^{-\lambda\tau_i}.$$

2.1.3 Transforming Index-1 Hessenberg DDAEs to DDEs

Section 2.1.2 have provided a way of transforming the non-index-1 Hessenberg DDAEs into DDEs, as illustrated by the right upper part of Fig. 2.1. It is noted that an alternative way of completing the task was presented in [6, 7]. The basic idea is shown by the left lower part of Fig. 2.1 and summarized as follows.

If $\boldsymbol{g}$ in (2.1) is memoryless, it is fully determined by the state and algebraic variables at the current time, i.e., $\boldsymbol{g}(\boldsymbol{x}, \boldsymbol{y}) = \boldsymbol{0}$. Given this, the algebraic equation in (2.1) can be decoupled into one equation at the current time and m equations at time $t - \tau_i$ $(i = 1, 2, \ldots, m)$. Combing $\boldsymbol{f}$ with $\boldsymbol{g}$ leads to the index-1 Hessenberg DDAEs of the retarded type:

$$\begin{cases} \dot{\boldsymbol{x}} = \boldsymbol{f}(\boldsymbol{x}, \boldsymbol{y}, \boldsymbol{x}_{\mathrm{d}1}, \boldsymbol{y}_{\mathrm{d}1}, \ldots, \boldsymbol{x}_{\mathrm{d}m}, \boldsymbol{y}_{\mathrm{d}m}) \\ \boldsymbol{0} = \boldsymbol{g}(\boldsymbol{x}, \boldsymbol{y}) \\ \boldsymbol{0} = \boldsymbol{g}(\boldsymbol{x}_{\mathrm{d}1}, \boldsymbol{y}_{\mathrm{d}1}) \\ \cdots \\ \boldsymbol{0} = \boldsymbol{g}(\boldsymbol{x}_{\mathrm{d}m}, \boldsymbol{y}_{\mathrm{d}m}) \end{cases} . \tag{2.22}$$

At the equilibrium point, linearizing (2.22) leads to

$$\begin{cases} \Delta\dot{\boldsymbol{x}} = \boldsymbol{A}_0\Delta\boldsymbol{x} + \boldsymbol{B}_0\Delta\boldsymbol{y} + \boldsymbol{A}_1\Delta\boldsymbol{x}_{\mathrm{d}1} + \boldsymbol{B}_1\Delta\boldsymbol{y}_{\mathrm{d}1} + \cdots + \boldsymbol{A}_m\Delta\boldsymbol{x}_{\mathrm{d}m} + \boldsymbol{B}_m\Delta\boldsymbol{y}_{\mathrm{d}m} \\ \boldsymbol{0} = \boldsymbol{C}_0\Delta\boldsymbol{x} + \boldsymbol{D}_0\Delta\boldsymbol{y} \\ \boldsymbol{0} = \boldsymbol{C}_0\Delta\boldsymbol{x}_{\mathrm{d}i} + \boldsymbol{D}_0\Delta\boldsymbol{y}_{\mathrm{d}i}, \quad i = 1, 2, \ldots, m \end{cases} . \tag{2.23}$$

Since $\boldsymbol{D}_i$ $(i = 0, 1, \ldots, m)$ is non-singular, (2.23) can be readily transformed into DDEs as given by (2.19) by eliminating algebraic variables $\Delta\boldsymbol{y}$ and $\Delta\boldsymbol{y}_{\mathrm{d}i}$ $(i = 1, 2, \ldots, m)$.

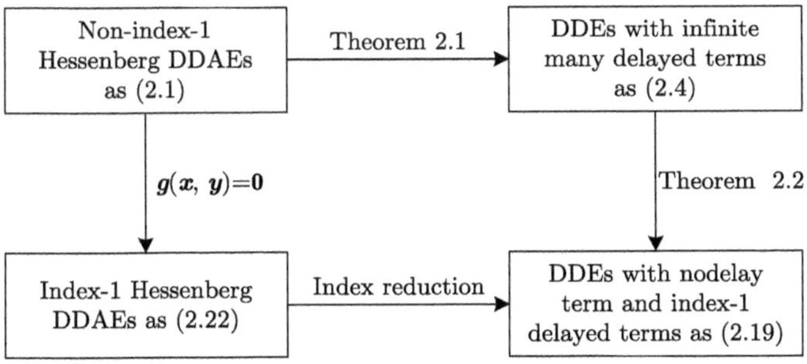

Fig. 2.1 Two ways of transforming non-index-1 Hessenberg DDAEs to DDEs

2.2 Eigen-Analysis

This section focuses on eigenvalue-based small-signal stability analysis of time-delay systems, including the characteristic equation and eigenvalue distribution of the system as well as eigenvalue sensitivities with respect to time delays and parameters.

2.2.1 Characteristic Equation

The small-signal model of time-delay system, i.e., (2.23), can be rewritten as

$$\begin{cases} \boldsymbol{E}\Delta\dot{\hat{\boldsymbol{x}}}(t) = \boldsymbol{J}_0\Delta\hat{\boldsymbol{x}}(t) + \sum\limits_{i=1}^{m} \boldsymbol{J}_i\Delta\hat{\boldsymbol{x}}(t-\tau_i), & t \geq 0 \\ \Delta\hat{\boldsymbol{x}}(t) = \hat{\varphi}(t), & t \in [-\tau_{\max},\ 0] \end{cases} \tag{2.24}$$

where $\Delta\hat{\boldsymbol{x}}(t) = [\Delta\boldsymbol{x}^{\mathrm{T}}(t),\ \Delta\boldsymbol{y}^{\mathrm{T}}(t)]^{\mathrm{T}} \in \mathbb{R}^{d\times 1}$ denotes the augmented system state variable. $\hat{\varphi} = [\varphi_x^{\mathrm{T}},\ \varphi_y^{\mathrm{T}}]^{\mathrm{T}} \in \mathbb{R}^{d\times 1}$ represents the augmented initial condition of the system, where $\varphi_x \in \mathbb{R}^{n\times 1}$ and $\varphi_y \in \mathbb{R}^{l\times 1}$ correspond to $\Delta\boldsymbol{x}$ and $\Delta\boldsymbol{y}$, respectively. $\tau_i > 0\,(i = 1,\ 2,\ \ldots,\ m)$ are time delays, and the maximum one is denoted as $\tau_{\max}$. $\boldsymbol{E} = \mathrm{diag}\{\boldsymbol{I}_n,\ \boldsymbol{0}_l\} \in \mathbb{R}^{d\times d}$. The augmented system state matrices $\boldsymbol{J}_i \in \mathbb{R}^{d\times d}$ can be described as

$$\boldsymbol{J}_0 = \begin{bmatrix} \boldsymbol{A}_0 & \boldsymbol{B}_0 \\ \boldsymbol{C}_0 & \boldsymbol{D}_0 \end{bmatrix},\ \boldsymbol{J}_i = \begin{bmatrix} \boldsymbol{A}_i & \boldsymbol{B}_i \\ \boldsymbol{0}_{l\times n} & \boldsymbol{0}_l \end{bmatrix},\ i = 1,\ 2,\ \ldots,\ m$$

with $\boldsymbol{A}_i \in \mathbb{R}^{n\times n}$, $\boldsymbol{B}_i \in \mathbb{R}^{n\times l}$, $\boldsymbol{C}_0 \in \mathbb{R}^{l\times n}$ and $\boldsymbol{D}_0 \in \mathbb{R}^{l\times l}$.

The characteristic equation corresponding to (2.24) is given by

$$\left(\boldsymbol{J}_0 + \sum_{i=1}^{m} \boldsymbol{J}_i \mathrm{e}^{-\lambda\tau_i}\right)\hat{\boldsymbol{v}} = \lambda\boldsymbol{E}\hat{\boldsymbol{v}} \tag{2.25}$$

where $\lambda \in \mathbb{C}$ and $\hat{\boldsymbol{v}} \in \mathbb{C}^{d\times 1}$ are eigenvalue and the augmented right eigenvector, respectively. Equation (2.25) can be further rewritten as

$$\begin{bmatrix} \boldsymbol{A}'(\lambda) & \boldsymbol{B}'(\lambda) \\ \boldsymbol{C}_0 & \boldsymbol{D}_0 \end{bmatrix} \begin{bmatrix} \boldsymbol{v} \\ \bar{\boldsymbol{v}} \end{bmatrix} = \boldsymbol{0} \tag{2.26}$$

where $\boldsymbol{v} \in \mathbb{C}^{n\times 1}$ is the right eigenvector, and $\bar{\boldsymbol{v}} = -\boldsymbol{D}_0^{-1}\boldsymbol{C}_0\boldsymbol{v} \in \mathbb{C}^{l\times 1}$ is related to $\Delta\boldsymbol{y}$. Note that $\hat{\boldsymbol{v}} = [\boldsymbol{v}^{\mathrm{T}},\ \bar{\boldsymbol{v}}^{\mathrm{T}}]^{\mathrm{T}}$. $\boldsymbol{A}'(\lambda) \in \mathbb{C}^{n\times n}$ and $\boldsymbol{B}'(\lambda) \in \mathbb{C}^{n\times l}$ can be respectively expressed as

$$A'(\lambda) = A_0 - \lambda I_n + \sum_{i=1}^{m} A_i e^{-\lambda\tau_i}, \quad B'(\lambda) = B_0 + \sum_{i=1}^{m} B_i e^{-\lambda\tau_i}.$$

Since $\boldsymbol{D}_0$ is non-singular, one can obtain

$$\begin{cases} \Delta y = -D_0^{-1} C_0 \Delta x \\ \Delta y(t-\tau_i) = -D_0^{-1} C_0 \Delta x(t-\tau_i), \quad i = 1, 2, \ldots, m \end{cases}. \tag{2.27}$$

according to the last item in (2.23). Then, by eliminating $\Delta \boldsymbol{y}$ and $\Delta \boldsymbol{y}(t-\tau_i)$ ($i = 1, 2, \ldots, m$), (2.24) can be reformulated into the following DDEs (also as (2.19)):

$$\begin{cases} \Delta \dot{x}(t) = \tilde{A}_0 \Delta x(t) + \sum_{i=1}^{m} \tilde{A}_i \Delta x(t-\tau_i), \ t \geq 0 \\ \Delta x(t) = \varphi(t), \qquad t \in [-\tau_{\max}, 0] \end{cases} \tag{2.28}$$

where φ is the alias of φ_x, and the system state matrix $\tilde{\boldsymbol{A}}_0 \in \mathbb{R}^{n\times n}$ and the delayed counterparts $\tilde{\boldsymbol{A}}_i \in \mathbb{R}^{n\times n}$ ($i = 1, 2, \ldots, m$) are computed as $\tilde{\boldsymbol{A}}_i = \boldsymbol{A}_i - \boldsymbol{B}_i \boldsymbol{D}_0^{-1} \boldsymbol{C}_0$.

Note that matrices $\boldsymbol{J}_i, \boldsymbol{A}_i, \boldsymbol{B}_i$ ($i = 0, 1, \ldots, m$), $\boldsymbol{C}_0$ and $\boldsymbol{D}_0$ as well as $\tilde{\boldsymbol{A}}_i$ ($i = 1, 2, \ldots, m$) are generally acknowledged to be highly sparse, while $\tilde{\boldsymbol{A}}_0$ is dense. This is widely appreciated in analyzing dynamic systems, such as power systems, mechanical systems and so on.

The characteristic equation of (2.28) can be formulated as

$$\left(\tilde{A}_0 + \sum_{i=1}^{m} \tilde{A}_i e^{-\lambda\tau_i}\right) v = \lambda v \tag{2.29}$$

or

$$u^{\mathrm{T}} \left(\tilde{A}_0 + \sum_{i=1}^{m} \tilde{A}_i e^{-\lambda\tau_i}\right) = \lambda u^{\mathrm{T}} \tag{2.30}$$

where $\boldsymbol{u} \in \mathbb{C}^{n\times 1}$ is the left eigenvector associated with λ.

To sum up, due to presence of time delays, the classical problem of finding roots of the state matrix of the system without delays at the equilibrium point has been transformed into the solution of a transcendental characteristic equation.

2.2.2 Eigenvalue Distribution

The necessary and sufficient condition for the asymptotic stability of time-delay system is that all its eigenvalues strictly lie on the left half s-plane. Compared with systems without time delay, the time-delay system has infinitely many eigenvalues due to the exponential terms in (2.25) and (2.29) [9–11]. To explain the "infinity", this subsection introduces the asymptotic distribution of eigenvalues of time-delay system described in [12].

The characteristic quasi-polynomial corresponding to (2.29) is

$$h(s) = \det\left(s\boldsymbol{I}_n - \tilde{\boldsymbol{A}}_0 - \sum_{i=1}^{m} \tilde{\boldsymbol{A}}_i \mathrm{e}^{-\lambda\tau_i} \right) = \sum_{j=0}^{N} p_j(s)\mathrm{e}^{-\beta_j s}$$

where $\det(\cdot)$ returns the determinant of a square matrix, and $\beta_0 > \beta_1 > \cdots > \beta_{N-1} > \beta_N = 0$ with $N \le 2^m$. $p_j(s)$ $(j = 0,\ 1,\ \ldots,\ N)$ are polynomials of degree m_j $(\le n)$ and can be expressed as

$$p_j(s) = \sum_{k=0}^{m_j} p_{j,k}s^k = p_{j,0} + p_{j,1}s + \cdots + p_{j,m_j}s^{m_j} = p_{j,m_j}s^{m_j}(1 + \varepsilon_j(s))$$

where $p_{j,k}$ $(k = 0,\ 1,\ \ldots,\ m_j)$ are coefficients, and the functions

$$\varepsilon_j(s) = \frac{\sum_{k=0}^{m_j-1} p_{j,k}s^k}{p_{j,m_j}s^{m_j}}$$

have the property $\lim_{|s|\to\infty}|\varepsilon_j(s)| = 0$.

Since $\mathrm{e}^{\beta_0 s}$ has no zeros on the s-plane, then

$$g(s) \triangleq h(s)\mathrm{e}^{\beta_0 s} = \sum_{j=0}^{N} p_j(s)\mathrm{e}^{\theta_j s} = \sum_{j=0}^{N} p_{j,m_j}s^{m_j}(1 + \varepsilon_j(s))\mathrm{e}^{\theta_j s} \tag{2.31}$$

has the same zeros as $h(s)$, where $\theta_j = \beta_0 - \beta_j$ $(j = 0,\ 1,\ \ldots,\ N)$ satisfy $0 = \theta_0 < \theta_1 < \cdots < \theta_{N-1} < \theta_N = \beta_0$.

Based on (2.31), the set $\boldsymbol{P} = \{P_j|(\theta_j,\ m_j),\ j = 0,\ 1,\ \ldots,\ N\}$ is established. Then, one can draw a polyline L connecting P_0 to P_N. It is endowed with the following features [12, 13]: has vertices only at points of the set $\boldsymbol{P}$, is convex upward in the sense that no points lie above it. Taking the quasi-polynomial

$$\begin{aligned} h(s) =& 51.7\mathrm{e}^{-24.99s} + (0.03s^3 - 0.1s + 1.5)\mathrm{e}^{-23.35s} + 0.5s^3\mathrm{e}^{-19.9s} \\ &+ (0.15s^5 + 0.2s^4)\mathrm{e}^{-18.52s} + 0.8s^6\mathrm{e}^{-13.52s} + (-8.7s^4 + 19.3s)\mathrm{e}^{-10.33s} \\ &+ (s^7 - 1.1s^5 + 6.7)\mathrm{e}^{-8.52s} + 29.1\mathrm{e}^{-4.61s} + (0.2s^8 - 1.8s) \end{aligned} \tag{2.32}$$

as an example, its polyline L is depicted in Fig. 2.2.

Then, let the successive segments of L be denoted by $L_1,\ L_2,\ \ldots,\ L_r$, numbered from left to right, where $r \le N - 1$. For each segment L_r $(r = 1,\ 2,\ \cdots)$, the zeros of $g(s)$ are asymptotic to those of the comparison function

$$g_r(s) = \sum_{Nr} p_{j,m_j} s^{m_j} \mathrm{e}^{\theta_j s} \tag{2.33}$$

where the sum is taken over all terms corresponding to the N_r points on the same L_r. Generally, $N_r = 2$.

Next, multiplying $g_r(s)$ by $s^{-m_{r,\min}} \mathrm{e}^{-\theta_{r,\min} s}$ yields

$$\bar{g}_r(s) = g_r(s) s^{-m_{r,\min}} \mathrm{e}^{-\theta_{r,\min} s} = \sum_{Nr} p_{j,m_j} s^{\bar{m}_j} \mathrm{e}^{\bar{\theta}_j s}$$

where $m_{r,\min} = \min_{N_r}\{m_j\}$, $\theta_{r,\min} = \min_{N_r}\{\theta_j\}$, $\bar{m}_j = m_j - m_{r,\min}$ and $\bar{\theta}_j = \theta_j - \theta_{r,\min}$.

Considering the slope $\gamma_r = \dfrac{\bar{m}_j}{\bar{\theta}_j}$ of L_r and $w = s + \gamma_r \ln s$, one can obtain

$$\bar{g}_r(s) = \sum_{Nr} p_{j,m_j} \mathrm{e}^{\bar{\theta}_j s + \bar{m}_j \ln s} = \sum_{Nr} p_{j,m_j} \mathrm{e}^{\bar{\theta}_j w}. \tag{2.34}$$

Let $z = \mathrm{e}^{w/\gamma_r}$, then (2.34) can be further rewritten as

$$f_r(z) = \sum_{Nr} p_{j,m_j} z^{\bar{m}_j}. \tag{2.35}$$

Obviously, the quasi-polynomial $g_r(s)$ in s as shown in (2.33) has been equivalently transformed into the general polynomial $f_r(z)$ in z as shown in (2.35). Furthermore, each zero of $f_r(z)$ corresponds a chain of zeros of $g_r(s)$ determined by

$$w = \gamma_r \mathrm{Ln} z = \gamma_r \ln|z| + \mathrm{j}\gamma_r(\arg(z) + 2k\pi), \ k = 0, \ \pm 1, \ \pm 2, \ \cdots. \tag{2.36}$$

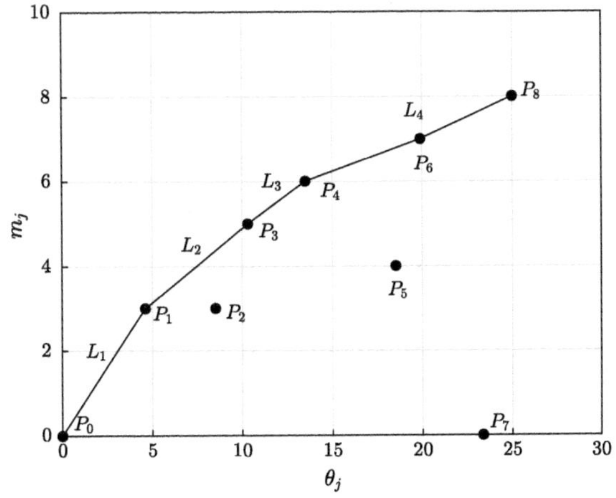

Fig. 2.2 The polyline L corresponding to (2.32)

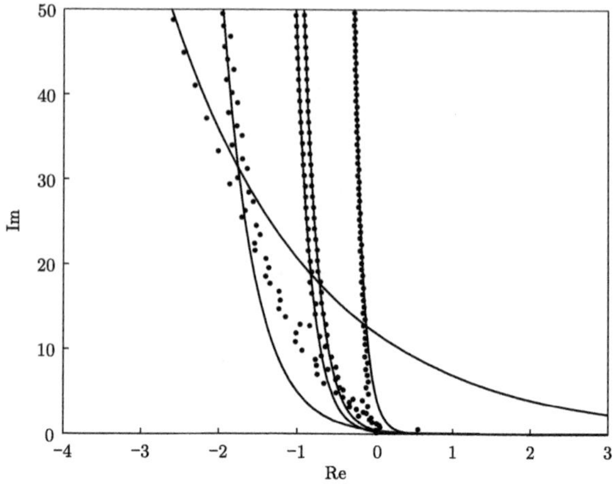

Fig. 2.3 The zeros and asymptotic exponential curves of (2.32)

Considering $w = s + \gamma_r \ln s$ and $s = \alpha + \mathrm{j}\omega$, one has

$$\mathrm{Re}(w) = \mathrm{Re}(s + \gamma_r \ln s) = \alpha + \gamma_r \ln|s|. \tag{2.37}$$

Substituting (2.36) into (2.37) and taking $\lim_{|s|\to\infty}|s| = |\omega|$ into account yield [12]

$$\alpha + \gamma_r \ln|\omega| = \gamma_r \ln|z|$$

whose solution is an exponential function on the s-plane, i.e.,

$$\omega = \pm \mathrm{e}^{-\frac{\alpha}{\gamma_r}+\ln|z|}.$$

That is to say, for each segment L_r $(r = 1,\ 2,\ \cdots)$ with the slope γ_r, zeros of $h(s)$, i.e., eigenvalues λ in (2.29), of large moduli lie asymptotically along a finite number (equal to the number of absolute values of zeros of $f_r(z)$) of exponential curves. Taking the quasi-polynomial $h(s)$ shown in (2.32) as an example, its zeros and asymptotic exponential curves are depicted in Fig. 2.3.

It should be noted that the asymptotic features are not apparent in the region near the origin, due to the prerequisite $|\varepsilon_j(s)| = 0$ $(j = 0,\ 1,\ \ldots,\ N)$.

2.2.3 *Eigenvalue Sensitivity*

To quantify the effect of time delays and system parameters on the system's small-signal stability, this subsection gives sensitivities of eigenvalues with respect to time delays and system parameters.

2.2.3.1 Sensitivity with Respect to Time Delays

Taking derivative of (2.29) with respect to τ_j $(j = 1, 2, \ldots, m)$, one can obtain

$$\frac{\partial\left(\left(\tilde{A}_0 + \sum_{i=1}^{m} \tilde{A}_i e^{-\lambda\tau_i}\right) v\right)}{\partial\tau_j} = \frac{\partial(\lambda v)}{\partial\tau_j}. \tag{2.38}$$

Since both λ and v are related to τ_j, the left and right-hand sides of (2.38) can be respectively expanded as

$$\begin{aligned}
&\left(\sum_{i=1}^{m} \tilde{A}_i \frac{\partial e^{-\lambda\tau_i}}{\partial\tau_j}\right) v + \left(\tilde{A}_0 + \sum_{i=1}^{m} \tilde{A}_i e^{-\lambda\tau_i}\right) \frac{\partial v}{\partial\tau_j} \\
&= \left(-\tilde{A}_j \lambda e^{-\lambda\tau_j} - \sum_{i=1}^{m} \tilde{A}_i \tau_i e^{-\lambda\tau_i} \frac{\partial\lambda}{\partial\tau_j}\right) v + \left(\tilde{A}_0 + \sum_{i=1}^{m} \tilde{A}_i e^{-\lambda\tau_i}\right) \frac{\partial v}{\partial\tau_j}
\end{aligned}$$

and

$$\frac{\partial\lambda}{\partial\tau_j} v + \lambda \frac{\partial v}{\partial\tau_j}.$$

Then, by pre-multiplying them by u^{T} and taking (2.30) into account, (2.38) becomes

$$u^{\mathrm{T}} \left(-\tilde{A}_j \lambda e^{-\lambda\tau_j} - \sum_{i=1}^{m} \tilde{A}_i \tau_i e^{-\lambda\tau_i} \frac{\partial\lambda}{\partial\tau_j}\right) v = \frac{\partial\lambda}{\partial\tau_j} u^{\mathrm{T}} v.$$

The sensitivity of λ with respect to τ_j $(j = 1, 2, \ldots, m)$ thus can be derived as

$$\frac{\partial\lambda}{\partial\tau_j} = -\frac{\lambda e^{-\lambda\tau_j} u^{\mathrm{T}} \tilde{A}_j v}{u^{\mathrm{T}} \left(I_n + \sum_{i=1}^{m} \tilde{A}_i \tau_i e^{-\lambda\tau_i}\right) v}.$$

2.2.3.2 Sensitivity with Respect to System Parameters

Taking derivative of (2.29) with respect to a certain system parameter p yields

$$\begin{aligned}
&\left(\frac{\partial\tilde{A}_0}{\partial p} + \sum_{i=1}^{m} \frac{\partial\tilde{A}_i}{\partial p} e^{-\lambda\tau_i} - \sum_{i=1}^{m} \tilde{A}_i \tau_i e^{-\lambda\tau_i} \frac{\partial\lambda}{\partial p}\right) v + \left(\tilde{A}_0 + \sum_{i=1}^{m} \tilde{A}_i e^{-\lambda\tau_i}\right) \frac{\partial v}{\partial p} \\
&= \frac{\partial\lambda}{\partial p} v + \lambda \frac{\partial v}{\partial p}.
\end{aligned} \tag{2.39}$$

By pre-multiplying (2.39) by u^{T} and taking (2.30) into account, one can obtain

$$\boldsymbol{u}^{\mathrm{T}}\left(\frac{\partial \tilde{\boldsymbol{A}}_0}{\partial p}+\sum_{i=1}^{m}\frac{\partial \tilde{\boldsymbol{A}}_i}{\partial p}\mathrm{e}^{-\lambda\tau_i}-\sum_{i=1}^{m}\tilde{\boldsymbol{A}}_i\tau_i\mathrm{e}^{-\lambda\tau_i}\frac{\partial\lambda}{\partial p}\right)\boldsymbol{v}=\frac{\partial\lambda}{\partial p}\boldsymbol{u}^{\mathrm{T}}\boldsymbol{v}.$$

The sensitivity of λ with respect to the parameter p is solved and given by

$$\frac{\partial\lambda}{\partial p}=\frac{\boldsymbol{u}^{\mathrm{T}}\left(\frac{\partial \tilde{\boldsymbol{A}}_0}{\partial p}+\sum\limits_{i=1}^{m}\frac{\partial \tilde{\boldsymbol{A}}_i}{\partial p}\mathrm{e}^{-\lambda\tau_i}\right)\boldsymbol{v}}{\boldsymbol{u}^{\mathrm{T}}\left(\boldsymbol{I}_n+\sum\limits_{i=1}^{m}\tilde{\boldsymbol{A}}_i\tau_i\mathrm{e}^{-\lambda\tau_i}\right)\boldsymbol{v}}.$$

2.2.3.3 Perturbation on Time Delays

Let the perturbation of the time delay τ_i $(i=1,\ 2,\ \ldots,\ m)$ be $\Delta\tau_i$. The perturbed time delay can be denoted as

$$\tau_i'=\tau_i+\varepsilon\Delta\tau_i \tag{2.40}$$

where the sufficiently small variable ε indicates the order of magnitude of the perturbation. Accordingly, the system's characteristic equation becomes

$$\left(\tilde{\boldsymbol{A}}_0+\sum_{i=1}^{m}\tilde{\boldsymbol{A}}_i\mathrm{e}^{-\lambda'\tau_i'}\right)\boldsymbol{v}'=\lambda'\boldsymbol{v}' \tag{2.41}$$

with

$$\begin{cases}\lambda'=\lambda+\varepsilon\lambda_1+\varepsilon^2\lambda_2+\cdots\\ \boldsymbol{v}'=\boldsymbol{v}+\varepsilon\boldsymbol{v}_1+\varepsilon^2\boldsymbol{v}_2+\cdots\end{cases} \tag{2.42}$$

where $\varepsilon^i\lambda_i$ and $\varepsilon^i\boldsymbol{v}_i$ are the i-th order perturbation of λ_i and $\boldsymbol{v}_i$ $(i=1,\ 2,\ \cdots)$, respectively.

The first order perturbation $\varepsilon\lambda_1$ can be derived as follows. First, (2.40) and (2.42) are substituted into (2.41). Second, the exponential terms $\mathrm{e}^{-\varepsilon(\lambda_1\tau_i+\lambda\Delta\tau_i)}$ $(i=1,\ 2,\ \ldots,\ m)$ can be expanded into the Taylor series. Then, one can obtain the first order perturbation of the characteristic equation

$$\left(\tilde{\boldsymbol{A}}_0+\sum_{i=1}^{m}\tilde{\boldsymbol{A}}_i\mathrm{e}^{-\lambda\tau_i}\right)\boldsymbol{v}_1-\sum_{i=1}^{m}\tilde{\boldsymbol{A}}_i\mathrm{e}^{-\lambda\tau_i}(\lambda_1\tau_i+\lambda\Delta\tau_i)\boldsymbol{v}=\lambda_1\boldsymbol{v}+\lambda\boldsymbol{v}_1. \tag{2.43}$$

Next, pre-multiplying (2.43) by $\boldsymbol{u}^{\mathrm{T}}$ yields

$$\boldsymbol{u}^{\mathrm{T}}\left(\tilde{\boldsymbol{A}}_0+\sum_{i=1}^{m}\tilde{\boldsymbol{A}}_i\mathrm{e}^{-\lambda\tau_i}\right)\boldsymbol{v}_1-\boldsymbol{u}^{\mathrm{T}}\sum_{i=1}^{m}\tilde{\boldsymbol{A}}_i\mathrm{e}^{-\lambda\tau_i}(\lambda_1\tau_i+\lambda\Delta\tau_i)\boldsymbol{v}=\lambda_1\boldsymbol{u}^{\mathrm{T}}\boldsymbol{v}+\lambda\boldsymbol{u}^{\mathrm{T}}\boldsymbol{v}_1.$$

Last, considering (2.30), $\varepsilon\lambda_1$ can be computed as

$$\varepsilon\lambda_1 = -\frac{\boldsymbol{u}^{\mathrm{T}}\left(\sum_{i=1}^{m}\tilde{\boldsymbol{A}}_i\mathrm{e}^{-\lambda\tau_i}\lambda\varepsilon\Delta\tau_i\right)\boldsymbol{v}}{\boldsymbol{u}^{\mathrm{T}}\left(\boldsymbol{I}_n+\sum_{i=1}^{m}\tilde{\boldsymbol{A}}_i\mathrm{e}^{-\lambda\tau_i}\tau_i\right)\boldsymbol{v}}.$$

2.2.3.4 Perturbation on System Parameters

Suppose that the perturbations of system state matrices $\tilde{\boldsymbol{A}}_i$ $(i = 0,\ 1,\ \ldots,\ m)$ led by system parameters are $\varepsilon\Delta\tilde{\boldsymbol{A}}_i$. The perturbed system state matrices can be denoted as

$$\tilde{\boldsymbol{A}}_i' = \tilde{\boldsymbol{A}}_i + \varepsilon\Delta\tilde{\boldsymbol{A}}_i. \tag{2.44}$$

Accordingly, the characteristic equation of the system changes to

$$\left(\tilde{\boldsymbol{A}}_0' + \sum_{i=1}^{m}\tilde{\boldsymbol{A}}_i'\mathrm{e}^{-\lambda'\tau_i}\right)\boldsymbol{v}' = \lambda'\boldsymbol{v}' \tag{2.45}$$

where the expressions of λ' and $\boldsymbol{v}'$ are identical to those in (2.42).

The first order perturbation $\varepsilon\lambda_1$ can be derived as follows. First, (2.44) and (2.42) are substituted into (2.45). Second, the exponential terms $\mathrm{e}^{-\varepsilon\lambda_1\tau_i}$ $(i = 1,\ 2,\ \ldots,\ m)$ can be expanded into the Taylor series. Then, one can obtain the first order perturbation of the characteristic equation

$$\begin{aligned}&\left(\tilde{\boldsymbol{A}}_0 + \sum_{i=1}^{m}\tilde{\boldsymbol{A}}_i\mathrm{e}^{-\lambda\tau_i}\right)\boldsymbol{v}_1 - \sum_{i=1}^{m}\tilde{\boldsymbol{A}}_i\mathrm{e}^{-\lambda\tau_i}\lambda_1\tau_i\boldsymbol{v} + \left(\Delta\tilde{\boldsymbol{A}}_0 + \sum_{i=1}^{m}\Delta\tilde{\boldsymbol{A}}_i\mathrm{e}^{-\lambda\tau_i}\right)\boldsymbol{v}\\ &\quad = \lambda_1\boldsymbol{v} + \lambda\boldsymbol{v}_1.\end{aligned} \tag{2.46}$$

By pre-multiplying (2.46) with $\boldsymbol{u}^{\mathrm{T}}$ and taking (2.30) into account, $\varepsilon\lambda_1$ can be solved as

$$\varepsilon\lambda_1 = \frac{\boldsymbol{u}^{\mathrm{T}}\left(\varepsilon\Delta\tilde{\boldsymbol{A}}_0 + \sum_{i=1}^{m}\varepsilon\Delta\tilde{\boldsymbol{A}}_i\mathrm{e}^{-\lambda\tau_i}\right)\boldsymbol{v}}{\boldsymbol{u}^{\mathrm{T}}\left(\boldsymbol{I}_n+\sum_{i=1}^{m}\tilde{\boldsymbol{A}}_i\mathrm{e}^{-\lambda\tau_i}\tau_i\right)\boldsymbol{v}}.$$

References

1. Milano, F., & Dassios, I. (2016). Small-signal stability analysis for non-index 1 Hessenberg form systems of delay differential-algebraic equations. *IEEE Transactions on Circuits and Systems-I: Regular Papers, 63*(9), 1521–1530.
2. Hairer, E., & Wanner, G. (1996). *Solving ordinary differential equations II. Stiff and differential-algebraic problems* (2nd ed.). Heidelberg: Springer.
3. Ye, H., Liu, K., Mou, Q., et al. (2019). Modeling and formulation of delayed cyber-physical power system for small signal stability analysis and control. *IEEE Transactions on Power Systems, 34*(3), 2419–2432.
4. Milano, F., & Anghel, M. (2012). Impact of time delays on power system stability. *IEEE Transactions on Circuit and Systems-I: Regular Papers, 59*(4), 889–900.
5. Milano, F. (2016). Small-signal stability analysis of large power systems with inclusion of multiple delays. *IEEE Transactions on Power Systems, 31*(4), 3257–3266.
6. Jia, H., Yu, X., Yu, Y., et al. (2008). Power system small signal stability region with time delay. *International Journal of Electrical Power and Energy Systems, 30*(1), 16–22.
7. Yu, X., Jia, H., & Wang, C. (2013). CTDAE & CTODE models and their applications to power system stability analysis with time delays. *Science China Technological Sciences, 56*(5), 1213–1223.
8. Ascher, U. M., & Petzold, L. R. (1995). The numerical solution of delay-differential-algebraic equations of retarded and neutral type. *SIAM Journal on Numerical Analysis, 32*(5), 1635–1657.
9. Hale, J. K., & Lunel, S. M. V. (1991). *Introduction to functional differential equations*. New York: Springer.
10. Breda, D. (2004). *Numerical computation of characteristic roots for delay differential equations*. Veneto, Italy: Universitá degli Studi di Padova.
11. Michiels, W., & Niculescu, S.-I. (2014). *Stability, control, and computation for time-delay systems: An eigenvalue-based approach* (2nd ed.). Philadelphia: SIAM.
12. Vyhlidal, T., & Zitek, P. (2009). Mapping based algorithm for large-scale computation of quasi-polynomial zeros. *IEEE Transactions on Automatic Control, 54*(1), 171–177.
13. Bellman, R., & Cooke, K. L. (1963). *Differential-difference equation*. New York: Academic Press.

Chapter 3
PSD-Based Eigenvalue Computation Framework for Large-Scale Time-Delay Systems

This chapter proposes the framework of PSD-based eigenvalue computation theory for time-delay systems, which consists of mapping, discretization, transform, estimation, and correction. The basic idea is that, the spectra of a time-delay system can be mapped to those of the infinite-dimensional semigroup operators; then eigenvalues of the finite-dimensional discretized matrices approximating semigroup operators are in turn used as the estimated eigenvalues of the time-delay system.

3.1 Semigroup Operators

Two types of semigroup operators, i.e., the integral operator and the differential operator [1], will be discussed in this book. They are also known as the solution operator and the infinitesimal generator, respectively.

3.1.1 Solution Operator

3.1.1.1 Retarded Functional Differential Equation (RFDE)

Due to the presence of time delays τ_i $(i = 1, 2, \ldots, m)$, the system as expressed in (2.28) exhibits the "memorability". In other words, the solution of the system, i.e., $\Delta \boldsymbol{x}(t)$ $(t \geq t_0)$, depends on the segment $\Delta \boldsymbol{x}(t_0 + \theta)$ with $\theta \in [-\tau_{\max}, 0]$, where t_0 is an arbitrary initial moment and usually taken as 0.

Let $\boldsymbol{X} := \mathrm{C}([-\tau_{\max}, 0], \mathbb{R}^{n \times 1})$ be the Banach space of continuous functions mapping the time interval $[-\tau_{\max}, 0]$ into an n-dimensional space $\mathbb{R}^{n \times 1}$ and endowed with the maximum norm $\sup_{\theta \in [-\tau_{\max}, 0]} |\varphi(\theta)|$. Then, the function segment [2] is defined as

H. Ye et al., *Spectral Discretization-Based Eigen-Analysis of Time-Delay Systems*,
https://doi.org/10.1007/978-981-96-0738-9_3

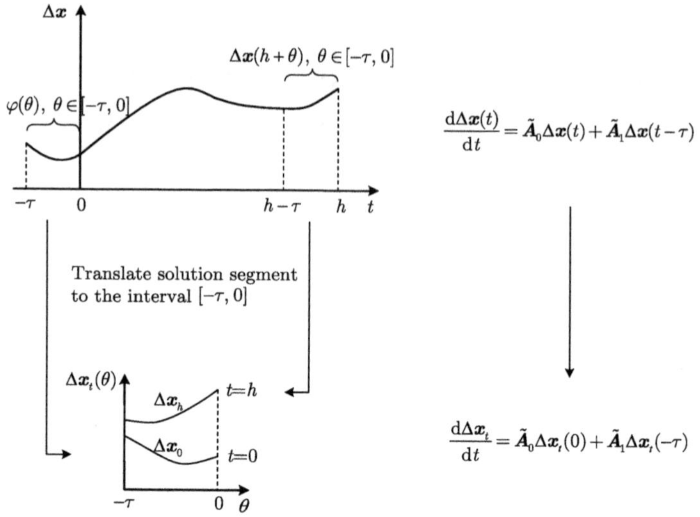

Fig. 3.1 Principle behind the transformation of (2.28) to (3.1) with a single delay

$$\Delta \boldsymbol{x}_t(\theta) \triangleq \Delta \boldsymbol{x}(t+\theta), \quad t \geq 0; \ \theta \in [-\tau_{\max}, \ 0].$$

By omitting the argument θ, it can be abbreviated as $\Delta \boldsymbol{x}_t$. Then, with the help of $\Delta \boldsymbol{x}_t(\theta)$, one can analyze $\Delta \boldsymbol{x}(t+\theta)$ with $\theta \in [-\tau_{\max}, \ 0]$ at different moments t on the same $\boldsymbol{X}$.

Next, DDEs (2.28) can be rewritten as the following RFDEs with discrete delays [3]:

$$\Delta \dot{\boldsymbol{x}}_t(0) = \tilde{\boldsymbol{A}}_0 \Delta \boldsymbol{x}_t(0) + \sum_{i=1}^{m} \tilde{\boldsymbol{A}}_i \Delta \boldsymbol{x}_t(-\tau_i). \tag{3.1}$$

To facilitate the understanding, Fig. 3.1 illustrates the principle behind the transformation of (2.28) to (3.1) with a single delay ($m = 1$ and $\tau = \tau_1$) [3]. In the figure, two function segments, namely $\Delta \boldsymbol{x}_0(\theta) = \varphi(\theta)$ and $\Delta \boldsymbol{x}_h(\theta) = \Delta \boldsymbol{x}(\theta + h)$ with $\theta \in [-\tau, \ 0]$, are depicted, where $h \geq 0$.

3.1.1.2 Definition

The solution operator $\mathcal{T}(h)$: $\boldsymbol{X} \rightarrow \boldsymbol{X}$ is the linear operator transforming the initial condition $\varphi(\theta)$ with $\theta \in [-\tau_{\max}, \ 0]$ to the function segment $\Delta \boldsymbol{x}_h(\theta)$ after a time step-length h, which can be expressed as

$$(\mathcal{T}(h)\varphi)(\theta) = \Delta \boldsymbol{x}_h(\theta), \quad h \geq 0; \; \theta \in [-\tau_{\max}, \; 0] \tag{3.2}$$

or in the reduced form:

$$\mathcal{T}(h)\varphi = \Delta \boldsymbol{x}_h, \quad h \geq 0. \tag{3.3}$$

The set of $\{\mathcal{T}(h)\}_{h\geq 0}$ is a strongly continuous semigroup [1, 2, 4] as the solution operator fulfills that

(1) $\mathcal{T}(0) = \boldsymbol{I}$, identity operator on $\boldsymbol{X}$,
(2) $\mathcal{T}(h + s) = \mathcal{T}(h)\mathcal{T}(s)$ for any $h, \; s \geq 0$,
(3) for any $\varphi \in \boldsymbol{X}, \; \lim_{h\to 0^+} \|\mathcal{T}(h)\varphi - \varphi\| = \boldsymbol{0}$.

3.1.1.3 Explicit Expression

The explicit expression of the solution operator $\mathcal{T}(h)$ is a combination of a shift and the solution of ordinary differential equations (ODEs).

(1) Shift: If $\theta \in [-\tau_{\max}, \; -h]$, $\Delta \boldsymbol{x}_h(\theta)$ is always the initial condition of the system since $\theta + h \leq 0$, namely

$$(\mathcal{T}(h)\varphi)(\theta) = \Delta \boldsymbol{x}_h(\theta) = \varphi(\theta + h). \tag{3.4}$$

(2) Time Integration: If $\theta \in (-h, \; 0]$, i.e., $\theta + h \in (0, \; h]$, $\Delta \boldsymbol{x}_h(\theta)$ is formulated as

$$\Delta \boldsymbol{x}(\theta + h) = \varphi(0) + \int_0^{\theta+h} \left(\tilde{\boldsymbol{A}}_0 \Delta \boldsymbol{x}(s) + \sum_{i=1}^{m} \tilde{\boldsymbol{A}}_i \Delta \boldsymbol{x}(s - \tau_i) \right) \mathrm{d}s, \;\; \theta \in [-h, \; 0] \tag{3.5}$$

which follows from the Picard's Existence and Uniqueness Theorem [5].

By combining the above two segments, one can obtain the explicit expression of the solution operator $\mathcal{T}(h)$ shown in (3.2) and (3.3) in terms of a piecewise function, i.e.,

$$\Delta \boldsymbol{x}_h(\theta) = (\mathcal{T}(h)\varphi)(\theta)$$
$$= \begin{cases} \varphi(0) + \displaystyle\int_{-h}^{\theta} \left(\tilde{\boldsymbol{A}}_0 \Delta \boldsymbol{x}_h(s) + \sum_{i=1}^{m} \tilde{\boldsymbol{A}}_i \Delta \boldsymbol{x}_h(s - \tau_i) \right) \mathrm{d}s, & \theta \in [-h, \; 0] \\ \varphi(\theta + h), & \theta \in [-\tau_{\max}, \; -h] \end{cases} . \tag{3.6}$$

3.1.2 Infinitesimal Generator

3.1.2.1 Definition

The infinitesimal generator $\mathcal{A} : \boldsymbol{X} \to \boldsymbol{X}$ of $\mathcal{T}(h)$, indeed the differential operator in θ-direction, is defined as

$$\mathcal{A}\varphi = \lim_{h \to 0^+} \frac{\mathcal{T}(h)\varphi - \varphi}{h} \tag{3.7}$$

with

$$\mathcal{D}(\mathcal{A}) = \left\{ \varphi \in \boldsymbol{X} | \lim_{h \to 0^+} \frac{\mathcal{T}(h)\varphi - \varphi}{h} \text{ exists} \right\}.$$

As shown in (3.7), $\mathcal{A}$ represents the (right-hand) derivative of $\mathcal{T}(h)$ at $h = 0$. It is a linear, closed, densely defined and unbounded operator for the strongly continuous semigroup corresponding to $\mathcal{T}(h)$.

By substituting (3.4) and (3.5) into (3.7), one can obtain

$$\mathcal{A}\varphi = \lim_{h \to 0^+} \frac{\mathcal{T}(h)\varphi - \varphi}{h} = \frac{\mathrm{d}\varphi(\theta)}{\mathrm{d}\theta} = \varphi' \tag{3.8}$$

for $h + \theta \leq 0$ and

$$\mathcal{A}\varphi(0) = \varphi'(0) = \tilde{\boldsymbol{A}}_0\varphi(0) + \sum_{i=1}^{m} \tilde{\boldsymbol{A}}_i\varphi(-\tau_i).$$

The latter is commonly referred to as the splicing condition or boundary condition corresponding to (3.1).

Thus, the infinitesimal generator $\mathcal{A}$ and its domain are explicitly expressed as

$$\mathcal{A}\varphi = \varphi'$$

and

$$\mathcal{D}(\mathcal{A}) = \left\{ \varphi \in \boldsymbol{X} | \varphi' \in \boldsymbol{X},\ \varphi'(0) = \tilde{\boldsymbol{A}}_0\varphi(0) + \sum_{i=1}^{m} \tilde{\boldsymbol{A}}_i\varphi(-\tau_i) \right\}.$$

3.1.2.2 Transformation of RFDEs into Abstract Cauchy Problem

RFDEs (3.1) can be first reformulated as hyperbolic partial differential equations (PDEs) [6–8]

$$\frac{\partial \boldsymbol{u}}{\partial t}(t,\ \theta) = \frac{\partial \boldsymbol{u}}{\partial \theta}(t\ ,\theta),\quad t \geq 0;\ \theta \in [-\tau_{\max},\ 0] \tag{3.9}$$

with the boundary condition

$$\frac{\partial \boldsymbol{u}}{\partial \theta}(t,\ 0) = \tilde{\boldsymbol{A}}_0 \boldsymbol{u}(t,\ 0) + \sum_{i=1}^{m} \tilde{\boldsymbol{A}}_i \boldsymbol{u}(t,\ -\tau_i),\ \ t \geq 0$$

and the initial value

$$\boldsymbol{u}(0, \theta) = \varphi(\theta),\ \theta \in [-\tau_{\max},\ 0]$$

whose solution $\boldsymbol{u}(t,\ \theta) \in C([0,\ +\infty) \times [-\tau_{\max},\ 0],\ \mathbb{R}^{n\times 1})$ is expressed as

$$\boldsymbol{u}(t, \theta) = \Delta \boldsymbol{x}(t + \theta) = \Delta \boldsymbol{x}_t(\theta),\ \ t \geq 0;\ \theta \in [-\tau_{\max},\ 0]. \tag{3.10}$$

Then, by substituting (3.8) into the right-hand side of (3.9), PDEs can be equivalently interpreted as the following abstract Cauchy problem on the Banach space [7, 9]:

$$\begin{cases} \dfrac{\mathrm{d}\boldsymbol{u}(t)}{\mathrm{d}t} = \mathcal{A}\boldsymbol{u}(t),\ \ t \geq 0 \\ \boldsymbol{u}(0) = \varphi \end{cases}. \tag{3.11}$$

By considering (3.10), an alternative formulation of the problem can be obtained:

$$\begin{cases} \dfrac{\mathrm{d}\Delta \boldsymbol{x}_t}{\mathrm{d}t} = \mathcal{A}\Delta \boldsymbol{x}_t,\ \ t \geq 0 \\ \Delta \boldsymbol{x}_0 = \varphi \end{cases}. \tag{3.12}$$

The interpretation of (3.1) as (3.11) and (3.12) associated with the infinitesimal generator $\mathcal{A}$, permits (abstract) application of many results and properties that hold in the field of PDEs and ODEs into DDEs.

3.1.3 Augmented Form of Semigroup Operators

For the time-delay system described by DDAEs and expressed as (2.24), this subsection reformulates the above two semigroup operators in an augmented form. Based on Sect. 3.1.1, the augmented Banach space $\hat{\boldsymbol{X}} := \mathrm{C}([-\tau_{\max},\ 0],\ \mathbb{R}^{d\times 1})$ and function segment $\Delta\hat{\boldsymbol{x}}_t(\theta)$ ($t \geq 0;\ \theta \in [-\tau_{\max},\ 0]$) can be redefined.

With the help of $\Delta\hat{\boldsymbol{x}}_t$, the augmented counterpart of (3.1) thus can be formulated as

$$\boldsymbol{E}\frac{\mathrm{d}\Delta\hat{\boldsymbol{x}}_t(0)}{\mathrm{d}t} = \boldsymbol{J}_0 \Delta\hat{\boldsymbol{x}}_t(0) + \sum_{i=1}^{m} \boldsymbol{J}_i \Delta\hat{\boldsymbol{x}}_t(-\tau_i) \tag{3.13}$$

or

$$\begin{cases} \boldsymbol{z}(t) = [\boldsymbol{A}_0 \ \ \boldsymbol{B}_0]\Delta\hat{\boldsymbol{x}}_t(0) + \sum_{i=1}^{m}[\boldsymbol{A}_i \ \ \boldsymbol{B}_i]\Delta\hat{\boldsymbol{x}}_t(-\tau_i) \\ \boldsymbol{w}(t) = -\boldsymbol{D}_0^{-1}\boldsymbol{C}_0\boldsymbol{z}(t) \end{cases} \tag{3.14}$$

where $\boldsymbol{z}(t) \triangleq \Delta\dot{\boldsymbol{x}}_t(0) = \Delta\dot{\boldsymbol{x}}(t) \in \mathbb{R}^{n\times 1}$ and $\boldsymbol{w}(t) \triangleq \Delta\dot{\boldsymbol{y}}_t(0) = \Delta\dot{\boldsymbol{y}}(t) \in \mathbb{R}^{l\times 1}$ for $t \geq 0$. The second item of (3.14) is derived by differentiating the first item of (2.27).

3.1.3.1 Augmented Solution Operator

The augmented solution operator $\mathcal{T}(h) : \hat{\boldsymbol{X}} \to \hat{\boldsymbol{X}}$ is defined as the linear operator that maps $\hat{\varphi}$ to $\Delta\hat{\boldsymbol{x}}_h$, which can be expressed as

$$\left(\mathcal{T}(h)\hat{\varphi}\right)(\theta) = \Delta\hat{\boldsymbol{x}}_h(\theta) = \begin{cases} \hat{\varphi}(0) + \int_0^{\theta+h} \begin{bmatrix} \boldsymbol{z}(s) \\ \boldsymbol{w}(s) \end{bmatrix} \mathrm{d}s, & \theta \in [-h, \ 0) \\ \hat{\varphi}(\theta + h), & \theta \in [-\tau_{\max}, \ -h] \end{cases} \tag{3.15}$$

or

$$\left(\mathcal{T}(h)\hat{\varphi}\right)(\theta) = \begin{cases} \begin{bmatrix} \boldsymbol{I}_n \\ -\boldsymbol{D}_0^{-1}\boldsymbol{C}_0 \end{bmatrix} \left(\varphi_x(0) + \int_0^{\theta+h} \boldsymbol{z}(s)\mathrm{d}s\right), & \theta \in [-h, \ 0) \\ \hat{\varphi}(\theta + h), & \theta \in [-\tau_{\max}, \ -h] \end{cases}. \tag{3.16}$$

3.1.3.2 Augmented Infinitesimal Generator

The augmented infinitesimal generator $\mathcal{A} : \hat{\boldsymbol{X}} \to \hat{\boldsymbol{X}}$ is defined as

$$\mathcal{A}\hat{\varphi} = \hat{\varphi}', \ \hat{\varphi} \in \mathcal{D}(\mathcal{A})$$

where

$$\mathcal{D}(\mathcal{A}) = \left\{ \hat{\varphi} \in \hat{\boldsymbol{X}} \mid \hat{\varphi}' \in \hat{\boldsymbol{X}}, \ \boldsymbol{E}\hat{\varphi}'(0) = \boldsymbol{J}_0\hat{\varphi}(0) + \sum_{i=1}^{m} \boldsymbol{J}_i\hat{\varphi}(-\tau_i) \right\}. \tag{3.17}$$

With $\mathcal{A}$, (3.13) can be transformed into the augmented abstract Cauchy problem

$$\begin{cases} \dfrac{\mathrm{d}\Delta\hat{\boldsymbol{x}}_t}{\mathrm{d}t} = \mathcal{A}\Delta\hat{\boldsymbol{x}}_t, \ \ t \geq 0 \\ \Delta\hat{\boldsymbol{x}}_0 = \hat{\varphi} \end{cases}. \tag{3.18}$$

3.2 Spectral Mapping

The eigenvalues of the time-delay system, denoted by λ, are fully characterized by the spectra of the infinitesimal generator $\mathcal{A}$ in the sense that

$$\lambda \in \sigma(\mathcal{A})$$

where $\sigma(\cdot)$ stands for the point spectrum operation. Meanwhile, it relates to the non-zero spectrum μ of the solution operator $\mathcal{T}(h)$ by

$$\mu = \mathrm{e}^{\lambda h}, \quad \lambda = \frac{\ln \mu}{h}, \quad \mu \in \sigma(\mathcal{T}(h)) \setminus \{0\} \tag{3.19}$$

for $h > 0$. μ is generally called the characteristic multiplier. μ and λ are distributed on the z-plane and s-plane, respectively. As shown in Fig. 3.2, the imaginary axis on the s-plane corresponds to the unit circle on the z-plane. λ_i, λ_j and λ_k are mapped to μ_i, μ_j and μ_k according to (3.19), respectively.

Note that, for an arbitrary μ of $\mathcal{T}(h)$, the imaginary part of the resultant λ, i.e., $\operatorname{Im}(\lambda) = \operatorname{Im}\left(\frac{\ln \mu}{h}\right)$, is restricted to $[-\frac{\pi}{h}, \frac{\pi}{h}]$. Therefore, to ensure that the imaginary part of the interested eigenvalue λ of time-delay systems can be correctly computed, the transfer step h of $\mathcal{T}(h)$ should not exceed $\left|\frac{\pi}{\operatorname{Im}(\lambda)}\right|$.

The above statements on spectral mapping are the cornerstone of the PSD-based methods to approximate eigenvalues of time-delay systems.

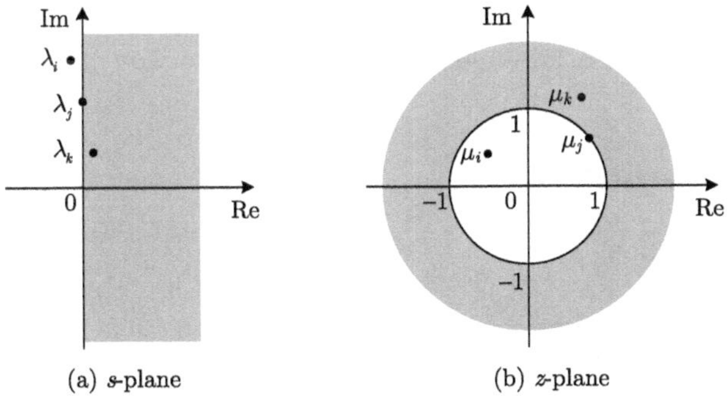

Fig. 3.2 Mapping between λ of the time-delay system/infinitesimal generator $\mathcal{A}$, and μ of the solution operator $\mathcal{T}(h)$

3.3 Spectral Discretization

3.3.1 *Classification*

It is known from the last section that eigenvalues of a time-delay system can be obtained from spectra of $\mathcal{A}$ and $\mathcal{T}(h)$. Given their infinite-dimensional nature, $\mathcal{A}$ and $\mathcal{T}(h)$ are required to be discretized and reduced into their approximation matrices with finite dimensions. This process is indicated as "spectral operator discretization" or "spectral discretization" for short.

In the field of computational mathematics and numerical analysis, there exist several spectral discretization schemes. They can be categorized into three groups according to spectral operators, spectral discretization ideas and numerical methods.

3.3.1.1 By Spectral Operators

IGD

Based on the augmented abstract Cauchy equation (3.18), different numerical methods can be applied to derive the expression between $\Delta\hat{\boldsymbol{x}}_0$ and its derivative at discrete points on the interval $[-\tau_{\max},\ 0]$, leading to the finite-dimensional discretized matrix approximating $\mathcal{A}$.

SOD

Based on the augmented piecewise function (3.15), the expression between $\Delta\hat{\boldsymbol{x}}_0$ and $\Delta\hat{\boldsymbol{x}}_h$ at discrete points on $[-\tau_{\max},\ 0]$ is derived by using different numerical methods. Its coefficient matrix is the finite-dimensional approximation matrix of $\mathcal{T}(h)$.

3.3.1.2 By Discretization Ideas

Spectral Discretization

The traditional spectral discretization idea requires all system state variables to be discretized over the delay interval $[-\tau_{\max},\ 0]$. The dimensions of the discretized matrices of $\mathcal{A}$ and $\mathcal{T}(h)$ are generally dozens of times that of the system. Massive computational burden is thus involved and prolonged CPU time is needed when analyzing real-life large-scale time-delay systems. In addition, the memory occupation and computational burden significantly increase as the number of discrete points on $[-\tau_{\max},\ 0]$.

PSD

Due to the fact that the delay-free state variables in the past time have nothing to do with system dynamics, only the retarded ones are required to be discretized over the delay interval $[-\tau_{\max}, 0]$. Since the number of retarded state variables is few, the dimensions of the resultant discretized matrices approximating the spectral operators $\mathcal{A}$ and $\mathcal{T}(h)$ associated to the time-delay system are close to the number of actual system state variables. Compared with the traditional spectral discretization idea, the discretized matrices generated by the PSD idea exhibit very low order but with identical accuracy.

3.3.1.3 By Numerical Methods

Numerical methods, such as pseudospectral (PS), implicit Runge-Kutta and linear multi-step methods, can be utilized to discretize $\mathcal{A}$ and $\mathcal{T}(h)$. Note that the characteristic equations of $\mathcal{A}$ and $\mathcal{T}(h)$ are analytic and infinitely regular [1]. By exploiting this nature, the PS method is superior in approximating the two operators with an error of $\mathcal{O}(N^{-N})$ (also called spectral accuracy). Here N is the number of discretized points on $[-\tau_{\max}, 0]$. The application of the PS method will be elaborated in Chaps. 4 and 5.

3.3.2 PSD

3.3.2.1 State Variable Partitioning of the Augmented System

The first step of PSD is to partition system state variables $\Delta\boldsymbol{x}$ into two parts, i.e., the delay-free state variables $\Delta\boldsymbol{x}^{(1)} \in \mathbb{R}^{n_1\times 1}$ and retarded state variables $\Delta\boldsymbol{x}^{(2)} \in \mathbb{R}^{n_2\times 1}$, satisfying $\Delta\boldsymbol{x} = [(\Delta\boldsymbol{x}^{(1)})^{\mathrm{T}}, (\Delta\boldsymbol{x}^{(2)})^{\mathrm{T}}]^{\mathrm{T}}$ and $n_1 + n_2 = n$. Similarly, system algebraic variables $\Delta\boldsymbol{y}$ can also be rewritten as $\Delta\boldsymbol{y} = [(\Delta\boldsymbol{y}^{(1)})^{\mathrm{T}}, (\Delta\boldsymbol{y}^{(2)})^{\mathrm{T}}]^{\mathrm{T}}$ with delay-free algebraic variables $\Delta\boldsymbol{y}^{(1)} \in \mathbb{R}^{l_1\times 1}$ and retarded algebraic variables $\Delta\boldsymbol{y}^{(2)} \in \mathbb{R}^{l_2\times 1}$, where $l_1 + l_2 = l$. Thus, by stacking $\Delta\boldsymbol{x}$ and $\Delta\boldsymbol{y}$, the state variables of the augmented system can be reformulated as

$$\Delta\hat{\boldsymbol{x}} = \left[\Delta\boldsymbol{x}^{\mathrm{T}},\ \Delta\boldsymbol{y}^{\mathrm{T}}\right]^{\mathrm{T}} = \left[(\Delta\boldsymbol{x}^{(1)})^{\mathrm{T}}\ (\Delta\boldsymbol{x}^{(2)})^{\mathrm{T}} \;\vdots\; (\Delta\boldsymbol{y}^{(1)})^{\mathrm{T}}\ (\Delta\boldsymbol{y}^{(2)})^{\mathrm{T}}\right]^{\mathrm{T}}$$

with the initial condition

$$\hat{\varphi} = \left[(\varphi_x^{(1)})^{\mathrm{T}}\ (\varphi_x^{(2)})^{\mathrm{T}} \;\vdots\; (\varphi_y^{(1)})^{\mathrm{T}}\ (\varphi_y^{(2)})^{\mathrm{T}}\right]^{\mathrm{T}}$$

where $\boldsymbol{\varphi}_x^{(1)} \in \mathbb{R}^{n_1 \times 1}$, $\boldsymbol{\varphi}_x^{(2)} \in \mathbb{R}^{n_2 \times 1}$, $\boldsymbol{\varphi}_y^{(1)} \in \mathbb{R}^{l_1 \times 1}$ and $\boldsymbol{\varphi}_y^{(2)} \in \mathbb{R}^{l_2 \times 1}$.

Following the partitioning, $\boldsymbol{E}$ and $\boldsymbol{J}_i$ $(i = 0, 1, \ldots, m)$ in (2.24) are reformulated in the block form, namely

$$\boldsymbol{E} = \left[\begin{array}{cc:cc} \boldsymbol{I}_{n_1} & & & \\ & \boldsymbol{I}_{n_2} & & \\ \hdashline & & \boldsymbol{0}_{l_1} & \\ & & & \boldsymbol{0}_{l_2} \end{array}\right]$$

$$\boldsymbol{J}_0 = \left[\begin{array}{cc:cc} \boldsymbol{A}_0^{(1)} & \boldsymbol{A}_0^{(2)} & \boldsymbol{B}_0^{(1)} & \boldsymbol{B}_0^{(2)} \\ \hdashline \boldsymbol{C}_0^{(1)} & \boldsymbol{C}_0^{(2)} & \boldsymbol{D}_0^{(1)} & \boldsymbol{D}_0^{(2)} \end{array}\right] = \left[\begin{array}{cc:cc} \boldsymbol{A}_{11,0} & \boldsymbol{A}_{12,0} & \boldsymbol{B}_{11,0} & \boldsymbol{B}_{12,0} \\ \boldsymbol{A}_{21,0} & \boldsymbol{A}_{22,0} & \boldsymbol{B}_{21,0} & \boldsymbol{B}_{22,0} \\ \hdashline \boldsymbol{C}_{11,0} & \boldsymbol{C}_{12,0} & \boldsymbol{D}_{11,0} & \boldsymbol{D}_{12,0} \\ \boldsymbol{C}_{21,0} & \boldsymbol{C}_{22,0} & \boldsymbol{D}_{21,0} & \boldsymbol{D}_{22,0} \end{array}\right]$$

$$\boldsymbol{J}_i = \left[\begin{array}{cc:cc} \boldsymbol{0}_{n \times n_1} & \boldsymbol{A}_i^{(2)} & \boldsymbol{0}_{n \times l_1} & \boldsymbol{B}_i^{(2)} \\ \hdashline \boldsymbol{0}_{l \times n_1} & \boldsymbol{0}_{l \times n_2} & \boldsymbol{0}_{l \times l_1} & \boldsymbol{0}_{l \times l_2} \end{array}\right] = \left[\begin{array}{cc:cc} \boldsymbol{0}_{n_1} & \boldsymbol{A}_{12,i} & \boldsymbol{0}_{n_1 \times l_1} & \boldsymbol{B}_{12,i} \\ \boldsymbol{0}_{n_2 \times n_1} & \boldsymbol{A}_{22,i} & \boldsymbol{0}_{n_2 \times l_1} & \boldsymbol{B}_{22,i} \\ \hdashline \boldsymbol{0}_{l \times n_1} & \boldsymbol{0}_{l \times n_2} & \boldsymbol{0}_{l \times l_1} & \boldsymbol{0}_{l \times l_2} \end{array}\right]$$

where $\boldsymbol{A}_{11,0} \in \mathbb{R}^{n_1 \times n_1}$, $\boldsymbol{A}_{12,i} \in \mathbb{R}^{n_1 \times n_2}$, $\boldsymbol{A}_{21,0} \in \mathbb{R}^{n_2 \times n_1}$, $\boldsymbol{A}_{22,i} \in \mathbb{R}^{n_2 \times n_2}$, $\boldsymbol{B}_{11,0} \in \mathbb{R}^{n_1 \times l_1}$, $\boldsymbol{B}_{12,i} \in \mathbb{R}^{n_1 \times l_2}$, $\boldsymbol{B}_{21,0} \in \mathbb{R}^{n_2 \times l_1}$, $\boldsymbol{B}_{22,i} \in \mathbb{R}^{n_2 \times l_2}$, $\boldsymbol{C}_{11,0} \in \mathbb{R}^{l_1 \times n_1}$, $\boldsymbol{C}_{12,0} \in \mathbb{R}^{l_1 \times n_2}$, $\boldsymbol{C}_{21,0} \in \mathbb{R}^{l_2 \times n_1}$, $\boldsymbol{C}_{22,0} \in \mathbb{R}^{l_2 \times n_2}$, $\boldsymbol{D}_{11,0} \in \mathbb{R}^{l_1 \times l_1}$, $\boldsymbol{D}_{12,0} \in \mathbb{R}^{l_1 \times l_2}$, $\boldsymbol{D}_{21,0} \in \mathbb{R}^{l_2 \times l_1}$, $\boldsymbol{D}_{22,0} \in \mathbb{R}^{l_2 \times l_2}$, $\boldsymbol{A}_0^{(1)} \in \mathbb{R}^{n \times n_1}$, $\boldsymbol{A}_i^{(2)} \in \mathbb{R}^{n \times n_2}$, $\boldsymbol{B}_0^{(1)} \in \mathbb{R}^{n \times l_1}$, $\boldsymbol{B}_i^{(2)} \in \mathbb{R}^{n \times l_2}$, $i = 0, 1, \ldots, m$.

Let $d_1 = n_1 + l_1$ and $d_2 = n_2 + l_2$, and then define $\Delta\hat{\boldsymbol{x}}^{(1)} \in \mathbb{R}^{d_1 \times 1}$, $\Delta\hat{\boldsymbol{x}}^{(2)} \in \mathbb{R}^{d_2 \times 1}$, $\hat{\boldsymbol{\varphi}}^{(1)} \in \mathbb{R}^{d_1 \times 1}$, $\hat{\boldsymbol{\varphi}}^{(2)} \in \mathbb{R}^{d_2 \times 1}$, $\boldsymbol{J}_i^{(2)} \in \mathbb{R}^{d \times d_2}$, $\boldsymbol{J}_{x,i}^{(2)} \in \mathbb{R}^{d \times n_2}$ and $\boldsymbol{J}_{y,i}^{(2)} \in \mathbb{R}^{d \times l_2}$ $(i = 1, 2, \ldots, m)$ as

$$\Delta\hat{\boldsymbol{x}}^{(1)} = \left[\left(\Delta\boldsymbol{x}^{(1)}\right)^{\mathrm{T}}, \left(\Delta\boldsymbol{y}^{(1)}\right)^{\mathrm{T}}\right]^{\mathrm{T}}, \ \Delta\hat{\boldsymbol{x}}^{(2)} = \left[\left(\Delta\boldsymbol{x}^{(2)}\right)^{\mathrm{T}}, \left(\Delta\boldsymbol{y}^{(2)}\right)^{\mathrm{T}}\right]^{\mathrm{T}}$$

$$\hat{\boldsymbol{\varphi}}^{(1)} = \left[\left(\boldsymbol{\varphi}_x^{(1)}\right)^{\mathrm{T}}, \left(\boldsymbol{\varphi}_y^{(1)}\right)^{\mathrm{T}}\right]^{\mathrm{T}}, \ \hat{\boldsymbol{\varphi}}^{(2)} = \left[\left(\boldsymbol{\varphi}_x^{(2)}\right)^{\mathrm{T}}, \left(\boldsymbol{\varphi}_y^{(2)}\right)^{\mathrm{T}}\right]^{\mathrm{T}}$$

$$\boldsymbol{J}_i^{(2)} = \left[\boldsymbol{J}_{x,i}^{(2)}, \ \boldsymbol{J}_{y,i}^{(2)}\right], \ \boldsymbol{J}_{x,i}^{(2)} = \begin{bmatrix} \boldsymbol{A}_i^{(2)} \\ \boldsymbol{0}_{l \times n_2} \end{bmatrix}, \ \boldsymbol{J}_{y,i}^{(2)} = \begin{bmatrix} \boldsymbol{B}_i^{(2)} \\ \boldsymbol{0}_{l \times l_2} \end{bmatrix}.$$

In practical applications, very few state and algebraic variables are retarded. In this sense, it can be concluded that that $d_2 \ll d_1$. This is the premise of high efficiency of PSD.

3.3.2.2 The PSD Idea

After the partition, (2.24) can be rewritten as

$$\begin{cases} \boldsymbol{E}\Delta\dot{\hat{\boldsymbol{x}}}(t) = \boldsymbol{J}_0\Delta\hat{\boldsymbol{x}}(t) + \sum_{i=1}^{m} \boldsymbol{J}_i^{(2)}\Delta\hat{\boldsymbol{x}}^{(2)}(t-\tau_i),\ t \geq 0 \\ \Delta\hat{\boldsymbol{x}}^{(1)}(0) = \hat{\varphi}^{(1)}(0) \\ \Delta\hat{\boldsymbol{x}}^{(2)}(t) = \hat{\varphi}^{(2)}(t), \qquad t \in [-\tau_{\max},\ 0] \end{cases} . \tag{3.20}$$

It is known from (3.20) that system dynamics have nothing to do with $\Delta\boldsymbol{x}^{(1)}$ in the past time. Based on this fact, the idea of PSD is proposed. The core is that only the retarded $\Delta\boldsymbol{x}^{(2)}$ instead of $\Delta\boldsymbol{x}$ are discretized.

3.3.2.3 PIGD

Considering (3.20), the abstract Cauchy problem (3.18) is rewritten as

$$\begin{cases} \begin{bmatrix} \mathrm{d}\Delta\hat{\boldsymbol{x}}_t^{(1)}(0)/\mathrm{d}t \\ \mathrm{d}\Delta\hat{\boldsymbol{x}}_t^{(2)}/\mathrm{d}t \end{bmatrix} = \mathcal{A}\begin{bmatrix} \Delta\hat{\boldsymbol{x}}_t^{(1)}(0) \\ \Delta\hat{\boldsymbol{x}}_t^{(2)} \end{bmatrix},\ t \geq 0 \\ \Delta\hat{\boldsymbol{x}}_0^{(1)}(0) = \hat{\varphi}^{(1)}(0) \\ \Delta\hat{\boldsymbol{x}}_0^{(2)} = \hat{\varphi}^{(2)} \end{cases} \tag{3.21}$$

and the corresponding splicing condition (3.17) becomes

$$\boldsymbol{E}\hat{\varphi}'(0) = \boldsymbol{J}_0\hat{\varphi}(0) + \sum_{i=1}^{m} \boldsymbol{J}_i^{(2)}\hat{\varphi}^{(2)}(-\tau_i) \tag{3.22}$$

where $\Delta\hat{\boldsymbol{x}}_t^{(1)}(0) = \Delta\hat{\boldsymbol{x}}^{(1)}(t)$, and $\Delta\hat{\boldsymbol{x}}_t^{(2)}$ represents the function segment associated to $\Delta\hat{\boldsymbol{x}}^{(2)}$ over the interval $[t-\tau_{\max},\ t]$ for $t \geq 0$.

Partial infinitesimal generator discretization (PIGD) can be achieved by evaluating (3.21) at a set of discrete points over the interval $[-\tau_{\max},\ 0]$, as illustrated in Fig. 3.3. In contrast to IGD, the discretization of $\Delta\hat{\boldsymbol{x}}_0^{(1)}$ over $[-\tau_{\max},\ 0]$ in PIGD is reduced to evaluate it at only one point, i.e., $\theta = 0$.

Specifically, the partial discretization of $\mathcal{A}$ can be summarized as follows.

- A set of discrete points over $[-\tau_{\max},\ 0]$ are determined.
- The derivative of $\Delta\hat{\boldsymbol{x}}_0$ at $\theta = 0$ (i.e., $\hat{\varphi}'(0)$) and that of $\Delta\hat{\boldsymbol{x}}_0^{(2)}$ at other nonzero discrete points are evaluated, leading to a generalized eigenvalue problem. It can be represented by a matrix pencil $(\boldsymbol{E}_N,\ \hat{\mathcal{A}}_N)$.

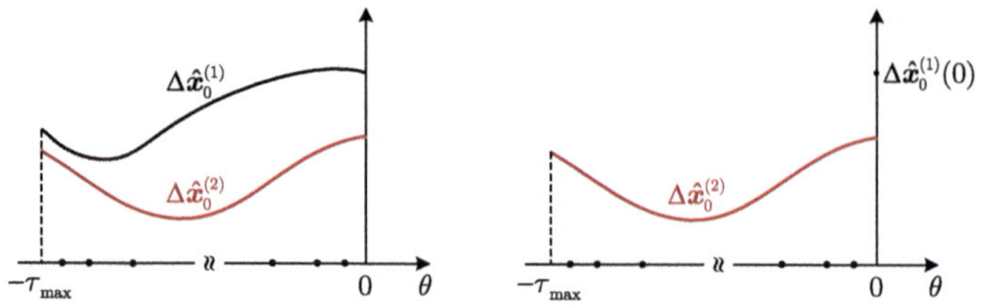

(a) IGD (b) PIGD

Fig. 3.3 Principle of PIGD

- The matrix pencil $(\boldsymbol{E}_N,\ \hat{\mathcal{A}}_N)$ is further transformed into a standard eigenvalue problem of the partial discretization matrix approximating $\mathcal{A}$, denoted as $\mathcal{A}_N$. The eigenvalues $\hat{\lambda}$ of $\mathcal{A}_N$ are good estimates to those λ of the time-delay system.

3.3.2.4 PSOD

Considering (3.20), the piecewise functions (3.15) and (3.16) of $\mathcal{T}(h)$ are rewritten as

$$\begin{cases} \Delta\hat{\boldsymbol{x}}_h(\theta) = \hat{\boldsymbol{\varphi}}(0) + \displaystyle\int_0^{\theta+h} \begin{bmatrix} \boldsymbol{z}(s) \\ \boldsymbol{w}(s) \end{bmatrix} \mathrm{d}s, & \theta \in [-h,\ 0) \\ \Delta\hat{\boldsymbol{x}}_h^{(2)}(\theta) = \hat{\boldsymbol{\varphi}}^{(2)}(\theta + h), & \theta \in [-\tau_{\max},\ -h] \end{cases} \tag{3.23}$$

and

$$\begin{cases} \Delta\hat{\boldsymbol{x}}_h(\theta) = \begin{bmatrix} \boldsymbol{I}_n \\ -\boldsymbol{D}_0^{-1}\boldsymbol{C}_0 \end{bmatrix} \left(\boldsymbol{\varphi}_x(0) + \displaystyle\int_0^{\theta+h} \boldsymbol{z}(s)\mathrm{d}s \right), & \theta \in [-h,\ 0) \\ \Delta\hat{\boldsymbol{x}}_h^{(2)}(\theta) = \hat{\boldsymbol{\varphi}}^{(2)}(\theta + h), & \theta \in [-\tau_{\max},\ -h] \end{cases} \tag{3.24}$$

where

$$\begin{cases} \boldsymbol{z}(t) = [\boldsymbol{A}_0\ \ \boldsymbol{B}_0]\Delta\hat{\boldsymbol{x}}(t) + \displaystyle\sum_{i=1}^{m} \left[\boldsymbol{A}_i^{(2)}\ \ \boldsymbol{B}_i^{(2)}\right] \Delta\hat{\boldsymbol{x}}^{(2)}(t - \tau_i) \\ \boldsymbol{C}_0\boldsymbol{z}(t) + \boldsymbol{D}_0\boldsymbol{w}(t) = \boldsymbol{0} \end{cases} \tag{3.25}$$

for $t \geq 0$.

Partial solution operator discretization (PSOD) turns out to be the estimation of $\Delta\hat{\boldsymbol{x}}_h^{(1)}(0)$ and $\Delta\hat{\boldsymbol{x}}_h^{(2)}$ at discrete points over $[-\tau_{\max},\ 0]$, as illustrated in Fig. 3.4. Different from SOD [5, 10], the discretization of $\Delta\hat{\boldsymbol{x}}^{(1)}$ at nonzero discrete points is unnecessary in PSOD.

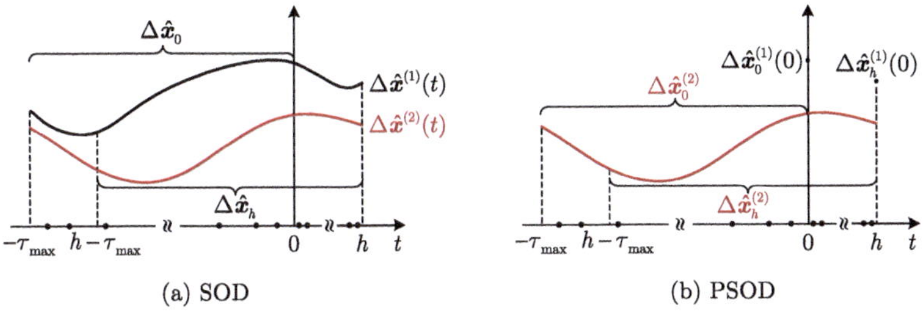

Fig. 3.4 Principle of PSOD

The process of partially discretizing $\mathcal{T}(h)$ has the following steps.

- The time step-length h and two sets of discrete points over intervals $[-\tau_{\max},\ 0]$ and $[0,\ h]$ are specified.
- The evaluation of $\Delta\hat{\boldsymbol{x}}_h^{(2)}$ at discrete points in $[-\tau_{\max},\ -h)$ are directly obtained according to the shift segment shown in (3.4).
- The evaluation of $\Delta\hat{\boldsymbol{x}}_h^{(1)}(0)$ and $\Delta\hat{\boldsymbol{x}}_h^{(2)}$ at discrete points in $[-h,\ 0]$ are obtained by solving (3.25).
- Either (3.23) or (3.24) is thus transformed into an explicit state transition equation. The coefficient matrix is the partial discretization matrix approximating $\mathcal{T}(h)$, denoted as $\hat{\boldsymbol{T}}_N$. Then, the eigenvalues λ of the time-delay system are estimated from those $\hat{\mu}$ of $\hat{\boldsymbol{T}}_N$.

3.3.3 PS Method

3.3.3.1 Overview

The PS method, also known as spectral collocation method, is a method for the numerical solution of ODEs, PDEs and integral equations. The core is to choose a finite-dimensional space of candidate solutions (usually, polynomials up to a certain degree) and a number of points in the domain (called collocation points), and to select that solution which satisfies the given equation at each collocation point [11]. Different from the finite difference and the finite element methods, the solution given by the PS method is an approximation for the whole computational domain with the so-called "spectral accuracy" (that is, the convergence accuracy is N^{-N}, where N is the number of collocation points). In addition, it should be noted that this method has nothing to do with pseudo-spectrum.

Consider the following ODE

$$\begin{cases} y'(t) = f(t,\ y(t)), & t = [t_0,\ t_0 + h] \\ y(t_0) = y_0 \end{cases}.$$

The collocation method approximates the solution y by the polynomial p of degree N which satisfies the initial condition $p(t_0) = y_0$, and the differential equation $p'(t) = f(t,\ p(t))$ at all collocation points $t_0 + c_k h$ $(k = 1, 2,\ \ldots,\ N)$ with $0 \leqslant c_1 < c_2 < \cdots < c_N \leqslant 1$. This gives $N + 1$ conditions, which matches $N + 1$ parameters needed to specify a polynomial of degree N.

Note that all these collocation methods are in fact implicit Runge-Kutta methods. The coefficients c_k $(k = 1,\ \ldots,\ N)$ in the Butcher tableau of a Runge-Kutta method are the collocation points. However, not all implicit Runge-Kutta methods are collocation methods [12].

Taking $N = 2$ as an example, the two collocation points $c_1 = 0$ and $c_2 = 1$ are specified. The collocation conditions are

$$\begin{cases} p(t_0) = y_0 \\ p'(t_0) = f(t_0,\ p(t_0)) \\ p'(t_0 + h) = f(t_0 + h,\ p(t_0 + h)) \end{cases}.$$

There are three conditions such that p should be a polynomial of degree 2. Write p in the following form to simplify computations:

$$p(t) = \alpha(t - t_0)^2 + \beta(t - t_0) + \gamma$$

where

$$\begin{cases} \alpha = \dfrac{1}{2h}\left(f(t_0 + h,\ p(t_0 + h)) - f(t_0,\ p(t_0))\right) \\ \beta = f(t_0,\ p(t_0)) \\ \gamma = y_0 \end{cases}.$$

The collocation method is now given (implicitly) by

$$y_1 = p(t_0 + h) = y_0 + \frac{1}{2}h\left(f(t_0 + h,\ y_1) + f(t_0,\ y_0)\right)$$

where $y_1 = p(t_0 + h)$ is the approximate solution at $t = t_0 + h$.

3.3.3.2 Trigonometric Definitions of Chebyshev Polynomials of the First and Second Kinds [13]

The First Kind Polynomial T_N

The Chebyshev polynomial $T_N(x)$ of the first kind is a polynomial in $x = \cos\theta$ of degree n, which is defined as

$$T_N(x) = \cos(N\theta) = \cos(N \arccos x) \tag{3.26}$$

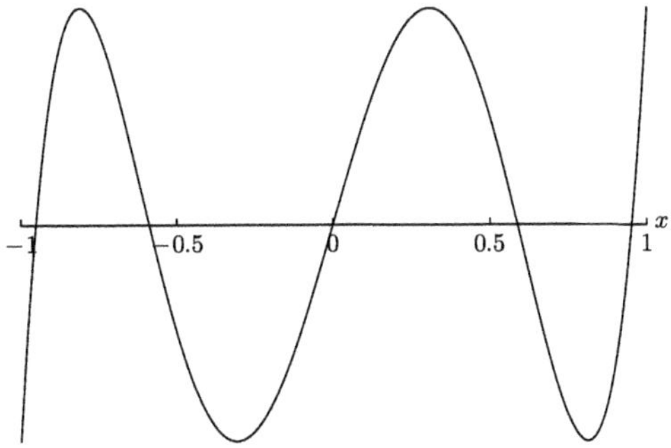

Fig. 3.5 $T_5(x)$ on range $[-1,\ 1]$

where the range of x is the interval $[-1,\ 1]$ and the range of values of the corresponding θ can be taken as $[0,\ \pi]$. These ranges are traversed in opposite directions, since $x = -1$ corresponds to $\theta = \pi$ and $x = 1$ corresponds to $\theta = 0$.

It is well known (as a consequence of de Moivre's Theorem) that $\cos(N\theta)$ is a polynomial of degree N in $\cos\theta$, and indeed one is familiar with the elementary formulae

$$\begin{aligned}&\cos(0\theta) = 1,\ \cos(1\theta) = \cos\theta,\ \cos(2\theta) = 2\cos^2\theta - 1,\ \cos(3\theta) = 4\cos^3\theta - 3\cos\theta,\\ &\cos(4\theta) = 8\cos^4\theta - 8\cos^2\theta + 1,\ \cos(5\theta) = 16\cos^5\theta - 20\cos^3\theta + 5\cos\theta,\ \ldots\end{aligned} \tag{3.27}$$

Substituting (3.27) into (3.26), one can obtain the explicit expression of T_N in terms of x

$$\begin{aligned}&T_0(x) = 1,\ T_1(x) = x,\ T_2(x) = 2x^2 - 1,\ T_3(x) = 4x^3 - 3x,\\ &T_4(x) = 8x^4 - 8x^2 + 1,\ T_5(x) = 16x^5 - 20x^3 + 5x,\ \ldots\end{aligned} \tag{3.28}$$

When $N = 5$, the graph of $T_5(x)$ on the interval $[-1,\ 1]$ is shown in Fig. 3.5.

The Second Kind Polynomial U_N

The N-th order Chebyshev polynomial $U_N(x)$ of the second kind is defined as

$$U_N(x) = \frac{\sin((N+1)\theta)}{\sin\theta} \tag{3.29}$$

where the ranges of x and θ are the same as for $T_N(x)$.

Elementary formulae give

$$\begin{aligned}&\sin(1\theta) = \sin,\ \sin(2\theta) = 2\sin\theta\cos\theta,\ \sin(3\theta) = \sin\theta\left(4\cos^2\theta - 1\right),\\ &\sin(4\theta) = \sin\theta\left(8\cos^3\theta - 4\cos\theta\right),\\ &\sin(5\theta) = \sin\theta\left(16\cos^4\theta - 8\cos^3\theta - 4\cos^2\theta + 1\right),\\ &\sin(6\theta) = \sin\theta\left(32\cos^5\theta - 8\cos^4\theta - 24\cos^3\theta + 6\cos\theta\right),\ \ldots\end{aligned}$$

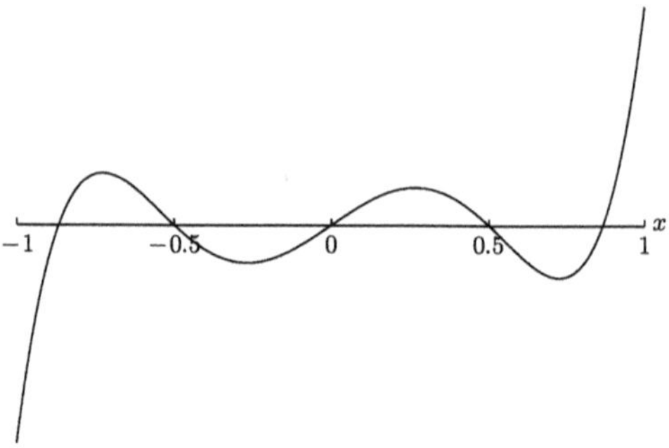

Fig. 3.6 $U_5(x)$ on range $[-1,\ 1]$

Accordingly, one can deduce the explicit expression of U_N in terms of x

$$\begin{aligned} &U_0(x) = 1,\ U_1(x) = 2x,\ U_2(x) = 4x^2 - 1,\ U_3(x) = 8x^3 - 4x, \\ &U_4(x) = 16x^4 - 12x^2 + 1,\ U_5(x) = 32x^5 - 32x^3 + 6x,\ \ldots \end{aligned} \tag{3.30}$$

When $N = 5$, the graph of $U_5(x)$ on the interval $[-1,\ 1]$ is shown in Fig. 3.6.

The functions chebyshevT(N, x) and chebyshevU(N, x) from MATLAB represent the N-th degree Chebyshev polynomial of the first kind at the point x, i.e., (3.28) and (3.30), respectively.

3.3.3.3 Zeros and Poles of Chebyshev Polynomials of the First and Second Kinds [13]

The Chebyshev polynomials of degree $n > 0$ of all four kinds have precisely n zeros and $n + 1$ local poles in the interval $[-1,\ 1]$. Note that $n - 1$ of these poles are interior to $[-1,\ 1]$, and are "true" alternate maxima and minima (in the sense that the gradient vanishes), the other two poles being at the end points ± 1 (where the gradient is non-zero). In general, the zeros of Chebyshev polynomial of the first kind are called Chebyshev points, which can be selected as the interpolation points to prevent the Runge phenomenon.

Zeros of T_N

It is known from (3.28) that the zeros for x of $T_N(x)$ on the interval $[-1,\ 1]$ correspond to those for θ of $\cos(N\theta)$ on the interval $[0,\ \pi]$, i.e.,

$$x_k = \cos\frac{(2k-1)\pi}{2N}$$

and

$$\theta_k = \frac{(2k-1)\pi}{2N},\quad k = 1,\ 2,\ \ldots,\ N.$$

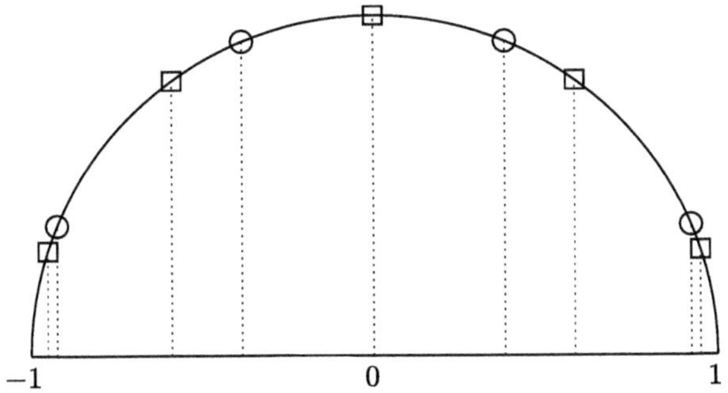

Fig. 3.7 Zeros of $T_N(x)$ when $N = 4$ and 5

Note that these zeros are in decreasing order in x (corresponding to increasing θ), and it is sometimes preferable to list them in their natural order as

$$x_k = \cos \frac{(2N - 2k + 1)\pi}{2N}, \quad k = 1,\ 2,\ \ldots,\ N.$$

Geometrically, the zeros of T_N represents the projection of the equidistant discrete points on the upper half-unit circle at a spacing of π/N onto the interval $[-1,\ 1]$. Note that $x = 0$ is a zero of $T_N(x)$ for all odd N, but not for even N, and that zeros are symmetrically placed in pairs on either side of $x = 0$. Taking $N = 4$ and 5 for example, the zeros of $T_N(x)$ correspond to the projections of "∘" and "□" on the interval $[-1,\ 1]$ in Fig. 3.7, respectively.

Zeros of U_N

The zeros of $U_N(x)$ (defined by (3.29)) are readily determined in a similar way from the zeros of $\sin((N + 1)\theta)$ as

$$x_k = \cos \frac{k\pi}{N + 1} \tag{3.31}$$

or in their natural order

$$x_k = \cos \frac{(N - k + 1)}{N + 1}\pi, \quad k = 1,\ 2,\ \ldots,\ N.$$

One is naturally tempted to extend the set of points (3.31) by including the further values $x_0 = 1$ and $x_{N+1} = -1$, giving the set

$$x_k = \cos \frac{k\pi}{N + 1}, \quad k = 0,\ 1,\ \ldots,\ N + 1. \tag{3.32}$$

In fact, these are zeros not of $U_N(x)$, but of the polynomial $(1 - x^2)U_N(x)$. However, these points are popular as nodes in applications to integration.

In addition, the zeros of $U_N(x)$ geometrically represent the projections of the equidistant discrete points on the upper half-unit circle at a spacing of $\pi/(N+1)$ onto the interval $[-1,\ 1]$.

Poles of T_N

The internal poles of $T_N(x)$ correspond to the poles of $\cos(N\theta)$, namely the zeros of $\sin(N\theta)$, since

$$\frac{\mathrm{d}}{\mathrm{d}x}T_N(x) = \frac{\mathrm{d}}{\mathrm{d}x}\cos(N\theta) = \frac{\mathrm{d}}{\mathrm{d}\theta}\cos(N\theta) \Big/ \frac{\mathrm{d}}{\mathrm{d}\theta}\cos\theta = \frac{N\sin(N\theta)}{\sin\theta}.$$

Hence, including those at $x = \pm 1$, the poles of $T_N(x)$ on $[-1,\ 1]$ are

$$x_k = \cos\frac{k\pi}{N}$$

or in their natural order

$$x_k = \cos\frac{(N-k)\pi}{N}, \quad k = 0,\ 1,\ \ldots,\ N. \tag{3.33}$$

These are precisely the zeros of $(1-x^2)U_{N-1}(x)$, namely the points (3.32) above (with N replaced by $N-1$). Note that the poles are all of equal magnitude (unity) and alternate in sign at the points (3.33) between -1 and 1. When $N = 5$, the poles and zeros of $T_N(x)$ correspond to the projections of "○" and "□" onto the interval $[-1,\ 1]$ as shown in Fig. 3.8, respectively.

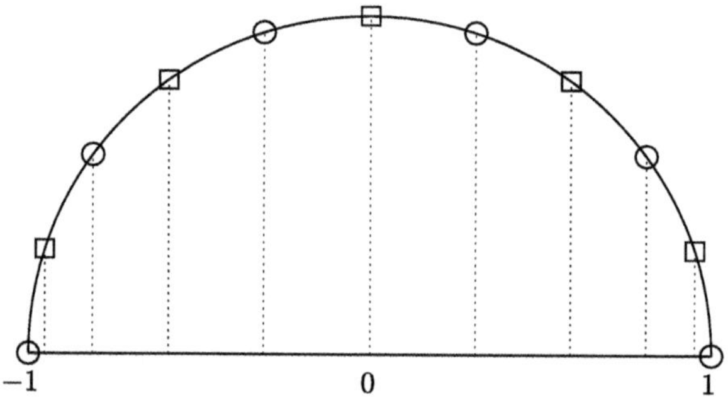

Fig. 3.8 Poles and zeros of $T_5(x)$

Poles of U_N

The poles of $U_N(x)$ are not in general as readily determined. Indeed, finding them involves the solution of transcendental equation

$$\frac{\mathrm{d}}{\mathrm{d}x}U_N(x) = \frac{\mathrm{d}}{\mathrm{d}x}\frac{\sin((N+1)\theta)}{\sin\theta} = \frac{-(N+1)\sin\theta\cos((N+1)\theta)+\cos\theta\sin((N+1)\theta)}{\sin^3\theta}.$$

The poles, therefore, correspond to values of θ satisfying the equation

$$\tan((N+1)\theta) = (N+1)\tan\theta \neq 0.$$

Unlike $T_N(x)$, the poles of $U_N(x)$ cannot be obtained through simple derivation. However, it is certain that the extreme values of $U_N(x)$ have magnitudes which increase monotonically as $|x|$ increases away from 0, until the largest magnitude of $N+1$ is achieved at $x = \pm 1$.

Based on (3.29), one can construct the weighted polynomial

$$\sqrt{1-x^2}U_N(x) = \sin((N+1)\theta)$$

whose poles can be explicitly determined, at

$$x_k = \cos\frac{(2k+1)\pi}{2(N+1)}, \quad k = 0,\ 1,\ \ldots,\ N.$$

When $N = 4$, the poles and zeros of $U_N(x)$ correspond to the projections of "∘" and "□" onto the interval $[-1,\ 1]$ in Fig. 3.9, respectively.

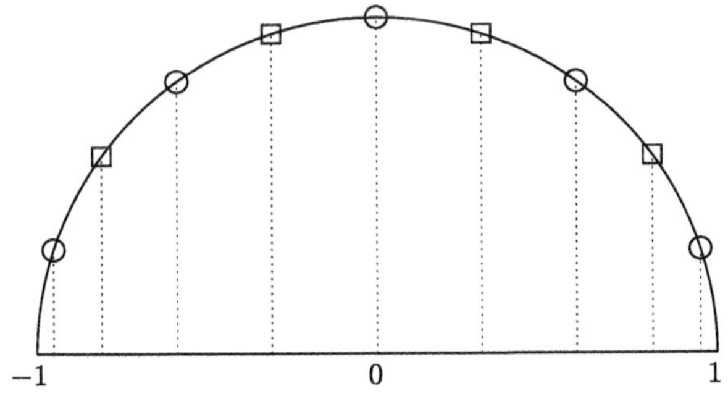

Fig. 3.9 Poles and zeros of $U_4(x)$

3.3.3.4 Connections Between T_N and U_N

The recurrence relation for the N-th degree Chebyshev polynomial $T_N(x)$ of the first kind is

$$T_{N+1}(x) = 2xT_N(x) - T_{N-1}(x), \quad N = 1,\ 2,\ \ldots$$

together with the initial conditions $T_0(x) = 1$ and $T_1(x) = x$. The recurrence relation for the N-th degree Chebyshev polynomial $U_N(x)$ of the second kind is

$$U_{N+1}(x) = 2xU_N(x) - U_{N-1}(x), \quad N = 1,\ 2,\ \ldots \tag{3.34}$$

whose initial conditions are $U_0(x) = 1$ and $U_1(x) = 2x$.

In addition, $T_N(x)$ and $U_N(x)$ have the following relation:

$$T_N(x) = U_N(x) - xU_{N-1}(x) \tag{3.35}$$

$$2T_N(x) = U_N(x) - U_{N-2}(x)$$

$$T_N'(x) = NU_{N-1}(x).$$

3.3.3.5 Chebyshev Differentiation Matrix [13]

Differentiation Matrix

Suppose that the values of any N-th degree polynomial $p(x)$ at $N+1$ points $x_0,\ x_1,\ \ldots,\ x_N$ is known. Then these values determine the polynomial uniquely, and so determine the values of the derivatives $p'(x) = \mathrm{d}p(x)/\mathrm{d}x$ at the same $N+1$ points. Each such derivative can, in fact, be expressed as a fixed linear combination of the given function values, and the whole relationship can be written in matrix form

$$\begin{bmatrix} p'(x_0) \\ \vdots \\ p'(x_N) \end{bmatrix} = \boldsymbol{D}_N \begin{bmatrix} p(x_0) \\ \vdots \\ p(x_N) \end{bmatrix} \tag{3.36}$$

where $\boldsymbol{D}_N \in \mathbb{R}^{(N+1)\times(N+1)}$ is called the differentiation matrix.

Suppose now that the points x_j are the $N+1$ zeros of some $(N+1)$-th degree polynomial $P_{N+1}(x)$. For $j = 0,\ 1,\ \ldots,\ N$, let $p_j(x) = P_{N+1}(x)/(x - x_j)$, which is an N-th degree polynomial since x_j is a zero of P_{N+1}, then one can obtain without much difficulty that

$$p_j(x_j) = P_{N+1}'(x_j)$$

$$p_j(x_i) = 0,\ i \neq j$$

$$p_j'(x_j) = \frac{1}{2} P_{N+1}''(x_j)$$

$$p_j'(x_i) = \frac{P_{N+1}'(x_i)}{x_i - x_j}, \ i \neq j.$$

From this, one can deduce (by setting $p(x) = p_j(x)$ in (3.36)) that the j-th column of the differentiation matrix $\boldsymbol{D}_N$ must have elements

$$\boldsymbol{D}_N(j, j) = \frac{P_{N+1}''(x_j)}{2P_{N+1}'(x_j)} \tag{3.37}$$

and

$$\boldsymbol{D}_N(i, j) = \frac{P_{N+1}'(x_i)}{(x_i - x_j)P_{N+1}'(x_j)}, \ i \neq j. \tag{3.38}$$

For instance, when $N = 1$, the interpolation points are $x_0 = 1$ and $x_1 = -1$, respectively, and the corresponding interpolation function values are $p(x_0) = u_0$ and $p(x_1) = u_1$, respectively. The interpolation polynomial on the interval $[-1, \ 1]$ is rewritten in the Lagrange form

$$p(x) = \frac{1}{2}(1 + x)u_0 + \frac{1}{2}(1 - x)u_1$$

whose derivative is

$$p'(x) = \frac{1}{2}u_0 - \frac{1}{2}u_1$$

and the corresponding differentiation matrix $\boldsymbol{D}_1$ is given by

$$\boldsymbol{D}_1 = \begin{bmatrix} 1/2 & -1/2 \\ 1/2 & -1/2 \end{bmatrix}.$$

When $N = 2$, the interpolation points are $x_0 = 1$, $x_1 = 0$ and $x_2 = -1$. Correspondingly, the interpolation function values are $p(x_0) = u_0$, $p(x_1) = u_1$ and $p(x_2) = u_2$, respectively. The interpolation polynomial on the interval $[-1, \ 1]$ is in the second order, that is

$$p(x) = \frac{1}{2}x(1 + x)u_0 + (1 + x)(1 - x)u_1 + \frac{1}{2}x(x - 1)u_2.$$

The derivative of $p(x)$ is a linear polynomial

$$p'(x) = \left(x + \frac{1}{2}\right)u_0 - 2xu_1 + \left(x - \frac{1}{2}\right)u_2$$

which corresponds to the differentiation matrix $\boldsymbol{D}_2$ described by

$$\boldsymbol{D}_2 = \begin{bmatrix} 3/2 & -2 & 1/2 \\ 1/2 & 0 & -1/2 \\ -1/2 & 2 & -3/2 \end{bmatrix}.$$

Notice that $\boldsymbol{D}_N$ is singular in each case, as expected.

Chebyshev Differentiation Matrix

In particular, suppose that the $N + 1$ points are $y_j = \cos(j\pi/N)$, which are the zeros of the polynomial $P_{N+1}(x) = (1 - x^2)U_{N-1}(x)$ and the poles in $[-1,\ 1]$ of $T_N(x)$. Making the usual substitution $x = \cos\theta$, one has

$$P_{N+1}(x) = \sin\theta \sin(N\theta).$$

Differentiating $P_{N+1}(x)$ with respect to x, one then obtains

$$P'_{N+1}(x) = -\frac{\cos\theta \sin(N\theta) + N \sin\theta \cos(N\theta)}{\sin\theta} \tag{3.39}$$

and

$$P''_{N+1}(x) = -\frac{\cos^2\theta \sin(N\theta) - N \sin\theta \cos\theta \cos(N\theta) + (1 + N^2)\sin^2\theta \sin(N\theta)}{\sin^3\theta}. \tag{3.40}$$

When $\theta_j = j\pi/N$, then $\sin(N\theta_j) = 0$ and $\cos(N\theta_j) = (-1)^j$. Therefore, (3.39) gives

$$P'_{N+1}(y_j) = \begin{cases} -(-1)^j N, & 0 < j < N \\ -2N, & j = 0 \\ -2(-1)^N N, & j = N \end{cases} \tag{3.41}$$

and (3.40) gives

$$P''_{N+1}(y_j) = \begin{cases} (-1)^j N \dfrac{y_j}{1 - y_j^2}, & 0 < j < N \\ -2N \dfrac{1 + 2N^2}{3}, & j = 0\,\cdot \\ 2(-1)^N N \dfrac{1 + 2N^2}{3}, & j = N \end{cases} \tag{3.42}$$

Substituting (3.41) and (3.42) into (3.37) and (3.38), the elements of Chebyshev differentiation matrix $\boldsymbol{D}_N$ can be expressed as

$$\boldsymbol{D}_N(i,\ j) = \frac{(-1)^{i-j}}{y_i - y_j},\ 0 < i \neq j < N,\ \boldsymbol{D}_N(j,\ j) = \frac{-y_j}{2(1-y_j^2)},\ 0 < j < N$$

$$\boldsymbol{D}_N(0,\ 0) = \frac{2N^2+1}{6},\ \boldsymbol{D}_N(N,\ N) = -\frac{2N^2+1}{6}$$

$$\boldsymbol{D}_N(0,\ N) = \frac{1}{2}(-1)^N,\ \boldsymbol{D}_N(N,\ 0) = -\frac{1}{2}(-1)^N$$

$$\boldsymbol{D}_N(0,\ j) = 2\frac{(-1)^j}{1-y_j},\ \boldsymbol{D}_N(N,\ j) = -2\frac{(-1)^{N-j}}{1+y_j},\ 0 < j < N$$

$$\boldsymbol{D}_N(i,\ 0) = -\frac{1}{2}\frac{(-1)^i}{1-y_i},\ \boldsymbol{D}_N(i,\ N) = \frac{1}{2}\frac{(-1)^{i-N}}{1+y_i},\ 0 < i < N$$

Notice that if $p(x_0) = p(x_1) = ... = p(x_N) = 1$, then $p(x)$ is a constant function, i.e., $p(x) \equiv 1$ and $p'(x) \equiv 0$. It follows that each row of the matrix $\boldsymbol{D}_N$ must sum to zero, so that the matrix is singular. In practice, to ensure the numerical stability of $\boldsymbol{D}_N$, the diagonal elements of $\boldsymbol{D}_N$ can be calculated as

$$\boldsymbol{D}_N(i,\ i) = -\sum_{j=0,\ j\neq i}^{N} \boldsymbol{D}_N(i,\ j).$$

3.4 Spectral Transform

To improve the convergence rate of sequential and Krylov subspace-based eigenvalue calculation methods, the critical eigenvalues of spectral discretization matrices are supposed to be transformed into the ones with the largest modulus and sparsely distributed. Then, the small-signal stability of time-delay systems can be reliably and efficiently determined by computing only a few eigenvalues. This section elaborates three kinds of spectral transforms for time-delay systems, including the shift-invert transform and Cayley transform for PIGD, and the rotation-and-multiplication preconditioning for PSOD.

3.4.1 Shift-Invert Transform

The shift-invert transform features in preferentially computing a subset of eigenvalues of the partial discretization matrix $\mathcal{A}_N$ that are close to the pre-specified shift points. The transform can be implemented by the following two schemes.

3.4.1.1 DS Scheme

In the discretization-to-shift (DS) scheme, both the shift and invert operations are directly executed on $\mathcal{A}_N$ [14]. The eigenvalues closer to the origin point can be obtained with higher accuracy [15].

Given a complex shift point λ_s, one has

$$\boldsymbol{S} = (\mathcal{A}_N - \lambda_s \boldsymbol{I})^{-1}.$$

Supposed that the eigenpair of $\mathcal{A}_N$ is denoted by $(\lambda_A, \ \boldsymbol{v}_A)$, one can obtain

$$(\mathcal{A}_N - \lambda_s \boldsymbol{I})^{-1} \boldsymbol{v}_A = \frac{1}{\lambda_A - \lambda_s} \boldsymbol{v}_A. \tag{3.43}$$

Thus, the relationship between the eigenpair of $\boldsymbol{S}$, i.e., $(\lambda_S, \ \boldsymbol{v}_S)$, and $(\lambda_A, \ \boldsymbol{v}_A)$ is that

$$\lambda_S = \frac{1}{\lambda_A - \lambda_s}, \quad \boldsymbol{v}_S = \boldsymbol{v}_A.$$

3.4.1.2 SD Scheme

In the shift-to-discretization (SD) scheme, the shift operation is executed on $\mathcal{A}$ and the invert operation is then carried out on the shifted discretization matrix $\mathcal{A}'_N$. This results in that the obtained eigenvalues closer to the shift point have higher accuracy.

Given a shift point λ_s, by replacing λ in the characteristic equation of the time-delay system with $\lambda' + \lambda_s$, one can obtain

$$\left(\tilde{\boldsymbol{A}}'_0 + \sum_{i=1}^{m} \tilde{\boldsymbol{A}}'_i \mathrm{e}^{-\lambda' \tau_i} \right) \boldsymbol{v} = \lambda' \boldsymbol{v}$$

or

$$\left(\boldsymbol{J}'_0 + \sum_{i=1}^{m} \boldsymbol{J}'_i \mathrm{e}^{-\lambda' \tau_i} \right) \hat{\boldsymbol{v}} = \lambda' \boldsymbol{E} \hat{\boldsymbol{v}}$$

where

$$\lambda' = \lambda - \lambda_s$$

$$\tilde{\boldsymbol{A}}'_0 = \boldsymbol{A}_0 - \lambda_s \boldsymbol{I}_n - \boldsymbol{B}_0 \boldsymbol{D}_0^{-1} \boldsymbol{C}_0$$

$$\tilde{\boldsymbol{A}}'_i = \boldsymbol{A}_i \mathrm{e}^{-\lambda_s \tau_i} - \boldsymbol{B}_i \mathrm{e}^{-\lambda_s \tau_i} \boldsymbol{D}_0^{-1} \boldsymbol{C}_0, \ i = 1, \ 2, \ \ldots, \ m$$

$$
J_0' = J_0 - \lambda_s E = \left[\begin{array}{cc|cc} A_{11,0} - \lambda_s I_{n_1} & A_{12,0} & B_{11,0} & B_{12,0} \\ A_{21,0} & A_{22,0} - \lambda_s I_{n_2} & B_{21,0} & B_{22,0} \\ \hline C_{11,0} & C_{12,0} & D_{11,0} & D_{12,0} \\ C_{21,0} & C_{22,0} & D_{21,0} & D_{22,0} \end{array}\right]
$$

$$
J_i' = J_i \mathrm{e}^{-\lambda_s \tau_i} = \left[\begin{array}{cc|cc} 0 & A_{12,i}\mathrm{e}^{-\lambda_s \tau_i} & 0 & B_{12,i}\mathrm{e}^{-\lambda_s \tau_i} \\ 0 & A_{22,i}\mathrm{e}^{-\lambda_s \tau_i} & 0 & B_{22,i}\mathrm{e}^{-\lambda_s \tau_i} \\ \hline \multicolumn{2}{c|}{0} & \multicolumn{2}{c}{0} \end{array}\right], \quad i = 1,\ 2,\ \ldots,\ m.
$$

For subsequent use, the following definitions are given:

$$
A_0' = A_0 - \lambda_s I_n = \left[\begin{array}{c|c} A_{11,0} - \lambda_s I_{n_1} & A_{12,0} \\ \hline A_{21,0} & A_{22,0} - \lambda_s I_{n_2} \end{array}\right] \triangleq \left[\begin{array}{c|c} A_0'^{(1)} & A_0'^{(2)} \end{array}\right] \tag{3.44}
$$

$$
A_i' = A_i \mathrm{e}^{-\lambda_s \tau_i} = \left[\begin{array}{c|c} 0_{n_1} & A_{12,i}\mathrm{e}^{-\lambda_s \tau_i} \\ \hline 0_{n_2 \times n_1} & A_{22,i}\mathrm{e}^{-\lambda_s \tau_i} \end{array}\right] \triangleq \left[0_{n \times n_1}\ \ A_i'^{(2)}\right], \quad i = 1,\ 2, \ldots, m \tag{3.45}
$$

$$
B_i' = B_i \mathrm{e}^{-\lambda_s \tau_i} = \left[\begin{array}{c|c} 0_{n_1 \times l_1} & B_{12,i}\mathrm{e}^{-\lambda_s \tau_i} \\ \hline 0_{n_2 \times l_1} & B_{22,i}\mathrm{e}^{-\lambda_s \tau_i} \end{array}\right] \triangleq \left[0_{n \times l_1}\ \ B_i'^{(2)}\right], \quad i = 1,\ 2, \ldots, m. \tag{3.46}
$$

Then, by directly replacing A_i $(i = 0,\ 1,\ \ldots,\ m)$ and B_i $(i = 1,\ 2,\ \ldots,\ m)$ as well as their submatrices in $\mathcal{A}_N$ with the corresponding shifted counterpart as shown in (3.44)–(3.46), one can obtain the shifted discretization matrix $\mathcal{A}_N'$ with eigenvalue being λ'.

Next, the invert operation yields $\left(\mathcal{A}_N'\right)^{-1}$, whose eigenvalue is $\lambda'' = 1/\lambda'$. It can be seen from Fig. 3.10 that the shift-invert transform maps eigenvalues of $\mathcal{A}_N$ closest to λ_s to those of $\left(\mathcal{A}_N'\right)^{-1}$ with the largest modulus, and the eigenvectors remain unchanged. Note that multiple shift points are required and multiple runs of the eigenvalue computation are executed in practical application, i.e., eigenvalue scanning, to compute all oscillation modes of interest.

3.4.2 Cayley Transform

In contrast to the shift-invert transform, the Cayley transform is used as an alternative to multiple shift-points, so that all critical eigenvalues can be obtained in one sequence.

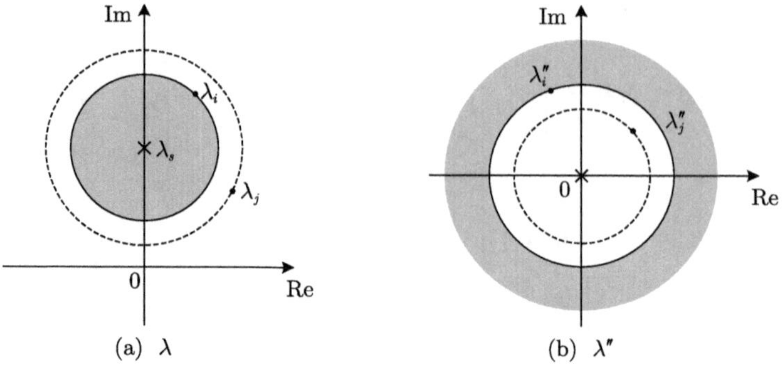

Fig. 3.10 Principle of the shift-invert transform

3.4.2.1 Real Cayley Transform

The Cayley transform is a particular linear-fractional transformation (also known as Mobius transformation), which is formulated as

$$\boldsymbol{S} = (\mathcal{A}_N - s_1\boldsymbol{I})(\mathcal{A}_N - s_2\boldsymbol{I})^{-1} \tag{3.47}$$

with complex shift points s_1 and s_2 ($s_1 \neq s_2$). The usual application, however, has real shift points.

According to (3.43), one has

$$(\mathcal{A}_N - s_1\boldsymbol{I})(\mathcal{A}_N - s_2\boldsymbol{I})^{-1}\boldsymbol{v}_A = \frac{\lambda_A - s_1}{\lambda_A - s_2}\boldsymbol{v}_A$$

and

$$\lambda_S = \frac{\lambda_A - s_1}{\lambda_A - s_2}, \ \lambda_A = \frac{s_2\lambda_S - s_1}{\lambda_S - 1}, \ \boldsymbol{v}_S = \boldsymbol{v}_A. \tag{3.48}$$

Let $\lambda_A = \sigma + \mathrm{j}\omega$, then the first item of (3.48) becomes

$$|\lambda_S| = \sqrt{\frac{(\sigma - s_1)^2 + \omega^2}{(\sigma - s_2)^2 + \omega^2}}$$

and can be further explicitly expressed as

$$|\lambda_S| \begin{cases} < 1, & \sigma < \dfrac{s_1 + s_2}{2} \\ = 1, & \sigma = \dfrac{s_1 + s_2}{2} \\ > 1, & \sigma > \dfrac{s_1 + s_2}{2} \end{cases}. \tag{3.49}$$

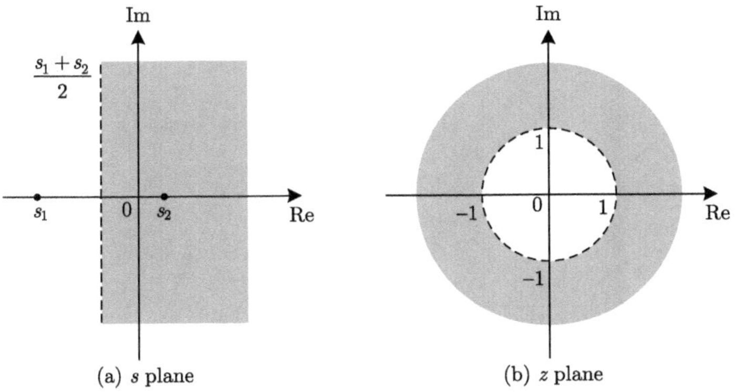

Fig. 3.11 Principle of the real Cayley transform

As shown in Fig. 3.11, the Cayley transform maps the symmetry axis (the vertical line with abscissa $\frac{s_1 + s_2}{2}$) of the s-plane to the unit circle of the z-plane. The right-half plane with respect to the symmetry axis is mapped outside the unit circle. Therefore, with a suitable selection of shift points, the symmetry axis can be placed in such a way that the critical (rightmost) eigenvalues of $\mathcal{A}_N$ are mapped to the dominant eigenvalues of $\boldsymbol{S}$ with the largest modulus.

Taking $s_1 = -s_2 = -3$ for example, two clusters of $|\lambda_S|$ isopleth curves in the s-plane are depicted in Fig. 3.12. It can be found that, an eigenvalue λ_A corresponds to λ_S with the larger modulus not necessarily located on the right of the one corresponds to λ_S with the smaller modulus.

3.4.2.2 Semi-Complex Cayley Transform

In the small-signal stability analysis, one generally focuses on the critical oscillation modes with damping ratio less than the specified threshold, which is denoted by ζ and formulated as $\zeta = \frac{-\sigma}{\sqrt{\sigma^2 + \omega^2}}$. A cluster of ζ isopleths in the s-plane are depicted in Fig. 3.13. The corresponding angle $\gamma = \arccos \zeta$ refers to the ζ isopleth subtends to the non-positive real axis of the s-plane.

Thus, the following describes a particular semi-complex Cayley transform of improved efficiency, by relaxing the requirement for the real shift points s_1 and s_2. As shown in Fig. 3.14, the symmetry axis is tilted and specified as the ζ_0 isopleth. Let $s_2 = a + \mathrm{j}b$, one can compute s_1 as

$$s_1 = (ac_1 - bc_2) - \mathrm{j}(ac_2 + bc_1)$$

with $c = \zeta_0,\ c_1 = 2c^2 - 1$ and $c_2 = 2c\sqrt{1 - c^2}$.

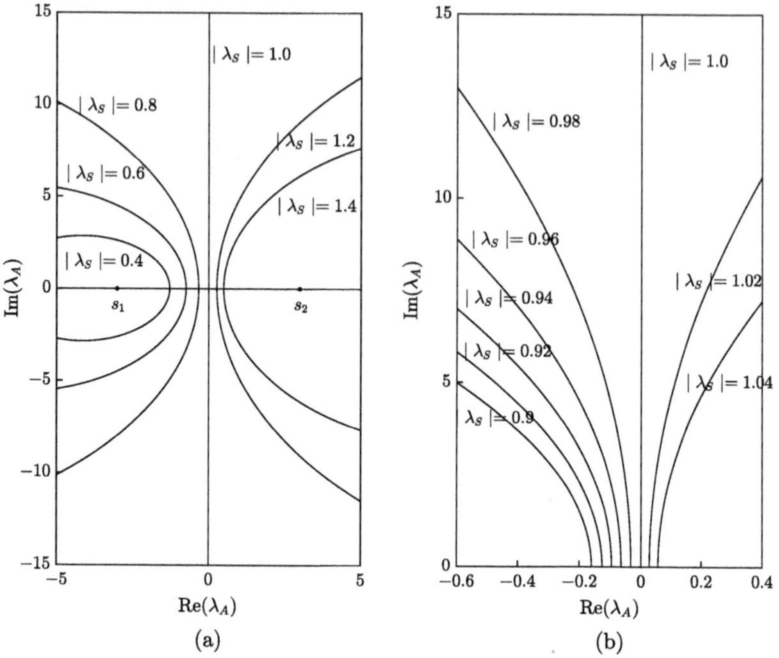

Fig. 3.12 Two clusters of $|\lambda_S|$ (as shown in (3.49)) isopleth curves

Fig. 3.13 A cluster of ζ isopleths on the s-plane

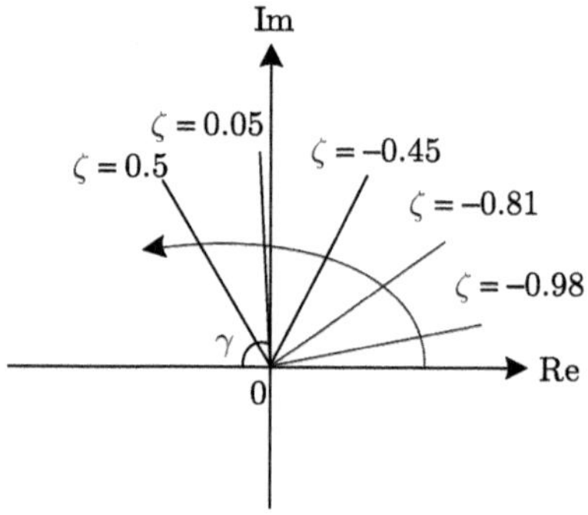

As shown in the shaded area of Fig. 3.14, critical eigenvalues λ_A with damping ratios less than ζ_0 in the s-plane is mapped to λ_S which are located outside the unit circle of the z-plane. Thus, those λ_S are computed by the sequential or Krylov subspace-based eigenvalue algorithms with priority.

Another common way to accomplish the same task is combining the real Cayley transform and the coordinate rotation [16]. As shown in Fig. 3.15, the coordinate can be rotated by $\theta = \arcsin \zeta$ radian in counterclockwise direction, leading to

$$\mathcal{A}'_N = \mathcal{A}_N \angle(-\theta) = \mathcal{A}_N \mathrm{e}^{-\mathrm{j}\theta}.$$

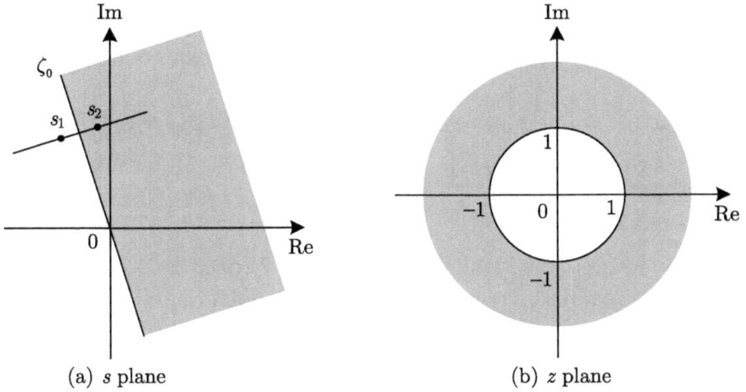

Fig. 3.14 Principle of the semi-complex Cayley transform

Fig. 3.15 Principle of the real Cayley transform combined with the coordinate rotation

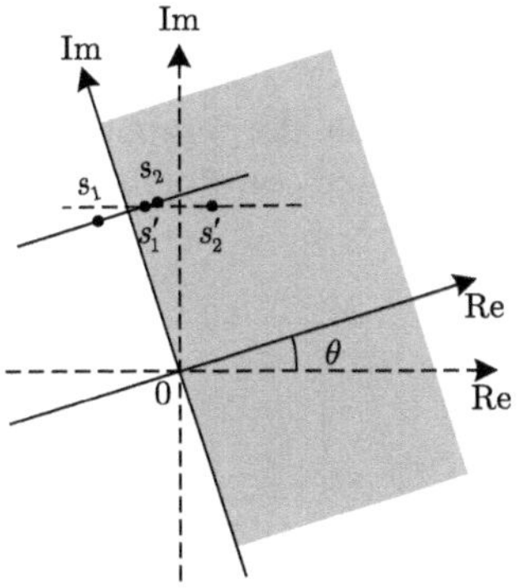

Supposed that the eigenpair of $\mathcal{A}_N'$ is denoted by $(\lambda_{A'},\ \boldsymbol{v}_{A'})$, then the relationship between it and $(\lambda_{\boldsymbol{A}},\ \boldsymbol{v}_{\boldsymbol{A}})$ is that

$$\lambda_{A'} = \lambda_A \angle(-\theta) = \lambda_A \mathrm{e}^{-\mathrm{j}\theta},\ \boldsymbol{v}_{A'} = \boldsymbol{v}_A.$$

Then, the right-half plane with respect to the rotated symmetry axis of the s-plane, i.e., the ζ_0 isopleth, is mapped outside the unit circle of the z-plane by implementing the real Cayley transform with shift points s_1' and s_2'

$$\left(\mathcal{A}_N' - s_1'\boldsymbol{I}\right)\left(\mathcal{A}_N' - s_2'\boldsymbol{I}\right)^{-1}\boldsymbol{v}_{A'} = \frac{\lambda_{A'} - s_1'}{\lambda_{A'} - s_2'}\boldsymbol{v}_{A'}$$

where the relationship between s_1', s_2', and s_1, s_2 in (3.47) satisfies

$$s_1 = s_1'\mathrm{e}^{-\mathrm{j}\theta},\ s_2 = s_2'\mathrm{e}^{-\mathrm{j}\theta}.$$

It should be noted that the semi-complex Cayley transform is equivalent to the combination of the real Cayley transform and the coordinate rotation, i.e.,

$$\begin{cases} \boldsymbol{S} = (\mathcal{A}_N - s_1\boldsymbol{I})(\mathcal{A}_N - s_2\boldsymbol{I})^{-1} = (\mathcal{A}'_N - s'_1\boldsymbol{I})(\mathcal{A}'_N - s'_2\boldsymbol{I})^{-1} \\ \lambda_S = \dfrac{\lambda_A - s_1}{\lambda_A - s_2} = \dfrac{\lambda_{A'} - s'_1}{\lambda_{A'} - s'_2}, \ \boldsymbol{v}_S = \boldsymbol{v}_{A'} \end{cases} . \tag{3.50}$$

Taking $s_1 = -3.2 + \mathrm{j}3.85$ and $s_2 = 2.8 + \mathrm{j}4.15$ (or $s'_1 = -3 + \mathrm{j}4$, $s'_2 = 3 + \mathrm{j}4$ and $\zeta_0 = 0.05$) for example, two clusters of $|\lambda_S|$ isopleth curves in the s-plane are depicted in Fig. 3.16.

3.4.3 Rotation-and-Multiplication Preconditioning

The rotation-and-multiplication preconditioning is applicable to the PSOD methods to facilitate the computation of critical electromechanical oscillation modes with an accelerated convergence rate [17].

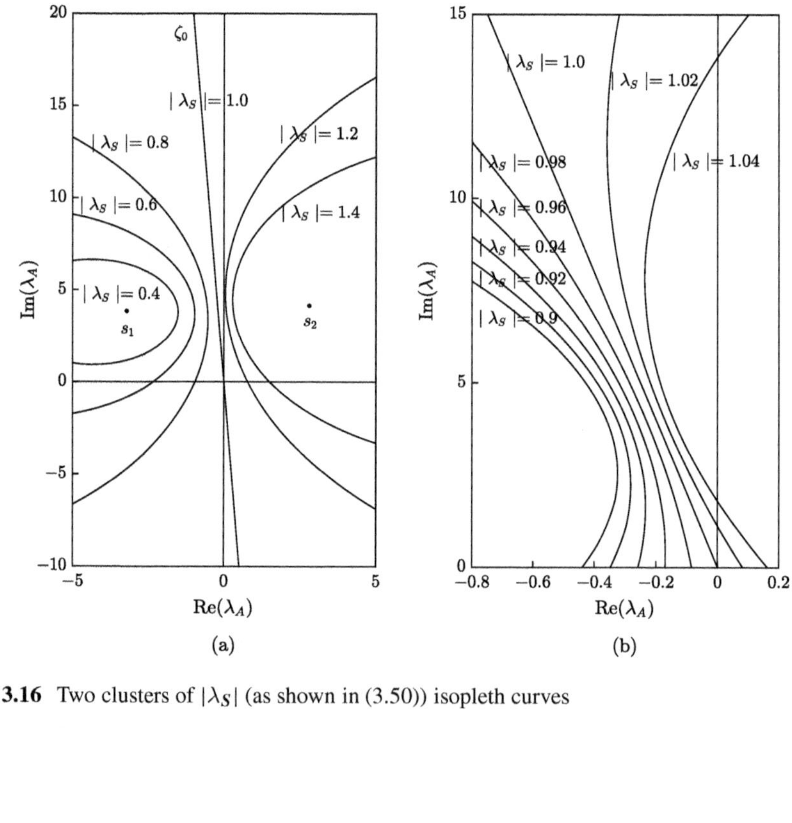

Fig. 3.16 Two clusters of $|\lambda_S|$ (as shown in (3.50)) isopleth curves

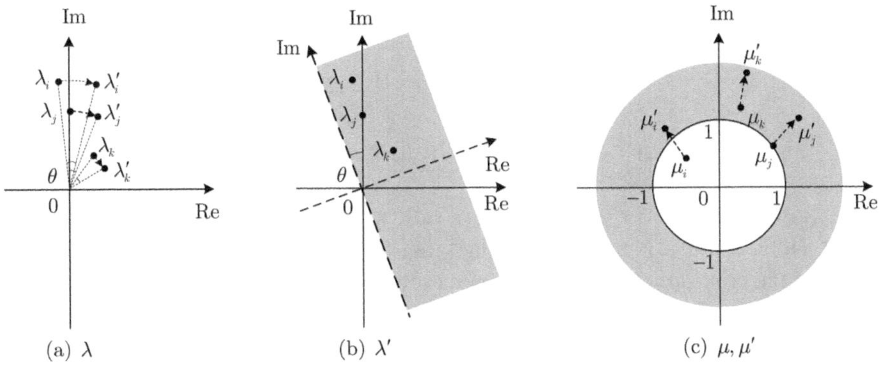

Fig. 3.17 Spectral mapping of the solution operator with coordinate rotation preconditioning

3.4.3.1 Inexact Coordinate Rotation

Suppose eigenvalues with damping ratios less than a threshold ζ are desired to be computed. To achieve this, one can rotate the coordinate by θ ($=\arcsin\zeta$) radian in counterclockwise direction [16]. According to the spectral mapping property, it is known that eigenvalues of the time-delay system located on the right of the rotated imaginary axis are transformed into those of the solution operator located outside the unit circle, as shown in Fig. 3.17.

Mathematically, the rotation operation can be accomplished by directly replacing λ with $\lambda' \mathrm{e}^{\mathrm{j}\theta}$ in the characteristic equation of the time-delay system, i.e.,

$$\left(\tilde{\boldsymbol{A}}_0' + \sum_{i=1}^{m} \tilde{\boldsymbol{A}}_i' \mathrm{e}^{-\lambda' \tau_i'}\right) \boldsymbol{v} = \lambda' \boldsymbol{v} \tag{3.51}$$

or

$$\left(\boldsymbol{J}_0' + \sum_{i=1}^{m} \boldsymbol{J}_i' \mathrm{e}^{-\lambda' \tau_i'}\right) \hat{\boldsymbol{v}} = \lambda' \boldsymbol{E} \hat{\boldsymbol{v}}$$

where

$$\begin{aligned}
\lambda' &= \lambda \mathrm{e}^{-\mathrm{j}\theta} \\
\tau_i' &= \tau_i \mathrm{e}^{\mathrm{j}\theta}, \quad i = 1, 2, \ldots, m \\
\tilde{\boldsymbol{A}}_i' &= \tilde{\boldsymbol{A}}_i \mathrm{e}^{-\mathrm{j}\theta}, \quad i = 0, 1, \ldots, m \\
\boldsymbol{J}_0' &\triangleq \begin{bmatrix} \boldsymbol{A}_0' & \boldsymbol{B}_0' \\ \boldsymbol{C}_0 & \boldsymbol{D}_0 \end{bmatrix} = \begin{bmatrix} \boldsymbol{A}_0 \mathrm{e}^{-\mathrm{j}\theta} & \boldsymbol{B}_0 \mathrm{e}^{-\mathrm{j}\theta} \\ \boldsymbol{C}_0 & \boldsymbol{D}_0 \end{bmatrix} \\
\boldsymbol{J}_i' &\triangleq \begin{bmatrix} \boldsymbol{A}_i' & \boldsymbol{B}_i' \\ \boldsymbol{0}_{l\times n} & \boldsymbol{0}_l \end{bmatrix} = \begin{bmatrix} \boldsymbol{A}_i \mathrm{e}^{-\mathrm{j}\theta} & \boldsymbol{B}_i \mathrm{e}^{-\mathrm{j}\theta} \\ \boldsymbol{0}_{l\times n} & \boldsymbol{0}_l \end{bmatrix}, \quad i = 1, 2, \ldots, m.
\end{aligned} \tag{3.52}$$

For subsequent use, the following definitions are given:

$$A_0'^{(1)} = A_0^{(1)}\mathrm{e}^{-\mathrm{j}\theta},\ \ B_0'^{(1)} = B_0^{(1)}\mathrm{e}^{-\mathrm{j}\theta}$$

$$A_i'^{(2)} = A_i^{(2)}\mathrm{e}^{-\mathrm{j}\theta},\ \ B_i'^{(2)} = B_i^{(2)}\mathrm{e}^{-\mathrm{j}\theta},\ \ i = 0,\ 1,\ \ldots,\ m.$$

It can be seen from (3.52) that τ_i' $(i = 1,\ 2,\ \ldots,\ m)$ become complex when $\theta \neq 0$. However, this contradicts the fact that the time step-length h should be real. To avoid the contradiction, the approximations

$$\tau_i' \approx \tau_i,\ \ i = 1,\ 2,\ \ldots,\ m$$

are necessary.

After this inexact coordinate rotation preconditioning, one has

$$\begin{cases} \mu' = \mathrm{e}^{\lambda' h} = \mathrm{e}^{\lambda \mathrm{e}^{-\mathrm{j}\theta} h} = \left(\mathrm{e}^{\lambda h}\right)^{\mathrm{e}^{-\mathrm{j}\theta}} = \mu^{\mathrm{e}^{-\mathrm{j}\theta}} \\ \lambda' = \dfrac{1}{h}\ln \mu' \end{cases}. \tag{3.53}$$

3.4.3.2 Effect of Delay Approximations

The time delay approximations in the coordinate rotation preconditioning can be expressed as

$$\Delta\tau_i = \tau_i - \tau_i' = \tau_i\left(1 - \mathrm{e}^{\mathrm{j}\theta}\right),\ \ i = 1,\ 2,\ \ldots,\ m. \tag{3.54}$$

Regarding $\Delta\tau_i$ as perturbations, the resulting error in estimating the rotated eigenvalue λ' can be approximated by applying the perturbation theory to the characteristic equation (3.51), i.e.,

$$\Delta\lambda' = -\frac{u^{\mathrm{T}}\left(\sum\limits_{i=1}^{m}\tilde{A}_i'\mathrm{e}^{-\lambda'\tau_i'}\lambda'\varepsilon\Delta\tau_i\right)v}{u^{\mathrm{T}}\left(I_n + \sum\limits_{i=1}^{m}\tilde{A}_i'\tau_i'\mathrm{e}^{-\lambda'\tau_i'}\right)v} = -\sum_{i=1}^{m}\frac{\partial\lambda}{\partial\tau_i}\tau_i\left(1 - \mathrm{e}^{\mathrm{j}\theta}\right). \tag{3.55}$$

It can be concluded from (3.55) that the bias $\Delta\lambda'$ resulted from delay approximations increases with the magnitudes of τ_i $(i = 1,\ 2,\ \ldots,\ m)$ and θ, and is largely affected by the sensitivity $\dfrac{\partial\lambda}{\partial\tau_i}$.

3.4.3.3 Multiplication Preconditioning

The spectral mapping (3.53) "compresses" λ' in the s-plane into μ' in the z-plane. The latter are closely distributed in the vicinity of the unit circle, which deteriorates the convergence rate of the sequential or Krylov subspace-based eigenvalue algorithms.

In theory, the convergence rate is determined by the dispersion $\dfrac{|\mu_i'| - |\mu_j'|}{|\mu_i'|}$, where μ_i' and μ_j' are two arbitrary eigenvalues. Therefore, nonlinear amplification of μ' is necessary which enhances its dispersion. Specifically, raising μ' to its α-th power (α is an integer and required to be >1) and using (3.53) yields

$$\mu'' \triangleq (\mu')^{\alpha} = \mathrm{e}^{\lambda'' h}$$

where

$$\lambda'' = \alpha \mathrm{e}^{-\mathrm{j}\theta} \lambda.$$

This implies that, the rotation-and-multiplication preconditioning indicates that the eigenvalues λ of the time-delay system are first rotated by θ radian and then amplified by α times.

3.4.3.4 Implementation of the Preconditioning

There are two implementations of the rotation-and-multiplication preconditioning.

(1) The first implementation can be summarized as: h is kept and

$$\tau_i'' = \frac{\tau_i \mathrm{e}^{\mathrm{j}\theta}}{\alpha} \approx \frac{\tau_i}{\alpha}, \ i = 1, 2, \ldots, m$$

$$\tilde{\boldsymbol{A}}_i'' = \alpha \tilde{\boldsymbol{A}}_i', \ i = 0, 1, \ldots, m$$

$$\boldsymbol{J}_0'' \triangleq \begin{bmatrix} \boldsymbol{A}_0'' & \boldsymbol{B}_0'' \\ \boldsymbol{C}_0 & \boldsymbol{D}_0 \end{bmatrix} = \begin{bmatrix} \alpha\boldsymbol{A}_0' & \alpha\boldsymbol{B}_0' \\ \boldsymbol{C}_0 & \boldsymbol{D}_0 \end{bmatrix}$$

$$\boldsymbol{J}_i'' \triangleq \begin{bmatrix} \boldsymbol{A}_i'' & \boldsymbol{B}_i'' \\ \boldsymbol{0}_{l\times n} & \boldsymbol{0}_l \end{bmatrix} = \begin{bmatrix} \alpha\boldsymbol{A}_i' & \alpha\boldsymbol{B}_i' \\ \boldsymbol{0}_{l\times n} & \boldsymbol{0}_l \end{bmatrix}, \ i = 1, 2, \ldots, m$$

$$\boldsymbol{A}_0''^{(1)} = \alpha \boldsymbol{A}_0'^{(1)}, \ \boldsymbol{B}_0''^{(1)} = \alpha \boldsymbol{B}_0'^{(1)}$$

$$\boldsymbol{A}_i''^{(2)} = \alpha \boldsymbol{A}_i'^{(i)}, \ \boldsymbol{B}_i''^{(2)} = \alpha \boldsymbol{B}_i'^{(i)}, \ i = 1, 2, \ldots, m.$$

(2) The second implementation can be summarized as: $\tilde{A}'_i$, J'_i, A'_i, B'_i ($i = 0, 1, \ldots, m$) and their submatrices as well as τ_i ($i = 1, 2, \ldots, m$) are kept, and h is amplified by α times.

3.4.3.5 Selection of α and θ

The parameters α and θ in the preconditioning should be carefully selected, because they greatly affect the convergence rate of the eigenvalue algorithms employed to compute μ'' from the preconditioned partial discretization matrices of the solution operator and hence deteriorate the efficiency and scalability of the PSOD methods.

(1) The factor $\alpha = 2$ or 3 is sufficient to improve the convergence of the PSOD methods from point of view of the authors. In the meanwhile, note that the selection of α should not violate the constraint $\alpha h \leq \tau_{\max}$.
(2) The selection of θ depends on different applications. On the one hand, if a majority of electromechanical oscillation modes are required to be computed, a larger value of θ is suggested. For example, $\theta = 5.74°$, which means $\zeta = 10\%$. On the other hand, if one wants to fast and reliably determine the system stability by computing few critical eigenvalues, a small value of θ can be specified, e.g., $\theta = 1.72°$ ($\zeta = 3\%$).

3.4.4 Summary

This subsection offers a full comparison of the above three kinds of spectral transforms for time-delay systems.

(1) The shift-invert transform features in computing eigenvalues close to the pre-specified shift points with high efficiency. However, multiple eigenvalue scannings with multiple initial shift points along the imaginary axis are required to ensure that all critical eigenvalues are reliably captured. The clusters of eigenvalues obtained by different runs are overlapped to some extent, indicating repeated and unnecessary computation. In addition, even if these eigenvalues have been accurately captured, one still cannot easily claim that all eigenvalues located on the right-half complex plane have been obtained. Hence, the system stability cannot be reliably evaluated.
(2) The Cayley transform is applied to the PIGD methods to capture rightmost or poorly damped eigenvalues of time-delay systems. In other words, a single duet of shift points and a single matrix factorization are sufficient. However, the advantages of the Cayley transform do not come without shortcomings. First, its convergence characteristics largely depends on the selection of shift points. Second, the transform reduces the relative differences in (the moduli of) the transformed eigenvalues, thereby reducing the convergence speed of eigen-analysis methods.

(3) The rotation-and-multiplication preconditioning provides a new way of efficiently computing the critical electromechanical oscillation modes with damping ratios less than a prespecified threshold by one run of the method. Thus, the small-signal stability of large-scale time-delay systems can be reliably determined. However, in contrast to the PIGD methods with the shift-invert transform, the PSOD methods with the rotation-and-multiplication preconditioning suffer from the problems of a large number of iterations and excessive computational time. The root cause is that the eigenvalues of the solution operator's discretization matrices are closely distributed in the vicinity of the unit circle, which is also encountered by the PIGD methods with the Cayley transform.

3.5 Spectral Estimation

Since real-life time-delay systems are generally large-scale and complex, enhancing the efficiency and scalability of PSD-based eigenvalue calculation methods is of critical importance. To resolve this, two critical techniques are discussed in this section.

3.5.1 Kronecker Product Transformation

It will be observed from the subsequent Chaps. 4 and 5 that, the blocks of the resultant discretized matrices of the infinitesimal generator and solution operator are essentially characterized by Kronecker products. Inspired by this, the unique property of Kronecker product is fully exploited in this subsection, which lays the basis for the highly efficient implementation of the PSD-based methods.

3.5.1.1 Definition

The Kronecker product (denoted by the symbol $\otimes$) is a special form of tensor product, named after the German mathematician Leopold Kronecker.

In essence, the Kronecker product is a binary matrix operator that maps two arbitrarily dimensioned matrices into a larger matrix with special block structure. Given matrices $\boldsymbol{A} \in \mathbb{C}^{m\times n}$ and $\boldsymbol{B} \in \mathbb{C}^{p\times q}$ described as

$$\boldsymbol{A} = \begin{bmatrix} a_{1,1} & a_{1,2} & \cdots & a_{1,n} \\ a_{2,1} & a_{2,2} & \cdots & a_{2,n} \\ \vdots & \vdots & & \vdots \\ a_{m,1} & a_{m,2} & \cdots & a_{m,n} \end{bmatrix}, \quad \boldsymbol{B} = \begin{bmatrix} b_{1,1} & b_{1,2} & \cdots & b_{1,q} \\ b_{2,1} & b_{2,2} & \cdots & b_{2,q} \\ \vdots & \vdots & & \vdots \\ b_{p,1} & b_{p,2} & \cdots & b_{p,q} \end{bmatrix}$$

their Kronecker product $\boldsymbol{A} \otimes \boldsymbol{B}$ is the $mp \times nq$ matrix with the following block structure:

$$\boldsymbol{A} \otimes \boldsymbol{B} = \begin{bmatrix} a_{1,1}\boldsymbol{B} & a_{1,2}\boldsymbol{B} & \cdots & a_{1,n}\boldsymbol{B} \\ a_{2,1}\boldsymbol{B} & a_{2,2}\boldsymbol{B} & \cdots & a_{2,n}\boldsymbol{B} \\ \vdots & \vdots & & \vdots \\ a_{m,1}\boldsymbol{B} & a_{m,2}\boldsymbol{B} & \cdots & a_{m,n}\boldsymbol{B} \end{bmatrix}.$$

3.5.1.2 Properties

Referring to [18], some properties of the Kronecker product are listed in the following. It is assumed that $\boldsymbol{A}$, $\boldsymbol{B}$, $\boldsymbol{C}$, and $\boldsymbol{D}$ are real valued matrices with appropriate dimensions.

(1) Scalar multiplication. Given $\alpha \in \mathbb{R}$,

$$(\alpha\boldsymbol{A}) \otimes \boldsymbol{B} = \boldsymbol{A} \otimes (\alpha\boldsymbol{B}) = \alpha(\boldsymbol{A} \otimes \boldsymbol{B}).$$

(2) Transpose and conjugate transpose:

$$(\boldsymbol{A} \otimes \boldsymbol{B})^{\mathrm{T}} = \boldsymbol{A}^{\mathrm{T}} \otimes \boldsymbol{B}^{\mathrm{T}}$$
$$(\boldsymbol{A} \otimes \boldsymbol{B})^{\mathrm{H}} = \boldsymbol{A}^{\mathrm{H}} \otimes \boldsymbol{B}^{\mathrm{H}}.$$

(3) Associative law:

$$(\boldsymbol{A} \otimes \boldsymbol{B}) \otimes \boldsymbol{C} = \boldsymbol{A} \otimes (\boldsymbol{B} \otimes \boldsymbol{C}).$$

(4) Distributive law:

$$(\boldsymbol{A} + \boldsymbol{C}) \otimes \boldsymbol{B} = \boldsymbol{A} \otimes \boldsymbol{B} + \boldsymbol{C} \otimes \boldsymbol{B}$$
$$\boldsymbol{A} \otimes (\boldsymbol{B} + \boldsymbol{C}) = \boldsymbol{A} \otimes \boldsymbol{B} + \boldsymbol{A} \otimes \boldsymbol{C}.$$

(5) Matrix multiplication:

$$(\boldsymbol{A} \otimes \boldsymbol{B})(\boldsymbol{C} \otimes \boldsymbol{D}) = (\boldsymbol{A}\boldsymbol{C}) \otimes (\boldsymbol{B}\boldsymbol{D})$$
$$(\boldsymbol{A} \otimes \boldsymbol{B})(\boldsymbol{C} \otimes \boldsymbol{D})(\boldsymbol{X} \otimes \boldsymbol{Y}) = (\boldsymbol{A}\boldsymbol{C}\boldsymbol{X}) \otimes (\boldsymbol{B}\boldsymbol{D}\boldsymbol{Y}).$$

(6) Trace:

$$\mathrm{trace}(\boldsymbol{A} \otimes \boldsymbol{B}) = \mathrm{trace}(\boldsymbol{B} \otimes \boldsymbol{A}) = \mathrm{trace}(\boldsymbol{A})\mathrm{trace}(\boldsymbol{B}).$$

(7) Determinant. For any square matrices $\boldsymbol{A} \in \mathbb{C}^{m \times m}$ and $\boldsymbol{B} \in \mathbb{C}^{n \times n}$,

$$\det(\boldsymbol{A} \otimes \boldsymbol{B}) = \det(\boldsymbol{B} \otimes \boldsymbol{A}) = (\det(\boldsymbol{A}))^{n} (\det(\boldsymbol{B}))^{m}.$$

(8) Inverse. For any non-singular square matrices $\boldsymbol{A} \in \mathbb{C}^{m\times m}$ and $\boldsymbol{B} \in \mathbb{C}^{n\times n}$,

$$(\boldsymbol{A} \otimes \boldsymbol{B})^{-1} = \boldsymbol{A}^{-1} \otimes \boldsymbol{B}^{-1}.$$

(9) Kronecker product transformation. The following equations

$$\boldsymbol{A}\boldsymbol{X} = \boldsymbol{B}$$

$$\boldsymbol{A}\boldsymbol{X} + \boldsymbol{X}\boldsymbol{B} = \boldsymbol{C}$$

$$\boldsymbol{A}\boldsymbol{X}\boldsymbol{B} = \boldsymbol{C}$$

$$\boldsymbol{A}\boldsymbol{X} + \boldsymbol{Y}\boldsymbol{B} = \boldsymbol{C}$$

can be equivalently transformed into

$$(\boldsymbol{I} \otimes \boldsymbol{A})\text{vec}(\boldsymbol{X}) = \text{vec}(\boldsymbol{B})$$

$$\left((\boldsymbol{I} \otimes \boldsymbol{A}) + \left(\boldsymbol{B}^{\text{T}} \otimes \boldsymbol{I}\right)\right)\text{vec}(\boldsymbol{X}) = \left(\boldsymbol{A} \oplus \boldsymbol{B}^{\text{T}}\right)\text{vec}(\boldsymbol{X}) = \text{vec}(\boldsymbol{C})$$

$$\left(\boldsymbol{B}^{\text{T}} \otimes \boldsymbol{A}\right)\text{vec}(\boldsymbol{X}) = \text{vec}(\boldsymbol{A}\boldsymbol{X}\boldsymbol{B}) = \text{vec}(\boldsymbol{C}) \tag{3.56}$$

$$(\boldsymbol{I} \otimes \boldsymbol{A})\text{vec}(\boldsymbol{X}) + \left(\boldsymbol{B}^{\text{T}} \otimes \boldsymbol{I}\right)\text{vec}(\boldsymbol{Y}) = \text{vec}(\boldsymbol{C})$$

respectively, where vec$(\cdot)$ means the vectorization of a matrix, i.e., stacking the columns of the matrix on top of one another:

$$\text{vec}(\boldsymbol{A}) = [a_{1,1},\ \ldots,\ a_{m,1},\ a_{1,2},\ \ldots,\ a_{m,2},\ \ldots,\ a_{1,n},\ \ldots,\ a_{m,n}]^{\text{T}}.$$

3.5.1.3 Application in Spectral Discretization Matrices

Taking the PSOD-PS method that will be given in Chap. 5 for example, the block matrix $\boldsymbol{M}_2 \in \mathbb{R}^{Nn\times((Q-1)M+1)n_2}$ can be reformulated as the sum of Kronecker products of the Lagrange coefficient vectors $\boldsymbol{L}^i_{M,N,2} \in \mathbb{R}^{N\times((Q-1)M+1)}$ and augmented system state matrices $\boldsymbol{A}^{(2)}_i$ $(i = 1,\ 2,\ \ldots,\ m)$, namely

$$\boldsymbol{M}_2 = \sum_{i=1}^{m} \boldsymbol{L}^i_{M,N,2} \otimes \boldsymbol{A}^{(2)}_i.$$

By utilizing the unique property (3.56), the product of $\boldsymbol{M}_2$ and the vector $\boldsymbol{q} \in \mathbb{C}^{((Q-1)M+1)n_2\times 1}$ can be efficiently computed as

$$\boldsymbol{M}_2\boldsymbol{q} = \sum_{i=1}^{m}\left(\boldsymbol{L}^{i}_{M,N,2} \otimes \boldsymbol{A}_i^{(2)}\right)\text{vec}(\boldsymbol{Q}) = \sum_{i=1}^{m}\boldsymbol{A}_i^{(2)}\left(\boldsymbol{Q}\left(\tilde{\boldsymbol{L}}^{i}_{M,N,2}\right)^{\text{T}}\right)$$

where $\boldsymbol{Q} \in \mathbb{C}^{n_2\times((Q-1)M+1)}$ satisfies $\text{vec}(\boldsymbol{Q}) = \boldsymbol{q}$.

3.5.2 *Sparse Eigenvalue Computation Using IRA Algorithm*

In the small-signal stability analysis and control of large-scale time-delay systems, it is necessary to preferentially compute a subset of critical eigenvalues with the smallest damping ratios or the largest real parts. Therefore, the implicitly restarted Arnoldi (IRA) algorithm [19–22], a kind of powerful Krylov subspace-based method, is generally applied. In the meanwhile, the inherent sparsity of the augmented system state matrix is required to be fully exploited to guarantee the capability of the PSD-based methods in analyzing large-scale time-delay systems.

3.5.2.1 MVP and MIVP

In the IRA algorithm, the most computationally expensive operation is the matrix-vector product (MVP) and matrix inversion-vector product (MIVP) associated with the system state matrix $\tilde{\boldsymbol{A}}$ to generate Krylov sequences $\boldsymbol{q}_k$ ($k = 0,\ 1,\ \ldots$), i.e.,

$$\boldsymbol{q}_{k+1} = \tilde{\boldsymbol{A}}\boldsymbol{q}_k \tag{3.57}$$

and

$$\boldsymbol{q}_{k+1} = \tilde{\boldsymbol{A}}^{-1}\boldsymbol{q}_k. \tag{3.58}$$

The general form of $\tilde{\boldsymbol{A}}$ is $\tilde{\boldsymbol{A}} = \boldsymbol{f}_A - \boldsymbol{f}_B\boldsymbol{D}_0^{-1}\boldsymbol{f}_C$, where $\boldsymbol{f}_A$, $\boldsymbol{f}_B$ and $\boldsymbol{f}_C$ are the simple functions of the augmented system state matrices $\boldsymbol{A}_i$, $\boldsymbol{B}_i$ ($i = 1,\ 2,\ \ldots,\ m$) and $\boldsymbol{C}_0$, respectively, and exhibit the high sparsity. Efficient implementations of (3.57) and (3.58) are given below.

3.5.2.2 Efficiently Computing (3.57)

Equation (3.57) can be decomposed into

$$\boldsymbol{D}_0\boldsymbol{z} = -\boldsymbol{f}_C\boldsymbol{q}_k$$

and

$$\boldsymbol{q}_{k+1} = \boldsymbol{f}_A \boldsymbol{q}_k + \boldsymbol{f}_B \boldsymbol{z}$$

which can be also expressed in the augmented form

$$\begin{bmatrix} \boldsymbol{q}_{k+1} \\ \boldsymbol{0} \end{bmatrix} = \begin{bmatrix} \boldsymbol{f}_A & \boldsymbol{f}_B \\ \boldsymbol{f}_C & \boldsymbol{D}_0 \end{bmatrix} \begin{bmatrix} \boldsymbol{q}_k \\ \boldsymbol{z} \end{bmatrix}.$$

In the special case where $\boldsymbol{f}_A$ and $\boldsymbol{f}_B$ are highly sparse, one can utilize

$$\tilde{\boldsymbol{A}} = \boldsymbol{f}_A - \left[\boldsymbol{f}_C^{\mathrm{T}} \left(\left(\boldsymbol{D}_0^{-1} \right)^{\mathrm{T}} \boldsymbol{f}_B^{\mathrm{T}} \right) \right]^{\mathrm{T}}$$

to explicitly obtain $\tilde{\boldsymbol{A}}$ and directly compute (3.57).

3.5.2.3 Efficiently Computing (3.58)

The Duncan-Guttman Formula is applied here to compute $\tilde{\boldsymbol{A}}^{-1}$ as

$$\begin{aligned} \tilde{\boldsymbol{A}}^{-1} &= \left(\boldsymbol{f}_A - \boldsymbol{f}_B \boldsymbol{D}_0^{-1} \boldsymbol{f}_C \right)^{-1} \\ &= \boldsymbol{f}_A^{-1} + \boldsymbol{f}_A^{-1} \boldsymbol{f}_B \left(\boldsymbol{D}_0 - \boldsymbol{f}_C \boldsymbol{f}_A^{-1} \boldsymbol{f}_B \right)^{-1} \boldsymbol{f}_C \boldsymbol{f}_A^{-1} \end{aligned}$$

whose core is to avoid explicitly forming the matrix $\tilde{\boldsymbol{A}}$. Thus, (3.58) can be decomposed into

$$\left(\boldsymbol{D}_0 - \boldsymbol{f}_C \boldsymbol{f}_A^{-1} \boldsymbol{f}_B \right) \boldsymbol{z} = \boldsymbol{f}_C \boldsymbol{f}_A^{-1} \boldsymbol{q}_k$$

and

$$\boldsymbol{q}_{k+1} = \boldsymbol{f}_A^{-1} \left(\boldsymbol{q}_k + \boldsymbol{f}_B \boldsymbol{z} \right)$$

which can be also expressed in the augmented form

$$\begin{bmatrix} \boldsymbol{f}_A & \boldsymbol{f}_B \\ \boldsymbol{f}_C & \boldsymbol{D}_0 \end{bmatrix} \begin{bmatrix} \boldsymbol{q}_{k+1} \\ \boldsymbol{z} \end{bmatrix} = \begin{bmatrix} \boldsymbol{q}_k \\ \boldsymbol{0} \end{bmatrix}.$$

It should be pointed out that, the matrix $\boldsymbol{D}^* = \boldsymbol{D}_0 - \boldsymbol{f}_C \boldsymbol{f}_A^{-1} \boldsymbol{f}_B$ exhibits the same sparsity as $\boldsymbol{D}_0$. Thus, the LU decomposition and inversion of $\boldsymbol{D}^*$ can be efficiently performed.

3.6 Spectral Correction

The obtained eigenvalues of the partial discretization matrices are the approximate counterparts of the time-delay system. Therefore, the Newton's method [23] is applied to efficiently search for the exact eigenvalues and the corresponding eigenvectors.

The augmented characteristic equation of the time-delay system, i.e., (2.26), can be linearized into

$$\begin{bmatrix} \boldsymbol{A}'(\lambda^{(k)}) & \boldsymbol{B}'(\lambda^{(k)}) \\ -\boldsymbol{C}_0 & -\boldsymbol{D}_0 \end{bmatrix} \begin{bmatrix} \Delta\boldsymbol{v}^{(k)} \\ \Delta\bar{\boldsymbol{v}}^{(k)} \end{bmatrix} + \begin{bmatrix} \boldsymbol{g}^{(k)} \\ \boldsymbol{0}_{l\times 1} \end{bmatrix} \Delta\lambda^{(k)} = -\boldsymbol{f}^{(k)} \tag{3.59}$$

where $\lambda^{(k)}$, $\boldsymbol{v}^{(k)}$ and $\bar{\boldsymbol{v}}^{(k)} = -\boldsymbol{D}_0^{-1}\boldsymbol{C}_0\boldsymbol{v}^{(k)}$ are estimates of eigenvalue, eigenvector and ancillary vector at the k-th iteration, respectively. The prefix Δ refers to the correction. $\boldsymbol{A}'(\lambda^{(k)}) \in \mathbb{C}^{n\times n}$, $\boldsymbol{B}'(\lambda^{(k)}) \in \mathbb{C}^{n\times L}$, $\boldsymbol{g}^{(k)} \in \mathbb{C}^{n\times 1}$ and the deviation of the augmented characteristic equation at the k-th iteration $\boldsymbol{f}^{(k)} \in \mathbb{C}^{(n+l)\times 1}$ are given by

$$\begin{cases} \boldsymbol{A}'(\lambda^{(k)}) \triangleq \lambda^{(k)}\boldsymbol{I}_n - \boldsymbol{A}_0 - \sum\limits_{i=1}^{m} \boldsymbol{A}_i \mathrm{e}^{-\lambda^{(k)}\tau_i} \\ \boldsymbol{B}'(\lambda^{(k)}) \triangleq -\boldsymbol{B}_0 - \sum\limits_{i=1}^{m} \boldsymbol{B}_i \mathrm{e}^{-\lambda^{(k)}\tau_i} \\ \boldsymbol{g}^{(k)} = \boldsymbol{v}^{(k)} + \sum\limits_{i=1}^{m} \tau_i \mathrm{e}^{-\lambda^{(k)}\tau_i} \left(\boldsymbol{A}_i\boldsymbol{v}^{(k)} + \boldsymbol{B}_i\bar{\boldsymbol{v}}^{(k)}\right) \\ \boldsymbol{f}^{(k)} = \begin{bmatrix} \boldsymbol{A}'(\lambda^{(k)}) & \boldsymbol{B}'(\lambda^{(k)}) \\ -\boldsymbol{C}_0 & -\boldsymbol{D}_0 \end{bmatrix} \begin{bmatrix} \boldsymbol{v}^{(k)} \\ \bar{\boldsymbol{v}}^{(k)} \end{bmatrix} \end{cases}.$$

An additional equation that needs to be considered is a normalization equation

$$\begin{bmatrix} \boldsymbol{v}_0^{\mathrm{H}} & \bar{\boldsymbol{v}}_0^{\mathrm{H}} \end{bmatrix} \begin{bmatrix} \boldsymbol{v}^{(k)} \\ \bar{\boldsymbol{v}}^{(k)} \end{bmatrix} - 1 = 0 \tag{3.60}$$

where the coefficient vectors $\boldsymbol{v}_0 = \boldsymbol{v}^{(0)}$ and $\bar{\boldsymbol{v}}_0 = \bar{\boldsymbol{v}}^{(0)}$ remain unchanged during the iteration.

Combining (3.59) and (3.60) yields

$$\begin{bmatrix} \boldsymbol{A}'(\lambda^{(k)}) & \boldsymbol{B}'(\lambda^{(k)}) & \boldsymbol{g}^{(k)} \\ -\boldsymbol{C}_0 & -\boldsymbol{D}_0 & \boldsymbol{0} \\ \boldsymbol{v}_0^{\mathrm{H}} & \bar{\boldsymbol{v}}_0^{\mathrm{H}} & 0 \end{bmatrix} \begin{bmatrix} \Delta\boldsymbol{v}^{(k)} \\ \Delta\bar{\boldsymbol{v}}^{(k)} \\ \Delta\lambda^{(k)} \end{bmatrix} = -\begin{bmatrix} \boldsymbol{f}^{(k)} \\ \begin{bmatrix} \boldsymbol{v}_0^{\mathrm{H}} & \bar{\boldsymbol{v}}_0^{\mathrm{H}} \end{bmatrix} \begin{bmatrix} \boldsymbol{v}^{(k)} \\ \bar{\boldsymbol{v}}^{(k)} \end{bmatrix} - 1 \end{bmatrix}. \tag{3.61}$$

For ease of calculation, (3.61) becomes

$$\left(\begin{bmatrix} \boldsymbol{A}'(\lambda^{(k)}) & \boldsymbol{B}'(\lambda^{(k)}) & \mathbf{0} \\ -\boldsymbol{C}_0 & -\boldsymbol{D}_0 & \mathbf{0} \\ \boldsymbol{v}_0^{\mathrm{H}} & \bar{\boldsymbol{v}}_0^{\mathrm{H}} & 1 \end{bmatrix} + \begin{bmatrix} \boldsymbol{g}^{(k)} \\ \mathbf{0} \\ -1 \end{bmatrix} \mathbf{e}_1^{\mathrm{T}}\right) \begin{bmatrix} \Delta\boldsymbol{v}^{(k)} \\ \Delta\bar{\boldsymbol{v}}^{(k)} \\ \Delta\lambda^{(k)} \end{bmatrix}$$

$$\triangleq \left(\bar{\boldsymbol{J}}^{(k)} + \bar{\boldsymbol{g}}^{(k)}\mathbf{e}_1^{\mathrm{T}}\right) \begin{bmatrix} \Delta\boldsymbol{v}^{(k)} \\ \Delta\bar{\boldsymbol{v}}^{(k)} \\ \Delta\lambda^{(k)} \end{bmatrix} = -\bar{\boldsymbol{f}}^{(k)}$$

where $\mathbf{e}_1^{\mathrm{T}} = [0, \ \ldots, \ 0, \ 1]^{\mathrm{T}} \in \mathbb{R}^{1\times(n+l+1)}$ and

$$\begin{cases} \bar{\boldsymbol{J}}^{(k)} = \begin{bmatrix} \boldsymbol{A}'(\lambda^{(k)}) & \boldsymbol{B}'(\lambda^{(k)}) & \mathbf{0} \\ -\boldsymbol{C}_0 & -\boldsymbol{D}_0 & \mathbf{0} \\ \boldsymbol{v}_0^{\mathrm{H}} & \bar{\boldsymbol{v}}_0^{\mathrm{H}} & 1 \end{bmatrix} \\ \bar{\boldsymbol{g}}^{(k)} = \begin{bmatrix} \boldsymbol{g}^{(k)} \\ \mathbf{0} \\ -1 \end{bmatrix}, \quad \bar{\boldsymbol{f}}^{(k)} = \begin{bmatrix} \boldsymbol{f}^{(k)} \\ \left[\boldsymbol{v}_0^{\mathrm{H}} \ \bar{\boldsymbol{v}}_0^{\mathrm{H}}\right] \begin{bmatrix} \boldsymbol{v}^{(k)} \\ \bar{\boldsymbol{v}}^{(k)} \end{bmatrix} - 1 \end{bmatrix} \end{cases}.$$

With the Sherman-Morrison formula, the corrections can be obtained as

$$\begin{bmatrix} \Delta\boldsymbol{v}^{(k)} \\ \Delta\bar{\boldsymbol{v}}^{(k)} \\ \Delta\lambda^{(k)} \end{bmatrix} = -\left(\bar{\boldsymbol{J}}^{(k)}\right)^{-1}\bar{\boldsymbol{f}}^{(k)} + \frac{\left(\bar{\boldsymbol{J}}^{(k)}\right)^{-1}\bar{\boldsymbol{g}}^{(k)}\mathbf{e}_1^{\mathrm{T}}\left(\bar{\boldsymbol{J}}^{(k)}\right)^{-1}\bar{\boldsymbol{f}}^{(k)}}{1 + \mathbf{e}_1^{\mathrm{T}}\left(\bar{\boldsymbol{J}}^{(k)}\right)^{-1}\bar{\boldsymbol{g}}^{(k)}}.$$

Once the correction is smaller than a given tolerance ε_1, namely

$$\max\left\{\left\|(\Delta\boldsymbol{v})^{(k)}\right\|, \ \left\|\Delta\bar{\boldsymbol{v}}^{(k)}\right\|, \ \left|\Delta\lambda^{(k)}\right|\right\} < \varepsilon_1 \text{ or } \left\|\bar{\boldsymbol{f}}^{(k)}\right\| < \varepsilon_1$$

the Newton iteration process converges.

Actually, the number of Newton iterations can be used as a good metric for evaluating the accuracy of the PSD-based eigenvalue computation methods.

References

1. Ahmed, N. U. (1991). *Semigroup theory with applications to systems and control*. Essex: Longman Scientific & Technical.
2. Diekmann, O., Gils, S. A. V., Lunel, S. M. V., et al. (1995). *Delay equations: functional, complex, and nonlinear analysis*. New York: Springer.
3. Michiels, W., Jarlebring, E., & Meerbergen, K. (2011). Krylov-based model order reduction of time-delay systems. *SIAM Journal on Matrix Analysis and Applications, 32*(4), 1399–1421.
4. Hale, J. K., & Lunel, S. M. V. (1991). *Introduction to functional differential equations*. New York: Springer.
5. Breda, D., Maset, S., & Vermiglio, R. (2015). *Stability of linear delay differential equations: A numerical approach with MATLAB*. New York: Springer.

6. Bellen, A., & Maset, S. (2000). Numerical solution of constant coefficient linear delay differential equations as abstract Cauchy problems. *Numerische Mathematik, 84*(3), 351–374.
7. Breda, D. (2004). *Numerical computation of characteristic roots for delay differential equations*. Veneto, Italy: Universitá degli Studi di Padova.
8. Jarlebring, E. (2008). *The spectrum of delay-differential equations: Numerical methods, stability and perturbation*. Braunschweig, Germany: TU Braunschweig.
9. Maset, S. (2003). Numerical solution of retarded functional differential equations as abstract Cauchy problems. *Journal of Computational and Applied Mathematics, 161*(2), 259–282.
10. Breda, D., Maset, S., & Vermiglio, R. (2005). Pseudospectral differencing methods for characteristic roots of delay differential equations. *Siam Journal on Scientific Computing, 27*(2), 482–495.
11. Trefethen, L. N. (2000). *Spectral methods in MATLAB*. Philadelphia: SIAM.
12. Ascher, U. M., & Petzold, L. R. (1998). *Computer methods for ordinary differential equations and differential equations and differential-algebraic equations*. Philadelphia: SIAM.
13. Mason, J. C., & Handscomb, D. (2003). *Chebyshev polynomials*. Boca Raton: CRC Press.
14. Meerbergen, K., Spence, A., & Roose, D. (1994). Shift-invert and Cayley transforms for detection of rightmost eigenvalues of nonsymmetric matrices. *BIT Numerical Mathematics, 34*(3), 409–423.
15. Ye, H., Li, T., & Liu, Y. (2020). Time integration-based IGD methods for eigen-analysis of large delayed cyber-physical power system. *IEEE Transactions on Power Systems, 35*(2), 1376–1388.
16. Yang, D., & Ajjarapu, V. (2007). Critical eigenvalues tracing for power system analysis via continuation of invariant subspaces and projected Arnoldi method. *IEEE Transactions on Power Systems, 22*(1), 324–332.
17. Ye, H., Mou, Q., & Liu, Y. (2017). Calculation of critical oscillation modes for large delayed cyber-physical power system using pseudo-spectral discretization of solution operator. *IEEE Transactions on Power Systems, 32*(6), 4464–4476.
18. Horn, R. A., & Johnson, C. R. (1991). *Topics in matrix analysis*. Cambridge: Cambridge University Press.
19. Wang, L., & Semlyen, A. (1990). Application of sparse eigenvalue techniques to the small signal stability analysis of large power systems. *IEEE Transactions on Power Systems, 5*(2), 635–642.
20. Angelidis, G., & Semlyen, A. (1996). Improved methodologies for the calculation of critical eigenvalues in small signal stability analysis. *IEEE Transactions on Power Systems, 11*(3), 1209–1217.
21. Bai, Z., Demmel, J., Dongarra, J., et al. (2000). *Templates for the solution of algebraic eigenvalue problems: A practical guide*. Philadelphia: SIAM.
22. Chabane, Y., & Hellal, A. (2014). An adaptive dynamic implicitly restarted Arnoldi method for the small signal stability eigen analysis of large power systems. *International Journal of Electrical Power and Energy Systems, 63*, 331–335.
23. Semlyen, A., & Wang, L. (1988). Sequential computation of the complete eigensystem for the study zone in small signal stability analysis of large power systems. *IEEE Transactions on Power Systems, 3*(2), 715–725.

Chapter 4
PIGD-PS Method

This chapter elaborates an eigenvalue computation method called "PIGD-PS" for time-delay systems. The core is to derive a finite-dimensional approximating matrix $\mathcal{A}_N$ to the infinitesimal generator $\mathcal{A}$ by applying the PS method. This allows one to compute the system's critical eigenvalues from $\mathcal{A}_N$ instead.

4.1 Partial Discretization Matrix of the Infinitesimal Generator

4.1.1 Fundamentals

4.1.1.1 Discrete Points

Given a positive integer N, the mesh $\Omega_N := \{\theta_{N,j},\ j = 1,\ 2,\ \ldots,\ N+1\}$ satisfying $-\tau_{\max} \leq \theta_{N,1} < \theta_{N,2} < \cdots < \theta_{N,N+1} = 0$ is established. The N non-zero points $\theta_{N,j}$ $(j = 1,\ 2,\ \ldots,\ N)$ are specified as the scaled and shifted zeros of an N-th order Chebyshev polynomial of the second kind $U_N(\cdot)$, i.e.,

$$\theta_{N,j} = \frac{\tau_{\max}}{2}(\alpha_j - 1)$$

where

$$\alpha_j = -\cos\left(\frac{\pi j}{N+1}\right),\quad j = 1,\ 2,\ \ldots,\ N.$$

H. Ye et al., *Spectral Discretization-Based Eigen-Analysis of Time-Delay Systems*,
https://doi.org/10.1007/978-981-96-0738-9_4

4.1.1.2 Interpolation Polynomial

At each point of Ω_N, $\varphi_x^{(2)}$ and $\varphi_y^{(2)}$ are evaluated, leading to block vectors $\boldsymbol{\Phi}_x^{(2)} \in \mathbb{R}^{(N+1)n_2\times 1}$ and $\boldsymbol{\Phi}_y^{(2)} \in \mathbb{R}^{(N+1)l_2\times 1}$, i.e.,

$$\boldsymbol{\Phi}_x^{(2)} = \left[(\boldsymbol{\varphi}_{x,1}^{(2)})^{\mathrm{T}},\ (\boldsymbol{\varphi}_{x,2}^{(2)})^{\mathrm{T}},\ \ldots,\ (\boldsymbol{\varphi}_{x,N+1}^{(2)})^{\mathrm{T}}\right]^{\mathrm{T}}$$

and

$$\boldsymbol{\Phi}_y^{(2)} = \left[(\boldsymbol{\varphi}_{y,1}^{(2)})^{\mathrm{T}},\ (\boldsymbol{\varphi}_{y,2}^{(2)})^{\mathrm{T}},\ \ldots,\ (\boldsymbol{\varphi}_{y,N+1}^{(2)})^{\mathrm{T}}\right]^{\mathrm{T}}$$

where $\boldsymbol{\varphi}_{x,k}^{(2)} = \varphi_x^{(2)}(\theta_{N,k}) \in \mathbb{R}^{n_2\times 1}$ and $\boldsymbol{\varphi}_{y,k}^{(2)} = \varphi_y^{(2)}(\theta_{N,k}) \in \mathbb{R}^{l_2\times 1}$, $k = 1,\ 2,\ \ldots,\ N+1$. Note that $\boldsymbol{\varphi}_{x,N+1}^{(2)} = \varphi_x^{(2)}(0)$ and $\boldsymbol{\varphi}_{y,N+1}^{(2)} = \varphi_y^{(2)}(0)$. Similarly, $\boldsymbol{\psi}_{x,k}^{(2)} \in \mathbb{R}^{n_2\times 1}$ and $\boldsymbol{\psi}_{y,k}^{(2)} \in \mathbb{R}^{l_2\times 1}$ denote the derivative of $\varphi_x^{(2)}$ and $\varphi_y^{(2)}$ at the discrete point $\theta_{N,k}$ ($k = 1,\ 2,\ \ldots,\ N+1$), respectively.

Furthermore, for the delay-free augmented system state variables $\varphi_x^{(1)}$ and $\varphi_y^{(1)}$, one has $\boldsymbol{\varphi}_{x,N+1}^{(1)} = \varphi_x^{(1)}(0) \in \mathbb{R}^{n_1\times 1}$, $\boldsymbol{\varphi}_{y,N+1}^{(1)} = \varphi_y^{(1)}(0) \in \mathbb{R}^{l_1\times 1}$, $\boldsymbol{\psi}_{x,N+1}^{(1)} = \varphi_x'^{(1)}(0) \in \mathbb{R}^{n_1\times 1}$ and $\boldsymbol{\psi}_{y,N+1}^{(1)} = \varphi_y'^{(1)}(0) \in \mathbb{R}^{l_1\times 1}$.

The unique interpolating polynomials $L_N\boldsymbol{\Phi}_x^{(2)}$ and $L_N\boldsymbol{\Phi}_y^{(2)}$ of degree smaller or equal to N are then introduced to approximate $\varphi_x^{(2)}$ and $\varphi_y^{(2)}$ during the interval $[-\tau_{\max},\ 0]$, namely

$$\begin{cases} \varphi_x^{(2)}(\theta) \approx (L_N\boldsymbol{\Phi}_x^{(2)})(\theta) = \displaystyle\sum_{j=0}^{N} \boldsymbol{c}_{x,j}^{(2)} T_j\left(\frac{2\theta}{\tau_{\max}}+1\right), \\ \varphi_y^{(2)}(\theta) \approx (L_N\boldsymbol{\Phi}_y^{(2)})(\theta) = \displaystyle\sum_{j=0}^{N} \boldsymbol{c}_{y,j}^{(2)} T_j\left(\frac{2\theta}{\tau_{\max}}+1\right), \quad \theta \in [-\tau_{\max},\ 0] \end{cases} \tag{4.1}$$

where $\boldsymbol{c}_{x,j}^{(2)} \in \mathbb{C}^{n_2\times 1}$ and $\boldsymbol{c}_{y,j}^{(2)} \in \mathbb{C}^{l_2\times 1}$ are constant vectors, and $T_j(\cdot)$ is the j-th order Chebyshev polynomial of the first kind, $j = 0,\ 1,\ \ldots,\ N$. In particular, one has $L_N\boldsymbol{\Phi}_x^{(2)}(\theta_{N,k}) = \boldsymbol{\varphi}_{x,k}^{(2)}$ and $L_N\boldsymbol{\Phi}_y^{(2)}(\theta_{N,k}) = \boldsymbol{\varphi}_{y,k}^{(2)}$, $k = 1,\ 2,\ \ldots,\ N+1$.

4.1.2 *Partial Discretization Matrix by PS Method*

The partial discretization of $\mathcal{A}$ using the PS method consists of two parts: at $\theta_{N,N+1} = 0$, the estimation of the derivatives of $\Delta\hat{\boldsymbol{x}}$ turns out to be that of the splicing condition (3.22); at other points $\theta_{N,k}$ ($k = 1,\ 2,\ \ldots,\ N$), the derivatives of the retarded ones, i.e., $\Delta\hat{\boldsymbol{x}}^{(2)}$, are estimated from the polynomials shown in (4.1). Thus, one can obtain

$$\begin{cases} \boldsymbol{E}\begin{bmatrix} \boldsymbol{\psi}_{x,N+1}^{(1)} \\ \boldsymbol{\psi}_{x,N+1}^{(2)} \\ \boldsymbol{\psi}_{y,N+1}^{(1)} \\ \boldsymbol{\psi}_{y,N+1}^{(2)} \end{bmatrix} = \lambda \boldsymbol{E}\begin{bmatrix} \boldsymbol{\varphi}_{x,N+1}^{(1)} \\ (L_N \boldsymbol{\Phi}_x^{(2)})(0) \\ \boldsymbol{\varphi}_{y,N+1}^{(1)} \\ (L_N \boldsymbol{\Phi}_y^{(2)})(0) \end{bmatrix} = \boldsymbol{J}_0 \begin{bmatrix} \boldsymbol{\varphi}_{x,N+1}^{(1)} \\ (L_N \boldsymbol{\Phi}_x^{(2)})(0) \\ \boldsymbol{\varphi}_{y,N+1}^{(1)} \\ (L_N \boldsymbol{\Phi}_y^{(2)})(0) \end{bmatrix} \\ + \sum_{i=1}^{m} \boldsymbol{J}_{x,i}^{(2)} (L_N \boldsymbol{\Phi}_x^{(2)})(-\tau_i) + \sum_{i=1}^{m} \boldsymbol{J}_{y,i}^{(2)} (L_N \boldsymbol{\Phi}_y^{(2)})(-\tau_i) \\ \boldsymbol{\psi}_{x,k}^{(2)} = \lambda (L_N \boldsymbol{\Phi}_x^{(2)})(\theta_{N,k}) = (L_N \boldsymbol{\Phi}_x^{(2)})'(\theta_{N,k}), \\ \boldsymbol{\psi}_{y,k}^{(2)} = \lambda (L_N \boldsymbol{\Phi}_y^{(2)})(\theta_{N,k}) = (L_N \boldsymbol{\Phi}_y^{(2)})'(\theta_{N,k}), \quad k = 1,\ 2,\ \ldots,\ N \end{cases} . \tag{4.2}$$

Substituting (4.1) into (4.2) and using the properties $T_j'(t) = jU_{j-1}(t)$ of Chebyshev polynomials yield

$$\begin{cases} (\boldsymbol{J}_0 - \lambda \boldsymbol{E}) \begin{bmatrix} \boldsymbol{c}_x^{(1)} \\ \sum\limits_{j=0}^{N} \boldsymbol{c}_{x,j}^{(2)} T_j(1) \\ \boldsymbol{c}_y^{(1)} \\ \sum\limits_{j=0}^{N} \boldsymbol{c}_{y,j}^{(2)} T_j(1) \end{bmatrix} \\ + \sum_{i=1}^{m} \boldsymbol{J}_{x,i}^{(2)} \sum_{j=0}^{N} \boldsymbol{c}_{x,j}^{(2)} T_j(\beta_i) + \sum_{i=1}^{m} \boldsymbol{J}_{y,i}^{(2)} \sum_{j=0}^{N} \boldsymbol{c}_{y,j}^{(2)} T_j(\beta_i) = \boldsymbol{0} \\ \sum_{j=0}^{N} \boldsymbol{c}_{x,j}^{(2)} \frac{2j}{\tau_{\max}} U_{j-1}(\alpha_k) = \lambda \sum_{j=0}^{N} \boldsymbol{c}_{x,j}^{(2)} T_j(\alpha_k) \\ \sum_{j=0}^{N} \boldsymbol{c}_{y,j}^{(2)} \frac{2j}{\tau_{\max}} U_{j-1}(\alpha_k), = \lambda \sum_{j=0}^{N} \boldsymbol{c}_{y,j}^{(2)} T_j(\alpha_k), \quad k = 1,\ 2,\ \ldots,\ N \end{cases} \tag{4.3}$$

where $\boldsymbol{c}_x^{(1)} = \boldsymbol{\varphi}_{x,N+1}^{(1)} \in \mathbb{C}^{n_1 \times 1}$, $\boldsymbol{c}_y^{(1)} = \boldsymbol{\varphi}_{y,N+1}^{(1)} \in \mathbb{C}^{l_1 \times 1}$, $\beta_i = -\frac{2\tau_i}{\tau_{\max}} + 1$, $i = 1,\ 2,\ \ldots,\ m$. $U_{j-1}(\cdot)$ denotes the $(j-1)$-th order Chebyshev polynomial of the second kind, $j = 0,\ 1,\ \ldots,\ N$.

Considering $T_j(1) = 1$, $T_0(t) = 1$ and $U_{-1}(t) = 0$, (4.3) is rewritten as

$$\begin{cases} \boldsymbol{J}_0 \begin{bmatrix} \boldsymbol{c}_x^{(1)} \\ \sum\limits_{j=0}^{N} \boldsymbol{c}_{x,j}^{(2)} \\ \boldsymbol{c}_y^{(1)} \\ \sum\limits_{j=0}^{N} \boldsymbol{c}_{y,j}^{(2)} \end{bmatrix} + \sum\limits_{i=1}^{m} \boldsymbol{J}_{x,i}^{(2)} \boldsymbol{c}_{x,0}^{(2)} + \sum\limits_{i=1}^{m} \boldsymbol{J}_{x,i}^{(2)} \sum\limits_{j=1}^{N} \boldsymbol{c}_{x,j}^{(2)} T_j(\beta_i) \\ + \sum\limits_{i=1}^{m} \boldsymbol{J}_{y,i}^{(2)} \boldsymbol{c}_{y,0}^{(2)} + \sum\limits_{i=1}^{m} \boldsymbol{J}_{y,i}^{(2)} \sum\limits_{j=1}^{N} \boldsymbol{c}_{y,j}^{(2)} T_j(\beta_i) = \lambda \boldsymbol{E} \begin{bmatrix} \boldsymbol{c}_x^{(1)} \\ \sum\limits_{j=0}^{N} \boldsymbol{c}_{x,j}^{(2)} \\ \boldsymbol{c}_y^{(1)} \\ \sum\limits_{j=0}^{N} \boldsymbol{c}_{y,j}^{(2)} \end{bmatrix} \\ \sum\limits_{j=1}^{N} \boldsymbol{c}_{x,j}^{(2)} \dfrac{2j}{\tau_{\max}} U_{j-1}(\alpha_k) = \lambda \sum\limits_{j=1}^{N} \boldsymbol{c}_{x,j}^{(2)} T_j(\alpha_k) + \lambda \boldsymbol{c}_{x,0}^{(2)}, \\ \sum\limits_{j=1}^{N} \boldsymbol{c}_{y,j}^{(2)} \dfrac{2j}{\tau_{\max}} U_{j-1}(\alpha_k), = \lambda \sum\limits_{j=1}^{N} \boldsymbol{c}_{y,j}^{(2)} T_j(\alpha_k) + \lambda \boldsymbol{c}_{y,0}^{(2)}, \quad k = 1,\ 2,\ \ldots,\ N \end{cases}$$

and can be further expressed in the matrix form

$$\left[\begin{array}{c:c:c:c} \boldsymbol{R}_x & \sum\limits_{i=0}^{m} \boldsymbol{\ell}_i^{\mathrm{T}} \otimes \boldsymbol{J}_{x,i}^{(2)} & \boldsymbol{R}_y & \sum\limits_{i=0}^{m} \boldsymbol{\ell}_i^{\mathrm{T}} \otimes \boldsymbol{J}_{y,i}^{(2)} \\ \hdashline & \boldsymbol{U} \otimes \boldsymbol{I}_{n_2} & & \\ \hdashline & & & \boldsymbol{U} \otimes \boldsymbol{I}_{l_2} \end{array}\right] \bar{\boldsymbol{c}} = \lambda \left[\begin{array}{c:c:c:c} \boldsymbol{I}_{n_1} & & & \\ & \boldsymbol{1}_{N+1}^{\mathrm{T}} \otimes \boldsymbol{I}_{n_2} & & \\ & \boldsymbol{0}_{l \times (N+1) n_2} & & \\ \hdashline & \boldsymbol{\Gamma} \otimes \boldsymbol{I}_{n_2} & \boldsymbol{0}_{N n_2 \times l_1} & \\ \hdashline & & & \boldsymbol{\Gamma} \otimes \boldsymbol{I}_{l_2} \end{array}\right] \bar{\boldsymbol{c}} \tag{4.4}$$

where $\boldsymbol{R}_x \in \mathbb{R}^{d \times n}$, $\boldsymbol{R}_y \in \mathbb{R}^{d \times l}$, $\boldsymbol{\ell}_i^{\mathrm{T}} \in \mathbb{R}^{1 \times N}$, $\bar{\boldsymbol{c}} \in \mathbb{C}^{(d+Nd_2) \times 1}$, $\boldsymbol{1}_{N+1} = [1,\ 1,\ \ldots,\ 1]^{\mathrm{T}} \in \mathbb{R}^{(N+1) \times 1}$, $\boldsymbol{U} \in \mathbb{R}^{N \times N}$ and $\boldsymbol{\Gamma} \in \mathbb{R}^{N \times (N+1)}$ are given by

$$\boldsymbol{R}_x = \begin{bmatrix} \boldsymbol{A}_{11,0} & \sum\limits_{0=1}^{m} \boldsymbol{A}_{12,i} \\ \boldsymbol{A}_{21,0} & \sum\limits_{i=0}^{m} \boldsymbol{A}_{22,i} \\ \boldsymbol{C}_{11,0} & \boldsymbol{C}_{12,0} \\ \boldsymbol{C}_{21,0} & \boldsymbol{C}_{22,0} \end{bmatrix} = \left[\boldsymbol{J}_{x,0}^{(1)} \ \sum\limits_{i=0}^{m} \boldsymbol{J}_{x,i}^{(2)} \right]$$

$$\boldsymbol{R}_y = \begin{bmatrix} \boldsymbol{B}_{11,0} & \sum_{i=0}^{m} \boldsymbol{B}_{12,i} \\ \boldsymbol{B}_{21,0} & \sum_{i=0}^{m} \boldsymbol{B}_{22,i} \\ \boldsymbol{D}_{11,0} & \boldsymbol{D}_{12,0} \\ \boldsymbol{D}_{21,0} & \boldsymbol{D}_{22,0} \end{bmatrix} = \begin{bmatrix} \boldsymbol{J}_{y,0}^{(1)} & \sum_{i=0}^{m} \boldsymbol{J}_{y,i}^{(2)} \end{bmatrix}$$

$$\begin{cases} \boldsymbol{\ell}_0^{\mathrm{T}} = \mathbf{1}_N^{\mathrm{T}} = [1,\ 1,\ \ldots,\ 1] \\ \boldsymbol{\ell}_i^{\mathrm{T}} = [T_1(\beta_i)\ T_2(\beta_i)\ \cdots\ T_N(\beta_i)],\quad i = 1,\ 2,\ \ldots,\ m \end{cases}$$

$$\bar{\boldsymbol{c}} = \Big[(\boldsymbol{c}_x^{(1)})^{\mathrm{T}}\ (\boldsymbol{c}_{x,0}^{(2)})^{\mathrm{T}} \,\vdots\, (\boldsymbol{c}_{x,1}^{(2)})^{\mathrm{T}}\ (\boldsymbol{c}_{x,2}^{(2)})^{\mathrm{T}} \cdots (\boldsymbol{c}_{x,N}^{(2)})^{\mathrm{T}} \,\vdots\, (\boldsymbol{c}_y^{(1)})^{\mathrm{T}}\ (\boldsymbol{c}_{y,0}^{(2)})^{\mathrm{T}} \,\vdots\, (\boldsymbol{c}_{y,1}^{(2)})^{\mathrm{T}}\ (\boldsymbol{c}_{y,2}^{(2)})^{\mathrm{T}} \cdots (\boldsymbol{c}_{y,N}^{(2)})^{\mathrm{T}} \Big]^{\mathrm{T}}$$

$$\boldsymbol{U} = \frac{2}{\tau_{\max}} \begin{bmatrix} U_0(\alpha_1) & 2U_1(\alpha_1) & \cdots & NU_{N-1}(\alpha_1) \\ U_0(\alpha_2) & 2U_1(\alpha_2) & \cdots & NU_{N-1}(\alpha_2) \\ \vdots & \vdots & & \vdots \\ U_0(\alpha_N) & 2U_1(\alpha_N) & \cdots & NU_{N-1}(\alpha_N) \end{bmatrix}$$

$$\boldsymbol{\Gamma} = \begin{bmatrix} T_0(\alpha_1) & T_1(\alpha_1) & \cdots & T_N(\alpha_1) \\ T_0(\alpha_2) & T_1(\alpha_2) & \cdots & T_N(\alpha_2) \\ \vdots & \vdots & & \vdots \\ T_0(\alpha_N) & T_1(\alpha_N) & \cdots & T_N(\alpha_N) \end{bmatrix}.$$

The relationship between $\boldsymbol{U}$ and $\boldsymbol{\Gamma}$ can be established as

$$\boldsymbol{\Gamma} = \boldsymbol{U}\boldsymbol{\Upsilon} \tag{4.5}$$

where $\boldsymbol{\Upsilon} \in \mathbb{R}^{N\times(N+1)}$ is given by

$$\boldsymbol{\Upsilon} = \frac{\tau_{\max}}{4} \begin{bmatrix} 2 & 0 & -1 & & & & \\ & \frac{1}{2} & 0 & -\frac{1}{2} & & & \\ & & \frac{1}{3} & 0 & \ddots & & \\ & & & \ddots & \ddots & -\frac{1}{N-2} & \\ & & & & \ddots & 0 & -\frac{1}{N-1} \\ & & & & & \frac{1}{N} & 0 \end{bmatrix}.$$

Proof It is known from (3.34) that

$$tU_{j-1}(t) = \frac{1}{2}\left(U_j(t) + U_{j-2}(t)\right). \tag{4.6}$$

Then, substituting (4.6) into (3.35) and taking account of $U_{-1}(t) = 0$ yield

$$T_0(t) = U_0(t),\ \ T_1(t) = \frac{1}{2}U_1(t),\ \ T_j(t) = \frac{1}{2}(U_j(t) - U_{j-2}(t)),\ \ j \geq 2. \tag{4.7}$$

With (4.7) and $U_0(t) = 1$, one has

$$\begin{bmatrix} T_0(\alpha_1) & T_1(\alpha_1) & \cdots & T_{N-1}(\alpha_1) \\ T_0(\alpha_2) & T_1(\alpha_2) & \cdots & T_{N-1}(\alpha_2) \\ \vdots & \vdots & & \vdots \\ T_0(\alpha_N) & T_1(\alpha_N) & \cdots & T_{N-1}(\alpha_N) \end{bmatrix}$$

$$= \frac{1}{2}\begin{bmatrix} U_0(\alpha_1) - U_{-2}(\alpha_1) & U_1(\alpha_1) - U_{-1}(\alpha_1) & \cdots & U_{N-1}(\alpha_1) - U_{N-3}(\alpha_1) \\ U_0(\alpha_2) - U_{-2}(\alpha_2) & U_1(\alpha_2) - U_{-1}(\alpha_2) & \cdots & U_{N-1}(\alpha_2) - U_{N-3}(\alpha_2) \\ \vdots & \vdots & & \vdots \\ U_0(\alpha_N) - U_{-2}(\alpha_N) & U_1(\alpha_N) - U_{-1}(\alpha_N) & \cdots & U_{N-1}(\alpha_N) - U_{N-3}(\alpha_N) \end{bmatrix}$$

$$= \frac{1}{2}\begin{bmatrix} 2U_0(\alpha_1) & U_1(\alpha_1) & U_2(\alpha_1) - U_0(\alpha_1) & \cdots & U_{N-1}(\alpha_1) - U_{N-3}(\alpha_1) \\ 2U_0(\alpha_2) & U_1(\alpha_2) & U_2(\alpha_2) - U_0(\alpha_2) & \cdots & U_{N-1}(\alpha_2) - U_{N-3}(\alpha_2) \\ \vdots & \vdots & \vdots & & \vdots \\ 2U_0(\alpha_N) & U_1(\alpha_N) & U_2(\alpha_N) - U_0(\alpha_N) & \cdots & U_{N-1}(\alpha_N) - U_{N-3}(\alpha_N) \end{bmatrix}$$

$$= \boldsymbol{U} \cdot \frac{\tau_{\max}}{4}\begin{bmatrix} 2 & 0 & -1 & & 0 \\ & \frac{1}{2} & 0 & \ddots & \\ & & \frac{1}{3} & \ddots & -\frac{1}{N-2} \\ & & & \ddots & 0 \\ & & & & \frac{1}{N} \end{bmatrix}.$$

According to the definition formula (3.29) of Chebyshev polynomial of the second kind, one can obtain

$$U_N(\alpha_k) = \frac{\sin((N+1)\pi + \pi k)}{\sin\left(\pi + \dfrac{\pi k}{N+1}\right)} = 0. \tag{4.8}$$

Substituting (4.8) into (4.7) yields

$$T_N(\alpha_k) = -\frac{1}{2}U_{N-2}(\alpha_k),\ \ k = 1,\ 2,\ \ldots,\ N;\ \ N \geq 2$$

which can be further expressed in the vector form, i.e.,

$$[T_N(\alpha_1),\ T_N(\alpha_2),\ \ldots,\ T_N(\alpha_N)]^{\mathrm{T}} = \boldsymbol{U}\left[0,\ 0,\ \ldots,\ 0,\ -\frac{\tau_{\max}}{4(N-1)},\ 0\right]^{\mathrm{T}}.$$

□

After substituting (4.5) into (4.4), the following generalized eigenvalue problem is obtained.

$$\begin{aligned}
&\left[\begin{array}{c:c:c:c}
\boldsymbol{R}_x & \sum\limits_{i=0}^{m} \boldsymbol{\ell}_i^{\mathrm{T}} \otimes \boldsymbol{J}_{x,i}^{(2)} & \boldsymbol{R}_y & \sum\limits_{i=0}^{m} \boldsymbol{\ell}_i^{\mathrm{T}} \otimes \boldsymbol{J}_{y,i}^{(2)} \\ \hdashline
 & \boldsymbol{I}_{Nn_2} & & \\ \hdashline
 & & & \boldsymbol{I}_{Nl_2}
\end{array}\right]\bar{\boldsymbol{c}} \\
&= \lambda \left[\begin{array}{cc:c:c:c}
\boldsymbol{I}_{n_1} & & & & \\
 & \boldsymbol{I}_{n_2} & \boldsymbol{1}_N^{\mathrm{T}} \otimes \boldsymbol{I}_{n_2} & & \\
 & \boldsymbol{0}_{l\times n_2} & \boldsymbol{0}_{l\times Nn_2} & & \\ \hdashline
 & \boldsymbol{\Upsilon}_1 \otimes \boldsymbol{I}_{n_2} & \boldsymbol{\Upsilon}_N \otimes \boldsymbol{I}_{n_2} & \boldsymbol{0}_{Nn_2\times l_1} & \\ \hdashline
 & & & \boldsymbol{\Upsilon}_1 \otimes \boldsymbol{I}_{l_2} & \boldsymbol{\Upsilon}_N \otimes \boldsymbol{I}_{l_2}
\end{array}\right]\bar{\boldsymbol{c}}
\end{aligned} \tag{4.9}$$

with $\boldsymbol{\Upsilon}_1 = \boldsymbol{\Upsilon}(:,\ 1) \in \mathbb{R}^{N\times 1}$ and $\boldsymbol{\Upsilon}_N = \boldsymbol{\Upsilon}(:, 2 : N+1) \in \mathbb{R}^{N\times N}$.
Equation (4.9) can be represented by the matrix pencil $(\boldsymbol{E}_N,\ \hat{\mathcal{A}}_N)$, where $\boldsymbol{E}_N \in \mathbb{R}^{(d+Nd_2)\times(d+Nd_2)}$ and $\hat{\mathcal{A}}_N \in \mathbb{R}^{(d+Nd_2)\times(d+Nd_2)}$ are given by

$$\boldsymbol{E}_N = \left[\begin{array}{c:c:c:cc}
\boldsymbol{I}_{n_1} & & & & \\
 & \boldsymbol{I}_{n_2} & \boldsymbol{1}_N^{\mathrm{T}} \otimes \boldsymbol{I}_{n_2} & & \\
 & \boldsymbol{0}_{l\times n_2} & \boldsymbol{0}_{l\times Nn_2} & & \\ \hdashline
 & \boldsymbol{\Upsilon}_1 \otimes \boldsymbol{I}_{n_2} & \boldsymbol{\Upsilon}_N \otimes \boldsymbol{I}_{n_2} \quad \boldsymbol{0}_{Nn_2\times l_1} & & \\ \hdashline
 & & & \boldsymbol{\Upsilon}_1 \otimes \boldsymbol{I}_{l_2} & \boldsymbol{\Upsilon}_N \otimes \boldsymbol{I}_{l_2}
\end{array}\right]$$

$$\hat{\mathcal{A}}_N = \left[\begin{array}{c:c:c:c}
\boldsymbol{R}_x & \sum\limits_{i=0}^{m} \boldsymbol{\ell}_i^{\mathrm{T}} \otimes \boldsymbol{J}_{x,i}^{(2)} & \boldsymbol{R}_y & \sum\limits_{i=0}^{m} \boldsymbol{\ell}_i^{\mathrm{T}} \otimes \boldsymbol{J}_{y,i}^{(2)} \\ \hdashline
 & \boldsymbol{I}_{Nn_2} & & \\ \hdashline
 & & & \boldsymbol{I}_{Nl_2}
\end{array}\right].$$

By means of the elementary transformation matrices

$$\boldsymbol{P}_1 = \left[\begin{array}{c:c:c}
\boldsymbol{I}_{n+Nn_2} & & \\ \hdashline
 & & \boldsymbol{I}_l \\ \hdashline
 & \boldsymbol{I}_{Nl_2} &
\end{array}\right] \in \mathbb{R}^{(d+Nd_2)\times(d+Nd_2)}$$

and

$$\boldsymbol{P}_2 = \left[\begin{array}{c:c:c}
\boldsymbol{I}_n & & \\ \hdashline
 & & \boldsymbol{I}_{Nd_2} \\ \hdashline
 & \boldsymbol{I}_l &
\end{array}\right] \in \mathbb{R}^{(d+Nd_2)\times(d+Nd_2)}$$

the matrix pencil $(\boldsymbol{E}_N, \hat{\mathcal{A}}_N)$ can be rewritten as

$$\begin{bmatrix} \boldsymbol{A}_N & \boldsymbol{B}_N \\ \boldsymbol{C}_N & \boldsymbol{D}_0 \end{bmatrix} \begin{bmatrix} \bar{\boldsymbol{c}}_x \\ \bar{\boldsymbol{c}}_y \end{bmatrix} = \lambda \begin{bmatrix} \boldsymbol{E}_1 & \boldsymbol{E}_2 \\ \boldsymbol{0} & \boldsymbol{0} \end{bmatrix} \begin{bmatrix} \bar{\boldsymbol{c}}_x \\ \bar{\boldsymbol{c}}_y \end{bmatrix} \tag{4.10}$$

where the upper triangular block matrix $\boldsymbol{A}_N \in \mathbb{R}^{(n+Nd_2)\times(n+Nd_2)}$, $\boldsymbol{B}_N \in \mathbb{R}^{(n+Nd_2)\times l}$, $\boldsymbol{C}_N \in \mathbb{R}^{l\times(n+Nd_2)}$, $\boldsymbol{E}_1 \in \mathbb{R}^{(n+Nd_2)\times(n+Nd_2)}$, $\boldsymbol{E}_2 \in \mathbb{R}^{(n+Nd_2)\times l}$, $\bar{\boldsymbol{c}}_x \in \mathbb{R}^{(n+Nd_2)\times 1}$ and $\bar{\boldsymbol{c}}_y \in \mathbb{R}^{l\times 1}$ are described as

$$\boldsymbol{A}_N = \left[\begin{array}{c|cc} \sum_{i=0}^{m} \boldsymbol{A}_i & \sum_{i=0}^{m} \boldsymbol{\ell}_i^{\mathrm{T}} \otimes \boldsymbol{A}_i^{(2)} & \sum_{i=0}^{m} \boldsymbol{\ell}_i^{\mathrm{T}} \otimes \boldsymbol{B}_i^{(2)} \\ \hline \boldsymbol{0}_{Nd_2\times n} & \multicolumn{2}{c}{\boldsymbol{I}_{Nd_2}} \end{array}\right]$$

$$\boldsymbol{B}_N = \left[\begin{array}{c} \sum_{i=0}^{m} \boldsymbol{B}_i \\ \hline \boldsymbol{0}_{Nd_2\times l} \end{array}\right]$$

$$\boldsymbol{C}_N = \left[\begin{array}{c|c|c} \boldsymbol{C}_0 & \boldsymbol{1}_N^{\mathrm{T}} \otimes \boldsymbol{C}_0^{(2)} & \boldsymbol{1}_N^{\mathrm{T}} \otimes \boldsymbol{D}_0^{(2)} \end{array}\right]$$

$$\boldsymbol{E}_1 = \left[\begin{array}{c|c|c} \boldsymbol{I}_{n_1} & & \\ & \boldsymbol{1}_{N+1}^{\mathrm{T}} \otimes \boldsymbol{I}_{n_2} & \\ \hline & \boldsymbol{\Upsilon} \otimes \boldsymbol{I}_{n_2} & \\ \hline & & \boldsymbol{\Upsilon}_N \otimes \boldsymbol{I}_{l_2} \end{array}\right]$$

$$\boldsymbol{E}_2 = \left[\begin{array}{cc} \multicolumn{2}{c}{\boldsymbol{0}_{(n+Nn_2)\times l}} \\ \hline \boldsymbol{0}_{Nl_2\times l_1} & \boldsymbol{\Upsilon}_1 \otimes \boldsymbol{I}_{l_2} \end{array}\right]$$

$$\bar{\boldsymbol{c}}_x = \Big[(\boldsymbol{c}_x^{(1)})^{\mathrm{T}} \; (\boldsymbol{c}_{x,0}^{(2)})^{\mathrm{T}} \,\big|\, (\boldsymbol{c}_{x,1}^{(2)})^{\mathrm{T}} \; (\boldsymbol{c}_{x,2}^{(2)})^{\mathrm{T}} \cdots (\boldsymbol{c}_{x,N}^{(2)})^{\mathrm{T}} \,\big| \\ (\boldsymbol{c}_{y,1}^{(2)})^{\mathrm{T}} \; (\boldsymbol{c}_{y,2}^{(2)})^{\mathrm{T}} \cdots (\boldsymbol{c}_{y,N}^{(2)})^{\mathrm{T}} \Big]^{\mathrm{T}}$$

$$\bar{\boldsymbol{c}}_y = \left[\left(\boldsymbol{c}_y^{(1)}\right)^{\mathrm{T}} \; \left(\boldsymbol{c}_{y,0}^{(2)}\right)^{\mathrm{T}} \right]^{\mathrm{T}}.$$

Then, by eliminating $\bar{\boldsymbol{c}}_y$, (4.10) can be further transformed into a standard eigenvalue problem

$$\mathcal{A}_N \bar{\boldsymbol{c}}_x = \lambda \bar{\boldsymbol{c}}_x$$

where $\mathcal{A}_N = \mathbf{\Pi}_N^{-1}\mathbf{\Sigma}_N \in \mathbb{R}^{(n+Nd_2)\times(n+Nd_2)}$ is the partial discretized matrix approximating to $\mathcal{A}$, with $\mathbf{\Sigma}_N \in \mathbb{R}^{(n+Nd_2)\times(n+Nd_2)}$ and $\mathbf{\Pi}_N \in \mathbb{R}^{(n+Nd_2)\times(n+Nd_2)}$ given by

$$\mathbf{\Sigma}_N = \boldsymbol{A}_N - \boldsymbol{B}_N\boldsymbol{D}_0^{-1}\boldsymbol{C}_N$$

and

$$\mathbf{\Pi}_N = \boldsymbol{E}_1 - \boldsymbol{E}_2\boldsymbol{D}_0^{-1}\boldsymbol{C}_N.$$

4.2 Eigenvalue Computation for Large-Scale Time-Delay Systems

4.2.1 Shift-Invert Transform

It has known from Sect. 3.4 that the shift-invert transform facilitates capturing a set of critical eigenvalues λ around a given point λ_s. After performing the transformation, one can obtain the transformed discretization matrix

$$(\mathcal{A}'_N)^{-1} = (\mathbf{\Sigma}'_N)^{-1}\mathbf{\Pi}_N$$

with the eigenvalue being $\lambda' = 1/(\lambda - \lambda_s)$, where $\mathbf{\Sigma}'_N \in \mathbb{R}^{(n+Nd_2)\times(n+Nd_2)}$ is given by

$$\mathbf{\Sigma}'_N = \boldsymbol{A}'_N - \boldsymbol{B}'_N\boldsymbol{D}_0^{-1}\boldsymbol{C}_N \tag{4.11}$$

with $\boldsymbol{A}'_N \in \mathbb{C}^{(n+Nd_2)\times(n+Nd_2)}$ and $\boldsymbol{B}'_N \in \mathbb{C}^{(n+Nd_2)\times l}$ described as

$$\boldsymbol{A}'_N = \left[\begin{array}{c|cc} \sum_{i=0}^{m}\boldsymbol{A}'_i & \sum_{i=0}^{m}\boldsymbol{\ell}_i^{\mathrm{T}}\otimes\boldsymbol{A}_i'^{(2)} & \sum_{i=0}^{m}\boldsymbol{\ell}_i^{\mathrm{T}}\otimes\boldsymbol{B}_i'^{(2)} \\ \hline \mathbf{0}_{Nd_2\times n} & \multicolumn{2}{c}{\boldsymbol{I}_{Nd_2}} \end{array}\right]$$

and

$$\boldsymbol{B}'_N = \left[\begin{array}{c} \sum_{i=0}^{m}\boldsymbol{B}'_i \\ \hline \mathbf{0}_{Nd_2\times l} \end{array}\right].$$

4.2.2 Sparse Eigenvalue Computation

The IRA algorithm is utilized to successively compute λ' from $\left(\mathcal{A}'_N\right)^{-1}$ in a descending order of modulus. In the process, the most critical and computationally expensive operation is to build the Krylov subspace by recursively solving the MIVP

$$\boldsymbol{p}_{k+1} = (\mathcal{A}'_N)^{-1}\boldsymbol{p}_k = (\boldsymbol{\Sigma}'_N)^{-1}\boldsymbol{\Pi}_N\boldsymbol{p}_k \tag{4.12}$$

where $\boldsymbol{p}_k$, $\boldsymbol{p}_{k+1} \in \mathbb{C}^{(n+Nd_2)\times 1}$ are the k-th and $(k+1)$-th orthogonalized Krylov vectors, respectively. Equation (4.12) can be achieved by the following two steps:

$$\boldsymbol{w} = \boldsymbol{\Pi}_N\boldsymbol{p}_j \tag{4.13}$$

and

$$\boldsymbol{p}_{j+1} = \left(\boldsymbol{\Sigma}'_N\right)^{-1}\boldsymbol{w} \tag{4.14}$$

where $\boldsymbol{w} \in \mathbb{C}^{(n+Nd_2)\times 1}$ is a temporary vector.

4.2.2.1 Implementation of (4.13)

The high sparsity of $\boldsymbol{E}_1$ and $\boldsymbol{E}_2$ can be applied to directly compute $\boldsymbol{\Pi}_N$ as

$$\begin{cases} \left[\, \boldsymbol{L}_1,\ \boldsymbol{U}_1,\ \boldsymbol{P}_1,\ \boldsymbol{Q}_1 \,\right] = \mathrm{lu}(\boldsymbol{D}_0) \\ \boldsymbol{\Pi}_N = \boldsymbol{E}_1 - \left(\boldsymbol{L}_1^{\mathrm{T}}\backslash\left(\boldsymbol{U}_1^{\mathrm{T}}\backslash\left(\boldsymbol{P}_1^{\mathrm{T}}\backslash\left(\boldsymbol{Q}_1^{\mathrm{T}}\backslash\boldsymbol{E}_2^{\mathrm{T}}\right)\right)\right)\right)^{\mathrm{T}}\boldsymbol{C}_N \\ \boldsymbol{w} = \boldsymbol{\Pi}_N\boldsymbol{p}_j \end{cases}$$

leading to the fast calculation of (4.13), where 'lu' stands for the LU decomposition operation.

4.2.2.2 Implementation of (4.14)

As shown in (4.11), $\boldsymbol{\Sigma}'_N$ can be expressed as the Schur complement of $\boldsymbol{D}_0$. Therefore, (4.14) can be decomposed into

$$\left[\boldsymbol{D}_0 - \boldsymbol{C}_N(\boldsymbol{A}'_N)^{-1}\boldsymbol{B}_N\right]\boldsymbol{z} = -\boldsymbol{C}_N(\boldsymbol{A}'_N)^{-1}\boldsymbol{p}_j \tag{4.15}$$

and

$$\boldsymbol{p}_{j+1} = (\boldsymbol{A}'_N)^{-1}(\boldsymbol{p}_j - \boldsymbol{B}_N\boldsymbol{z}) \tag{4.16}$$

to avoid explicitly computing $\boldsymbol{\Sigma}'_N$ and its inverse, where $\boldsymbol{z} \in \mathbb{R}^{l\times 1}$ is the intermediate vector. In the syntax of MATLAB, the specific implementation of solutions to (4.15) and (4.16) is listed as follows:

$$\begin{cases} [\boldsymbol{L}_2,\ \boldsymbol{U}_2,\ \boldsymbol{P}_2,\ \boldsymbol{Q}_2] = \mathrm{lu}(\boldsymbol{A}_N) \\ \boldsymbol{D}^* = \boldsymbol{D}_0 - \boldsymbol{C}_N \boldsymbol{Q}_2(\boldsymbol{U}_2\backslash(\boldsymbol{L}_2\backslash(\boldsymbol{P}_2\boldsymbol{B}_N))) \\ [\boldsymbol{L}_3,\ \boldsymbol{U}_3,\ \boldsymbol{P}_3,\ \boldsymbol{Q}_3] = \mathrm{lu}(\boldsymbol{D}^*) \\ \boldsymbol{y}_1 = -\boldsymbol{C}_N \boldsymbol{Q}_2(\boldsymbol{U}_2\backslash(\boldsymbol{L}_2\backslash(\boldsymbol{P}_2\boldsymbol{w}))) \\ \boldsymbol{z} = \boldsymbol{Q}_3(\boldsymbol{U}_3\backslash(\boldsymbol{L}_3\backslash(\boldsymbol{P}_3\boldsymbol{y}_1))) \\ \boldsymbol{y}_2 = \boldsymbol{w} - \boldsymbol{B}_N\boldsymbol{z} \\ \boldsymbol{p}_{j+1} = \boldsymbol{Q}_2(\boldsymbol{U}_2\backslash(\boldsymbol{L}_2\backslash(\boldsymbol{P}_2\boldsymbol{y}_2))) \end{cases} \tag{4.17}$$

where $\boldsymbol{y}_1 \in \mathbb{C}^{l\times 1}$ and $\boldsymbol{y}_2 \in \mathbb{C}^{(n+Nd_2)\times 1}$ are temporary vectors.

In the whole process of eigenvalue computation, the inverses of $\boldsymbol{A}_N$ and $\boldsymbol{D}^*$ need to be computed only once by LU decomposition. In addition, the sparsity of $\boldsymbol{A}_N$, $\boldsymbol{B}_N$, $\boldsymbol{C}_N$ and $\boldsymbol{D}_0$ can be fully exploited to reduce computational burden and save memory occupation.

4.2.3 *Eigenvalue Correction*

Once λ' of $(\mathcal{A}'_N)^{-1}$ with the largest moduli are computed by IRA, the eigenvalues $\hat{\lambda}$ of $\mathcal{A}_N$ approximating λ of $\mathcal{A}$ can be restored from

$$\hat{\lambda} = \lambda_s + \frac{1}{\lambda'}. \tag{4.18}$$

The first n elements of $\hat{\boldsymbol{v}}$ that corresponds to $\hat{\lambda}$ is a good estimate of the exact eigenvector $\boldsymbol{v}$. By taking $\hat{\lambda}$ and $\hat{\boldsymbol{v}}$ as initial values of the Newton's method as given in Sect. 3.6, the exact λ and $\boldsymbol{v}$ can be readily obtained due to the quadratic convergence of the method.

4.2.4 *Characteristic Analyses*

(1) **The low dimension**. Given that $d_2 \ll n$, the dimension of the discretization matrix $\mathcal{A}_N$ gets very close to the total number of system state variables.
(2) **The matrix structure**. The matrix $\boldsymbol{\Sigma}_N$ is an upper-triangular block matrix, as shown in Fig. 4.1. In the figure, the elements colored in light gray and in dark gray are respectively related to delay-free state variables $\Delta\boldsymbol{x}^{(1)}$ and retarded state variables $\Delta\hat{\boldsymbol{x}}^{(2)}$, and those colored in white are zero.

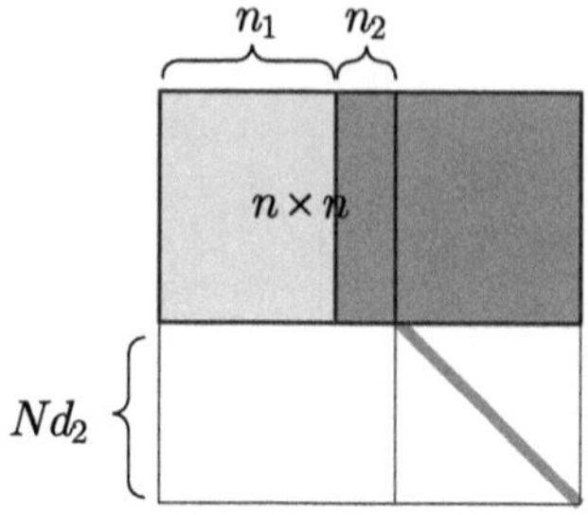

Fig. 4.1 Illustration of $\mathbf{\Sigma}_N$

(3) **The computational complexity**. Considering that the computational burden of (4.13) is negligible, the MIVP associated with $\mathbf{\Sigma}_N$ as shown in (4.14) dominates this PIGD-PS method. Thus, the computational burden of the method is almost identical to the traditional eigen-analysis of a delay-free system whose state matrix is $\mathbf{\Sigma}_N = \boldsymbol{A}_N - \boldsymbol{B}_N \boldsymbol{D}_0^{-1} \boldsymbol{C}_N$.

Chapter 5
PSOD-PS Method

This chapter addresses the PSOD-PS method for the eigen-analysis of time-delay systems. Compared with the PIGD-PS method, the PSOD-PS method shows higher efficiency on computing the critical eigenvalues, whose damping ratios are less than a given threshold, of time delayed systems.

5.1 Partial Discretization Matrix of the Solution Operator

5.1.1 Fundamentals

5.1.1.1 Discrete Points and Block Vectors over $[-\tau_{\max},\ 0]$

To establish the mesh Ω_M employed in the discretization of the solution operator $\mathcal{T}(h)$, the delay interval $[-\tau_{\max},\ 0]$ is first divided into Q sub-intervals denoted by $[\theta_1,\ \theta_0],\ [\theta_2,\ \theta_1],\ \ldots,\ [\theta_Q,\ \theta_{Q-1}]$, where $Q = \min\{q\,|qh > \tau_{\max},\ q \in \mathbb{N}\}$. Each sub-interval is further discretized by $M+1$ scaled and shifted poles of the M-th order Chebyshev polynomial of the first kind, leading to submeshes $\Omega_{M,i}$ with components $\theta_{M,i,j},\ (i = 1,\ 2,\ \ldots,\ Q;\ j = 0,\ 1,\ \ldots,\ M)$. Since $\theta_{M,i,M} = \theta_i = \theta_{M,i+1,0}$ $(i = 1,\ 2,\ \ldots,\ Q-1)$, assembling all $QM+1$ nodes on the interval $[-\tau_{\max},\ 0]$ yields [1, 2]

$$\Omega_M := \bigcup_{i=1}^{Q}\{\Omega_{M,i}\}$$

where

$$\Omega_{M,i} = \left\{\theta_{M,i,j},\ j = 0,\ 1,\ \ldots,\ M:\ \theta_{M,i,j} = \frac{h}{2}\left[\cos\left(\frac{j\pi}{M}\right) - 2i + 1\right]\right\},\quad i = 1,\ 2,\ \ldots,\ Q-1$$

H. Ye et al., *Spectral Discretization-Based Eigen-Analysis of Time-Delay Systems*,
https://doi.org/10.1007/978-981-96-0738-9_5

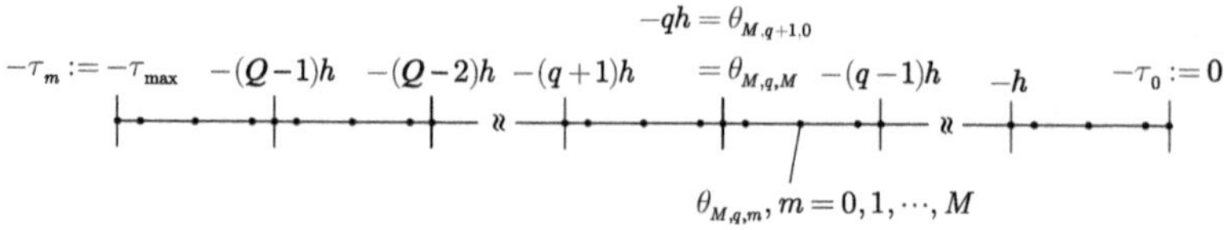

Fig. 5.1 The mesh Ω_M for PSOD-PS

and

$$\Omega_{M,Q} = \left\{ \theta_{M,Q,j},\ j = 0,\ 1,\ \ldots,\ M :\ \theta_{M,Q,j} \right. \\ \left. = \frac{\tau_{\max} - (Q-1)h}{2} \cos\left(\frac{j\pi}{M}\right) - \frac{\tau_{\max} + (Q-1)h}{2} \right\}.$$

The schematic is illustrated in Fig. 5.1.

At each point of Ω_M, $\varphi_x^{(2)}$ and $\varphi_y^{(2)}$ are evaluated, leading to block vectors $\mathbf{\Phi}_x^{(2)} \in \mathbb{R}^{(QM+1)n_2 \times 1}$ and $\mathbf{\Phi}_y^{(2)} \in \mathbb{R}^{(QM+1)l_2 \times 1}$, i.e.,

$$\begin{cases} \mathbf{\Phi}_x^{(2)} = \left[\left(\varphi_{x,1,0}^{(2)}\right)^{\mathrm{T}},\ \left(\varphi_{x,1,1}^{(2)}\right)^{\mathrm{T}},\ \ldots,\ \left(\varphi_{x,Q,M}^{(2)}\right)^{\mathrm{T}} \right]^{\mathrm{T}} \\ \mathbf{\Phi}_y^{(2)} = \left[\left(\varphi_{y,1,0}^{(2)}\right)^{\mathrm{T}},\ \left(\varphi_{y,1,1}^{(2)}\right)^{\mathrm{T}},\ \ldots,\ \left(\varphi_{y,Q,M}^{(2)}\right)^{\mathrm{T}} \right]^{\mathrm{T}} \end{cases}$$

where $\varphi_{x,i,j}^{(2)} = \varphi_x^{(2)}(\theta_{M,i,j}) \in \mathbb{R}^{n_2 \times 1}$ and $\varphi_{y,i,j}^{(2)} = \varphi_y^{(2)}(\theta_{M,i,j}) \in \mathbb{R}^{l_2 \times 1}$, $i = 1,\ 2,\ \ldots,\ Q;\ j = 0,\ 1,\ \ldots,\ M$. For $\varphi_x^{(1)}$ and $\varphi_y^{(1)}$, one has $\varphi_{x,1,0}^{(1)} = \varphi_x^{(1)}(\theta_{M,1,0}) \in \mathbb{R}^{n_1 \times 1}$ and $\varphi_{y,1,0}^{(1)} = \varphi_y^{(1)}(\theta_{M,1,0}) \in \mathbb{R}^{l_1 \times 1}$. Thus, $\varphi_{x,1,0} \in \mathbb{R}^{n \times 1}$ and $\varphi_{y,1,0} \in \mathbb{R}^{l \times 1}$ are expressed as

$$\varphi_{x,1,0} = \left[\left(\varphi_{x,1,0}^{(1)}\right)^{\mathrm{T}},\ \left(\varphi_{x,1,0}^{(2)}\right)^{\mathrm{T}} \right]^{\mathrm{T}},\ \varphi_{y,1,0} = \left[\left(\varphi_{y,1,0}^{(1)}\right)^{\mathrm{T}},\ \left(\varphi_{y,1,0}^{(2)}\right)^{\mathrm{T}} \right]^{\mathrm{T}}.$$

Furthermore, the block vector $\mathbf{\Phi} \in \mathbb{R}^{(n_1+(QM+1)d_2)\times 1}$ is defined as

$$\mathbf{\Phi} = \left[\left(\varphi_{x,1,0}^{(1)}\right)^{\mathrm{T}},\ \left(\varphi_{x,1,0}^{(2)}\right)^{\mathrm{T}},\ \ldots,\ \left(\varphi_{x,1,M-1}^{(2)}\right)^{\mathrm{T}},\ \left(\varphi_{y,1,0}^{(2)}\right)^{\mathrm{T}},\ \ldots,\ \left(\varphi_{y,1,M-1}^{(2)}\right)^{\mathrm{T}}, \right. \\ \left. \left(\varphi_{x,2,0}^{(2)}\right)^{\mathrm{T}},\ \ldots,\ \left(\varphi_{x,Q,M}^{(2)}\right)^{\mathrm{T}},\ \left(\varphi_{y,2,0}^{(2)}\right)^{\mathrm{T}},\ \ldots,\ \left(\varphi_{y,Q,M}^{(2)}\right)^{\mathrm{T}} \right]^{\mathrm{T}}. \tag{5.1}$$

The corresponding restriction operator R_M satisfies

$$R_M \hat{\varphi} = \mathbf{\Phi}.$$

The prolongation operator P_M is then introduced to approximate $\varphi_x^{(2)}$ and $\varphi_y^{(2)}$ on each sub-interval, i.e.,

$$\varphi_x^{(2)}(\theta) = \left(P_M \mathbf{\Phi}_x^{(2)}\right)(\theta) = \sum_{j=0}^{M} \ell_{M,i,j}(\theta)\varphi_{x,i,j}^{(2)},$$
$$\varphi_y^{(2)}(\theta) = \left(P_M \mathbf{\Phi}_y^{(2)}\right)(\theta) = \sum_{j=0}^{M} \ell_{M,i,j}(\theta)\varphi_{y,i,j}^{(2)}, \ \theta \in [\theta_i, \ \theta_{i-1}]; \ i = 1, \ 2, \ \ldots, \ Q$$

where

$$\ell_{M,i,j}(\theta) = \prod_{k=0,k\neq j}^{M} \frac{\theta - \theta_{M,i,k}}{\theta_{M,i,j} - \theta_{M,i,k}}, \quad i = 1, \ 2, \ \ldots, \ Q; \ j = 0, \ 1, \ \ldots, \ M$$

denote the Lagrange coefficients corresponding to $\theta_{M,i,j}$.

5.1.1.2 Discrete Points and Block Vectors over $[0, \ h]$

Given a positive integer N, the mesh Ω_N^+ over the interval $[0, \ h]$ is established, which consists of N scaled and shifted zeros of the N-th order Chebyshev polynomial of the first kind, i.e.,

$$\Omega_N^+ := \left\{t_{N,i}, \ i = 1, \ 2, \ \ldots, \ N : t_{N,i} = \frac{h}{2}\left[1 - \cos\left(\frac{(2i-1)\pi}{2N}\right)\right]\right\}.$$

At each point of Ω_N^+, the derivative z of $\Delta \boldsymbol{x}$ is discretized into the block vector $\boldsymbol{Z} \in \mathbb{R}^{Nn\times 1}$ as

$$\boldsymbol{Z} = \left[z_1^{\mathrm{T}}, \ z_2^{\mathrm{T}}, \ \ldots, \ z_N^{\mathrm{T}}\right]^{\mathrm{T}}$$

where $z_j = \left[\left(z_j^{(1)}\right)^{\mathrm{T}}, \ \left(z_j^{(2)}\right)^{\mathrm{T}}\right]^{\mathrm{T}} = z(t_{N,j}) \in \mathbb{R}^{n\times 1}$ with $z_j^{(1)} \in \mathbb{R}^{n_1\times 1}$ and $z_j^{(2)} \in \mathbb{R}^{n_2\times 1}$, $j = 1, \ 2, \ \ldots, \ N$. Similarly, the derivatives $\boldsymbol{w}$ of $\Delta \boldsymbol{y}$ during the interval $t \in [0, \ h]$ can be discretized into the block vector $\boldsymbol{W} \in \mathbb{R}^{Nl\times 1}$ as

$$\boldsymbol{W} = \left[\boldsymbol{w}_1^{\mathrm{T}}, \ \boldsymbol{w}_2^{\mathrm{T}}, \ \ldots, \ \boldsymbol{w}_N^{\mathrm{T}}\right]^{\mathrm{T}}$$

where $\boldsymbol{w}_j = \left[\left(\boldsymbol{w}_j^{(1)}\right)^{\mathrm{T}}, \ \left(\boldsymbol{w}_j^{(2)}\right)^{\mathrm{T}}\right]^{\mathrm{T}} = \boldsymbol{w}(t_{N,j}) \in \mathbb{R}^{l\times 1}$ with $\boldsymbol{w}_j^{(1)} \in \mathbb{R}^{l_1\times 1}$ and $\boldsymbol{w}_j^{(2)} \in \mathbb{R}^{l_2\times 1}$, $j = 1, \ 2, \ \ldots, \ N$.

The prolongation operator P_N^+ and restriction operator R_N^+ over the interval $[0, \ h]$ can be respectively expressed as

$$R_N^+ z = \mathbf{Z},\ \left(P_N^+ \mathbf{Z}\right)(t) = \sum_{j=1}^{N} \ell_{N,j}^+(t) z_j,\ t \in [0,\ h] \tag{5.2}$$

and

$$R_N^+ \boldsymbol{w} = \boldsymbol{W},\ \left(P_N^+ \boldsymbol{W}\right)(t) = \sum_{j=1}^{N} \ell_{N,j}^+(t) \boldsymbol{w}_j,\ t \in [0,\ h] \tag{5.3}$$

where

$$\ell_{N,j}^+(t) = \prod_{i=1, i\neq j}^{N} \frac{t - t_{N,i}}{t_{N,j} - t_{N,i}},\quad j = 1,\ 2,\ \ldots,\ N$$

denote the Lagrange coefficients corresponding to $t_{N,j}$.

5.1.2 Explicit Expression of the Solution Operator

This section elaborates the explicit expression of the solution operator $\mathcal{T}(h)$ by introducing two linear operators V_1 and V_2, which lays the groundwork for the subsequent derivation of its PS partial discretization matrix.

5.1.2.1 Mapping

The solution operator $\mathcal{T}(h)$ can be interpreted as the mapping operator V, that is

$$(V(\hat{\varphi},\ z,\ \boldsymbol{w}))(\theta) = \mathcal{T}(h)\hat{\varphi} = \begin{cases} \hat{\varphi}(0) + \displaystyle\int_0^\theta \begin{bmatrix} z(s) \\ \boldsymbol{w}(s) \end{bmatrix} \mathrm{d}s, & \theta \in [0,\ h] \\ \hat{\varphi}(\theta), & \theta \in [-\tau_{\max},\ 0] \end{cases}. \tag{5.4}$$

Let $z = \boldsymbol{w} = \boldsymbol{0}$, one can obtain

$$(V_1\hat{\varphi})(\theta) = (V(\hat{\varphi},\ \boldsymbol{0},\ \boldsymbol{0}))(\theta) = \begin{cases} \hat{\varphi}(0), & \theta \in [0,\ h] \\ \hat{\varphi}(\theta), & \theta \in [-\tau_{\max},\ 0] \end{cases}. \tag{5.5}$$

Similarly, $\hat{\varphi} = \boldsymbol{0}$ yields

$$\left(V_2 \begin{bmatrix} z \\ \boldsymbol{w} \end{bmatrix}\right)(\theta) = (V(\boldsymbol{0},\ z,\ \boldsymbol{w}))(\theta) = \begin{cases} \displaystyle\int_0^\theta \begin{bmatrix} z(s) \\ \boldsymbol{w}(s) \end{bmatrix} \mathrm{d}s, & \theta \in [0,\ h] \\ \boldsymbol{0}, & \theta \in [-\tau_{\max},\ 0] \end{cases}. \tag{5.6}$$

The mapping $V(\hat{\varphi},\ z,\ \boldsymbol{w})$ can be thus expressed in terms of V_1 and V_2, i.e.,

$$V(\hat{\varphi},\ z,\ \boldsymbol{w}) = V_1\hat{\varphi} + V_2\begin{bmatrix} z \\ \boldsymbol{w} \end{bmatrix}. \tag{5.7}$$

5.1.2.2 Application to Time-Delay Systems

Define the derivative operator F as

$$\begin{bmatrix} z \\ \boldsymbol{w} \end{bmatrix} = FV\left(\hat{\varphi},\ z,\ \boldsymbol{w}\right) \tag{5.8}$$

where

$$\begin{aligned} z(t) &= F\Delta\boldsymbol{x}(t) \\ &= [\boldsymbol{A}_0\ \ \boldsymbol{B}_0]\begin{bmatrix} \Delta\boldsymbol{x}(t) \\ \Delta\boldsymbol{y}(t) \end{bmatrix} + \sum_{i=1}^{m}\left[\boldsymbol{A}_i^{(2)}\ \ \boldsymbol{B}_i^{(2)}\right]\begin{bmatrix} \Delta\boldsymbol{x}^{(2)}(t-\tau_i) \\ \Delta\boldsymbol{y}^{(2)}(t-\tau_i) \end{bmatrix},\quad t\in[0,\ h] \end{aligned} \tag{5.9}$$

and

$$\boldsymbol{w}(t) = F\Delta\boldsymbol{y}(t) = -\boldsymbol{D}_0^{-1}\boldsymbol{C}_0 z(t),\quad t\in[0,\ h]. \tag{5.10}$$

Substituting (5.7) into (5.8), one has

$$z = FV_1\varphi_x + FV_2 z. \tag{5.11}$$

Combining (5.10) and (5.11) yields

$$\begin{bmatrix} \boldsymbol{I}_n & \\ \boldsymbol{C}_0 & \boldsymbol{D}_0 \end{bmatrix}\begin{bmatrix} z \\ \boldsymbol{w} \end{bmatrix} = \begin{bmatrix} FV_1\varphi_x + FV_2\boldsymbol{Z} \\ \boldsymbol{0} \end{bmatrix}.$$

$\left[z^{\mathrm{T}},\ \boldsymbol{w}^{\mathrm{T}}\right]^{\mathrm{T}}$, therefore, can be explicitly expressed as

$$\begin{bmatrix} z \\ \boldsymbol{w} \end{bmatrix} = \begin{bmatrix} \boldsymbol{I}_n - FV_2 & \\ \boldsymbol{C}_0 & \boldsymbol{D}_0 \end{bmatrix}^{-1}\begin{bmatrix} FV_1\varphi_x \\ \boldsymbol{0} \end{bmatrix}. \tag{5.12}$$

Substituting (5.12) into (5.7) and considering (5.4), one can obtain the explicit expression of $\mathcal{T}(h)\hat{\varphi}$ in terms of $\hat{\varphi}$:

$$\mathcal{T}(h)\hat{\varphi} = V_1\hat{\varphi} + V_2\begin{bmatrix} \boldsymbol{I}_n - FV_2 & \\ \boldsymbol{C}_0 & \boldsymbol{D}_0 \end{bmatrix}^{-1}\begin{bmatrix} FV_1\varphi_x \\ \boldsymbol{0} \end{bmatrix}. \tag{5.13}$$

5.1.2.3 Partial Discretization by PS Method

Given positive integers M and N, the restriction operators R_M, R_N^+ and the prolongation operators P_M, P_N^+ can be used to discretize (5.13) into $\boldsymbol{T}_{M,N}\boldsymbol{\Phi}$, whose coefficient matrix $\boldsymbol{T}_{M,N} \in \mathbb{R}^{(n_1+(QM+1)d_2)\times(n_1+(QM+1)d_2)}$ is PS partial discretization matrix of the solution operator and can be expressed as

$$\boldsymbol{T}_{M,N} = \boldsymbol{\Pi}_M + \boldsymbol{\Pi}_{M,N}\left(\begin{bmatrix} & \boldsymbol{I}_{Nn} \\ \boldsymbol{I}_N \otimes \boldsymbol{C}_0 & \boldsymbol{I}_N \otimes \boldsymbol{D}_0 \end{bmatrix} - \boldsymbol{\Sigma}_N\right)^{-1} \boldsymbol{\Sigma}_{M,N} \tag{5.14}$$

where $\boldsymbol{\Pi}_M \in \mathbb{R}^{(n_1+(QM+1)d_2)\times(n_1+(QM+1)d_2)}$, $\boldsymbol{\Pi}_{M,N} \in \mathbb{R}^{(n_1+(QM+1)d_2)\times Nd}$, $\boldsymbol{\Sigma}_N \in \mathbb{R}^{Nd\times Nd}$ and $\boldsymbol{\Sigma}_{M,N} \in \mathbb{R}^{Nd\times(n_1+(QM+1)d_2)}$ are given by

$$\boldsymbol{\Pi}_M\boldsymbol{\Phi} = R_M V_1 \begin{bmatrix} \varphi_{x,1,0}^{(1)} \\ P_M\boldsymbol{\Phi}_x^{(2)} \\ P_M\boldsymbol{\Phi}_y^{(2)} \end{bmatrix} \tag{5.15}$$

$$\boldsymbol{\Pi}_{M,N}\begin{bmatrix} \boldsymbol{Z} \\ \boldsymbol{W} \end{bmatrix} = R_M V_2 P_N^+ \begin{bmatrix} \boldsymbol{Z} \\ \boldsymbol{W} \end{bmatrix} \tag{5.16}$$

$$\boldsymbol{\Sigma}_{M,N}\boldsymbol{\Phi} = \begin{bmatrix} R_N^+ F V_1 \begin{bmatrix} \varphi_{x,1,0}^{(1)} \\ P_M\boldsymbol{\Phi}_x^{(2)} \end{bmatrix} \\ \boldsymbol{0}_{Nl\times 1} \end{bmatrix} \tag{5.17}$$

$$\boldsymbol{\Sigma}_N\begin{bmatrix} \boldsymbol{Z} \\ \boldsymbol{W} \end{bmatrix} = \begin{bmatrix} R_N^+ F V_2 P_N^+ & \\ & \boldsymbol{0}_{Nl} \end{bmatrix}\begin{bmatrix} \boldsymbol{Z} \\ \boldsymbol{W} \end{bmatrix}. \tag{5.18}$$

The subsequent subsections will give the detailed mathematical derivation of the partial discretization submatrices $\boldsymbol{\Pi}_M$, $\boldsymbol{\Pi}_{M,N}$, $\boldsymbol{\Sigma}_{M,N}$ and $\boldsymbol{\Sigma}_N$.

5.1.3 Matrix $\boldsymbol{\Pi}_{M,N}$

According to (5.5), the right-hand of (5.15) can be further written as

$$\left(V_1\begin{bmatrix} \varphi_{x,1,0}^{(1)} \\ P_M\boldsymbol{\Phi}_x^{(2)} \\ P_M\boldsymbol{\Phi}_y^{(2)} \end{bmatrix}\right)(\theta) = \begin{cases} \begin{bmatrix} \varphi_{x,1,0}^{(1)} \\ \left(P_M\boldsymbol{\Phi}_x^{(2)}\right)(0) \\ \left(P_M\boldsymbol{\Phi}_y^{(2)}\right)(0) \end{bmatrix}, & \theta \in (-h,\ 0] \\ \begin{bmatrix} \varphi_{x,1,0}^{(1)} \\ \left(P_M\boldsymbol{\Phi}_x^{(2)}\right)(h+\theta) \\ \left(P_M\boldsymbol{\Phi}_y^{(2)}\right)(h+\theta) \end{bmatrix}, & \theta \in [-\tau_{\max},\ -h] \end{cases}. \tag{5.19}$$

After implementing R_M, the first item of (5.19) becomes

$$\begin{cases} \left[R_M V_1 \varphi_{x,1,0}^{(1)}\right]_{1,0} = \varphi_{x,1,0}^{(1)} \\ \left[R_M V_1 P_M \Phi_x^{(2)}\right]_{1,k} = \varphi_{x,1,0}^{(2)}, \; k = 1,\, 2,\, \ldots,\, M-1 \\ \left[R_M V_1 P_M \Phi_y^{(2)}\right]_{1,k} = \varphi_{y,1,0}^{(2)} = -\left[\mathbf{0}_{l_2\times l_1}\; \boldsymbol{I}_{l_2}\right] \boldsymbol{D}_0^{-1} \boldsymbol{C}_0 \varphi_{x,1,0}, \\ \qquad\qquad\qquad\qquad\qquad k = 1,\, 2,\, \ldots,\, M-1 \end{cases}$$

where the subscript in the left-hand side has the same meaning as that specified in the discrete point $\theta_{M,1,k}$ ($k = 0,\, 1,\, \ldots,\, M-1$).

The second item of (5.19) becomes

$$\begin{cases} \left[R_M V_1 P_M \Phi_x^{(2)}\right]_{i,k} = \left(P_M \Phi_x^{(2)}\right)(h + \theta_{M,i,k}) \\ \left[R_M V_1 P_M \Phi_y^{(2)}\right]_{i,k} = \left(P_M \Phi_y^{(2)}\right)(h + \theta_{M,i,k}) \end{cases} \tag{5.20}$$

where the subscript in the left-hand side has the same meaning as those specified in $\theta_{M,i,k}$, $i = 2,\, 3,\, \ldots,\, Q$; $k = 0,\, 1,\, \ldots,\, M$. Considering that $h + \theta_{M,i,k}$ must lie in the $(i-1)$-th sub-interval, i.e., $h + \theta_{M,i,k} \in [-(i-1)h,\; -(i-2)h]$ ($i = 2,\, 3,\, \ldots,\, Q$), (5.20) can be rewritten as

$$\begin{cases} \left(P_M \Phi_x^{(2)}\right)(h + \theta_{M,i,k}) = \sum_{j=0}^{M} \ell_{M,i-1,j}(h + \theta_{M,i,k}) \varphi_{x,i-1,j}^{(2)} \\ \left(P_M \Phi_y^{(2)}\right)(h + \theta_{M,i,k}) = \sum_{j=0}^{M} \ell_{M,i-1,j}(h + \theta_{M,i,k}) \varphi_{y,i-1,j}^{(2)} \end{cases} \tag{5.21}$$

which is further analyzed in the following two aspects:

(1) If $i = 2,\, 3,\, \ldots,\, Q-1$, $h + \theta_{M,i,k} = \theta_{M,i-1,k}$ ($k = 0,\, 1,\, \ldots,\, M$) and thus

$$\begin{cases} \left(P_M \Phi_x^{(2)}\right)(h + \theta_{M,i,k}) = \sum_{j=0}^{M} \ell_{M,i-1,j}(\theta_{M,i-1,k}) \varphi_{x,i-1,j}^{(2)} \\ \left(P_M \Phi_y^{(2)}\right)(h + \theta_{M,i,k}) = \sum_{j=0}^{M} \ell_{M,i-1,j}(\theta_{M,i-1,k}) \varphi_{y,i-1,j}^{(2)} \end{cases}.$$

(2) If $i = Q$, $h + \theta_{M,i,k}$ ($k = 0,\, 1,\, \ldots,\, M-1$) falls into the $(Q-1)$-th sub-interval, i.e., $h + \theta_{M,i,k} \in [-(Q-1)h,\; -(Q-2)h]$. Considering that the Q-th sub-interval is generally less than the $(Q-1)$-th sub-interval, $h + \theta_{M,Q,k}$ does not coincide with $\theta_{M,Q-1,k}$ ($k = 0,\, 1,\, \ldots,\, M$). In this case, it is necessary to compute the coefficient $\ell_{M,Q-1,j}(h + \theta_{M,Q,k})$ ($j = 0,\, 1,\, \ldots,\, M$; $k = 0,\, 1,\, \ldots,\, M$) in (5.21) by using the barycentric Lagrange interpolation method [2].

Taken together, the right-hand of (5.15) can be explicitly expressed in terms of $\mathbf{\Phi}$, whose coefficient matrix is

$$\mathbf{\Pi}_M = \left[\begin{array}{c} \tilde{\mathbf{\Pi}}_M \\ \hdashline \begin{bmatrix} \mathbf{0}_{Mn_2\times n_1} & \boldsymbol{I}_{Mn_2} & & & & & \\ & & \boldsymbol{I}_{(Q-3)Mn_2} & & & & \\ & & & \boldsymbol{U}_{Mn} & & & \\ & & \boldsymbol{I}_{Ml_2} & & & & \\ & & & & & \boldsymbol{I}_{(Q-3)Ml_2} & \\ & & & & & & \boldsymbol{U}_{Ml} \end{bmatrix} \end{array}\right]$$

where $\tilde{\mathbf{\Pi}}_M \in \mathbb{R}^{(n_1+Md_2)\times(n_1+(QM+1)d_2)}$, $\boldsymbol{U}_{Mn} \in \mathbb{R}^{(M+1)n_2\times(2M+1)n_2}$ and $\boldsymbol{U}_{Ml} \in \mathbb{R}^{(M+1)l_2\times(2M+1)l_2}$ are given by

$$\tilde{\mathbf{\Pi}}_M = \left[\left[\begin{array}{c} \boldsymbol{I}_n \\ \mathbf{1}_{M-1}\otimes\left[\mathbf{0}_{n_2\times n_1}\ \ \boldsymbol{I}_{n_2}\right] \\ -\mathbf{1}_M\otimes\left[\mathbf{0}_{l_2\times l_1}\ \ \boldsymbol{I}_{l_2}\right]\boldsymbol{D}_0^{-1}\boldsymbol{C}_0 \end{array}\right] \begin{array}{:c} \mathbf{0}_{(n_1+Md_2)\times(l_2+QMd_2)} \end{array}\right] \tag{5.22}$$

$$\begin{cases} \boldsymbol{U}_{Mn} = \left[\boldsymbol{U}'_M\otimes\boldsymbol{I}_{n_2}\ \ \mathbf{0}_{(M+1)n_2\times Mn_2}\right] \\ \boldsymbol{U}_{Ml} = \left[\boldsymbol{U}'_M\otimes\boldsymbol{I}_{l_2}\ \ \mathbf{0}_{(M+1)l_2\times Ml_2}\right] \end{cases}$$

with $\mathbf{1}_M = [1,\ 1,\ \ldots,\ 1]^{\mathrm{T}} \in \mathbb{R}^{M\times 1}$, $\mathbf{1}_{M-1} = [1,\ 1,\ \ldots,\ 1]^{\mathrm{T}} \in \mathbb{R}^{(M-1)\times 1}$ and $\boldsymbol{U}'_M \in \mathbb{R}^{(M+1)\times(M+1)}$ described as

$$\boldsymbol{U}'_M = \begin{bmatrix} t_{Q,0,0} & \cdots & t_{Q,0,M-1} & t_{Q,0,M} \\ \vdots & & \vdots & \vdots \\ t_{Q,M-1,0} & \cdots & t_{Q,M-1,M-1} & t_{Q,M-1,M} \\ t_{Q,M,0} & \cdots & t_{Q,M,M-1} & t_{Q,M,M} \end{bmatrix}$$

where $t_{Q,k,j} = \ell_{M,Q-1,j}(\theta_{M,Q-1,k}),\ k,\ j = 0,\ 1,\ \ldots,\ M.$

5.1.4 Matrix $\boldsymbol{\Pi}_{M,N}$

According to (5.6), the right-hand of (5.16) can be further written as

$$\left(\mathcal{V}_2 P_N^+\begin{bmatrix}\boldsymbol{Z}\\ \boldsymbol{W}\end{bmatrix}\right)(\theta) = \begin{cases} \displaystyle\int_0^{h+\theta}\left(P_N^+\begin{bmatrix}\boldsymbol{Z}\\ \boldsymbol{W}\end{bmatrix}\right)(s)\mathrm{d}s, & \theta\in(-h,\ 0] \\ \mathbf{0}, & \theta\in[-\tau_{\max},\ -h] \end{cases}. \tag{5.23}$$

(1) For the discrete points $\theta_{M,1,k}$ $(k = 0,\ 1,\ \ldots,\ M-1)$ on the first sub-interval $(-h,\ 0]$, after implementing R_M, (5.23) becomes

$$\begin{cases} \left[R_M V_2 P_N^+ \boldsymbol{Z}\right]_{1,0} = \sum_{j=1}^{N} \int_0^{h+\theta_{M,1,0}} \ell_{N,j}^+(s) \boldsymbol{z}_j^* \mathrm{d}s \\ \left[R_M V_2 P_N^+ \boldsymbol{Z}\right]_{1,k} = \sum_{j=1}^{N} \int_0^{h+\theta_{M,1,k}} \ell_{N,j}^+(s) \left(\boldsymbol{z}_j^{(2)}\right)^* \mathrm{d}s,\ k = 1,\ 2,\ \ldots,\ M-1 \\ \left[R_M V_2 P_N^+ \boldsymbol{W}\right]_{1,k} = \sum_{j=1}^{N} \int_0^{h+\theta_{M,1,k}} \ell_{N,j}^+(s) \left(\boldsymbol{w}_j^{(2)}\right)^* \mathrm{d}s,\ k = 0,\ 1,\ \ldots,\ M-1 \end{cases}. \tag{5.24}$$

(2) For the discrete points $\theta_{M,i,k}$ $(i = 2,\ 3,\ \ldots,\ Q;\ k = 0,\ 1,\ \ldots,\ M-1)$ on the interval $[-\tau_{\max},\ -h]$, after implementing R_M, (5.23) becomes

$$\left[R_M V_2 P_N^+ \boldsymbol{Z}\right]_{i,k} = \boldsymbol{0}_{n_2 \times n},\quad \left[R_M V_2 P_N^+ \boldsymbol{W}\right]_{i,k} = \boldsymbol{0}_{l_2 \times l}. \tag{5.25}$$

By combining (5.24) and (5.25), the right-hand of (5.16) can be explicitly expressed in terms of $[\boldsymbol{Z}^{\mathrm{T}},\ \boldsymbol{W}^{\mathrm{T}}]^{\mathrm{T}}$, whose coefficient matrix is

$$\boldsymbol{\Pi}_{M,N} = \begin{bmatrix} \tilde{\boldsymbol{\Pi}}_{M,N} \\ \hdashline \boldsymbol{0}_{((Q-1)M+1)d_2 \times Nd} \end{bmatrix}$$

where $\tilde{\boldsymbol{\Pi}}_{M,N} \in \mathbb{R}^{(n_1+Md_2)\times Nd}$ is given by

$$\tilde{\boldsymbol{\Pi}}_{M,N} = \begin{bmatrix} \bar{\boldsymbol{U}}_{M,N} \otimes \boldsymbol{I}_n & \\ \tilde{\boldsymbol{U}}_{M,N} \otimes \left[\boldsymbol{0}_{n_2 \times n_1}\ \boldsymbol{I}_{n_2}\right] & \\ & \boldsymbol{U}_{M,N} \otimes \left[\boldsymbol{0}_{l_2 \times l_1}\ \boldsymbol{I}_{l_2}\right] \end{bmatrix} \tag{5.26}$$

with $\bar{\boldsymbol{U}}_{M,N} = \boldsymbol{U}_{M,N}(1,\ :)$, $\tilde{\boldsymbol{U}}_{M,N} = \boldsymbol{U}_{M,N}(2:M,\ :)$, and $\boldsymbol{U}_{M,N} \in \mathbb{R}^{M\times N}$ expressed as

$$\boldsymbol{U}_{M,N} = \begin{bmatrix} E_{1,1} & E_{1,2} & \cdots & E_{1,N} \\ E_{2,1} & E_{2,2} & \cdots & E_{2,N} \\ \vdots & \vdots & & \vdots \\ E_{M,1} & E_{M,2} & \cdots & E_{M,N} \end{bmatrix}$$

where

$$E_{k+1,j} = \int_0^{h+\theta_{M,1,k}} \ell_{N,j}^+(s)\mathrm{d}s,\quad k = 0,\ 1,\ \ldots,\ M-1;\ j = 1,\ 2,\ \ldots,\ N \tag{5.27}$$

are related to the integral of Lagrange interpolation coefficients $\ell_{N,j}^+$.

5.1.5 Matrix $\Sigma_{M,N}$

According to (5.9), $R_N^+ F V_1\left[\left(\varphi_{x,1,0}^{(1)}\right)^{\mathrm{T}},\ \left(P_M \mathbf{\Phi}_x^{(2)}\right)^{\mathrm{T}}\right]^{\mathrm{T}}$ in the right-hand of (5.17) can be further written as

$$\begin{aligned}&\left[R_N^+ F V_1\begin{bmatrix}\varphi_{x,1,0}^{(1)}\\ P_M\mathbf{\Phi}_x^{(2)}\end{bmatrix}\right]_j\\ &=[\boldsymbol{A}_0\ \ \boldsymbol{B}_0]\left(V_1\begin{bmatrix}\varphi_{x,1,0}^{(1)}\\ P_M\mathbf{\Phi}_x^{(2)}\\ \varphi_{y,1,0}^{(1)}\\ P_M\mathbf{\Phi}_y^{(2)}\end{bmatrix}\right)(t_{N,j})+\sum_{i=1}^{m}\left[\boldsymbol{A}_i^{(2)}\ \ \boldsymbol{B}_i^{(2)}\right]\left(V_1 P_M\begin{bmatrix}\mathbf{\Phi}_x^{(2)}\\ \mathbf{\Phi}_y^{(2)}\end{bmatrix}\right)(t_{N,j}-\tau_i)\end{aligned} \tag{5.28}$$

for $j = 1,\ 2,\ \ldots,\ N$. Substituting (5.5) into (5.28) yields

$$\begin{aligned}&\left[R_N^+ F V_1\begin{bmatrix}\varphi_{x,1,0}^{(1)}\\ P_M\mathbf{\Phi}_x^{(2)}\end{bmatrix}\right]_j\\ &=[\boldsymbol{A}_0\ \ \boldsymbol{B}_0]\,\hat{\varphi}_{1,0}+\sum_{i=1}^{m}\left[\boldsymbol{A}_i^{(2)}\ \ \boldsymbol{B}_i^{(2)}\right]\left(V_1 P_M\begin{bmatrix}\mathbf{\Phi}_x^{(2)}\\ \mathbf{\Phi}_y^{(2)}\end{bmatrix}\right)(t_{N,j}-\tau_i)\\ &=\tilde{\boldsymbol{A}}_0\varphi_{x,1,0}+\sum_{i=1}^{m}\left[\boldsymbol{A}_i^{(2)}\ \ \boldsymbol{B}_i^{(2)}\right]\left(V_1 P_M\begin{bmatrix}\mathbf{\Phi}_x^{(2)}\\ \mathbf{\Phi}_y^{(2)}\end{bmatrix}\right)(t_{N,j}-\tau_i),\quad j=1,\ 2,\ \ldots,\ N.\end{aligned} \tag{5.29}$$

For the subsequent derivation, the block vector $\bar{\mathbf{\Phi}} \in \mathbb{R}^{(n_1+(QM+1)d_2)\times 1}$, a counterpart of $\mathbf{\Phi}$, is defined as

$$\bar{\mathbf{\Phi}}=\left[\left(\varphi_{x,1,0}^{(1)}\right)^{\mathrm{T}},\ \left(\hat{\varphi}_{1,0}^{(2)}\right)^{\mathrm{T}},\ \left(\hat{\varphi}_{2,0}^{(2)}\right)^{\mathrm{T}},\ \ldots,\ \left(\hat{\varphi}_{Q,M}^{(2)}\right)^{\mathrm{T}}\right]^{\mathrm{T}} \tag{5.30}$$

where $\hat{\varphi}_{i,j}^{(2)}=\left[\left(\varphi_{x,i,j}^{(2)}\right)^{\mathrm{T}},\ \ \left(\varphi_{y,i,j}^{(2)}\right)^{\mathrm{T}}\right]^{\mathrm{T}}\in\mathbb{R}^{d_2\times 1},\quad i=1,\ \ 2,\ \ \ldots,\ \ Q;\ \ j=0,\ 1,\ \ldots,\ M$.

Based on (5.29), the right-hand of (5.17) is first required to be explicitly expressed in terms of $\bar{\mathbf{\Phi}}$, whose coefficient matrix $\bar{\mathbf{\Sigma}}_{M,N}\in\mathbb{R}^{Nn\times(n_1+(QM+1)d_2)}$ is expressed as

$$\bar{\mathbf{\Sigma}}_{M,N}=\begin{bmatrix}\mathbf{\Sigma}_{M,N}^{(1,1)} & \mathbf{\Sigma}_{M,N}^{(1,2)}\\ \mathbf{\Sigma}_{M,N}^{(2,1)} & \mathbf{\Sigma}_{M,N}^{(2,2)}\end{bmatrix}$$

with $\mathbf{\Sigma}_{M,N}^{(1,1)}\in\mathbb{R}^{\hat{N}n\times(n_1+((Q-1)M+1)d_2)}$, $\mathbf{\Sigma}_{M,N}^{(1,2)}\in\mathbb{R}^{\hat{N}n\times Md_2}$, $\mathbf{\Sigma}_{M,N}^{(2,2)}\in\mathbb{R}^{(N-\hat{N})n\times Md_2}$, and $\mathbf{\Sigma}_{M,N}^{(2,1)}\in\mathbb{R}^{(N-\hat{N})n\times(n_1+((Q-1)M+1)d_2)}$ given by

$$\boldsymbol{\Sigma}_{M,N}^{(1,1)} = \begin{bmatrix} \boldsymbol{F}_{1,1,0} & \cdots & \boldsymbol{F}_{1,1,M-1} & \boldsymbol{F}_{1,2,0} & \cdots & \boldsymbol{F}_{1,Q,0} \\ \vdots & & \vdots & \vdots & & \vdots \\ \boldsymbol{F}_{\hat{N},1,0} & \cdots & \boldsymbol{F}_{\hat{N},1,M-1} & \boldsymbol{F}_{\hat{N},2,0} & \cdots & \boldsymbol{F}_{\hat{N},Q,0} \end{bmatrix} \tag{5.31}$$

$$\boldsymbol{\Sigma}_{M,N}^{(1,2)} = \begin{bmatrix} \boldsymbol{F}_{1,Q,1} & \cdots & \boldsymbol{F}_{1,Q,M-1} & \boldsymbol{F}_{1,Q,M} \\ \vdots & & \vdots & \vdots \\ \boldsymbol{F}_{\hat{N},Q,1} & \cdots & \boldsymbol{F}_{\hat{N},Q,M-1} & \boldsymbol{F}_{\hat{N},Q,M} \end{bmatrix} \tag{5.32}$$

$$\boldsymbol{\Sigma}_{M,N}^{(2,1)} = \begin{bmatrix} \boldsymbol{F}_{\hat{N}+1,1,0} & \cdots & \boldsymbol{F}_{\hat{N}+1,1,M-1} & \boldsymbol{F}_{\hat{N}+1,2,0} & \cdots & \boldsymbol{F}_{\hat{N}+1,Q,0} \\ \vdots & & \vdots & \vdots & & \vdots \\ \boldsymbol{F}_{N,1,0} & \cdots & \boldsymbol{F}_{N,1,M-1} & \boldsymbol{F}_{N,2,0} & \cdots & \boldsymbol{F}_{N,Q,0} \end{bmatrix} \tag{5.33}$$

$$\boldsymbol{\Sigma}_{M,N}^{(2,2)} = \begin{bmatrix} \mathbf{0} & \cdots & \mathbf{0} & \mathbf{0} \\ \vdots & & \vdots & \vdots \\ \mathbf{0} & \cdots & \mathbf{0} & \mathbf{0} \end{bmatrix}. \tag{5.34}$$

Next, the expressions of the submatrices $\boldsymbol{F}_{j,1,0} \in \mathbb{R}^{n\times(n_1+d_2)}$ $(j = 1, 2, \ldots, N)$ and $\boldsymbol{F}_{j,k,l} \in \mathbb{R}^{n\times d_2}$ $(j = 1, 2, \ldots, N;\ k = 1, 2, \ldots, Q;\ l = 0, 1, \ldots, M)$ in (5.31)–(5.34) are deduced in detail.

5.1.5.1 Variable Declaration

The key to solving (5.29) comes in determining whether $t_{N,j} - \tau_i$ is positive for $i = 1, 2, \ldots, m$ and $j = 1, 2, \ldots, N$.

(1) If $t_{N,j} - \tau_i$ lies in the interval $[0,\ h]$, $\left(V_1 P_M\left[\left(\boldsymbol{\Phi}_x^{(2)}\right)^{\mathrm{T}},\ \left(\boldsymbol{\Phi}_y^{(2)}\right)^{\mathrm{T}}\right]^{\mathrm{T}}\right)(t_{N,j} - \tau_i) = \hat{\varphi}_{1,0}^{(2)}$ holds.

(2) If $t_{N,j} - \tau_i$ lies in the interval $[-\tau_{\max},\ 0]$, it is necessary to further determine the sub-interval on which it locates. Then, $\left(V_1 P_M\left[\left(\boldsymbol{\Phi}_x^{(2)}\right)^{\mathrm{T}},\ \left(\boldsymbol{\Phi}_y^{(2)}\right)^{\mathrm{T}}\right]^{\mathrm{T}}\right)(t_{N,j} - \tau_i)$ can be estimated by (5.2) defined on the sub-interval.

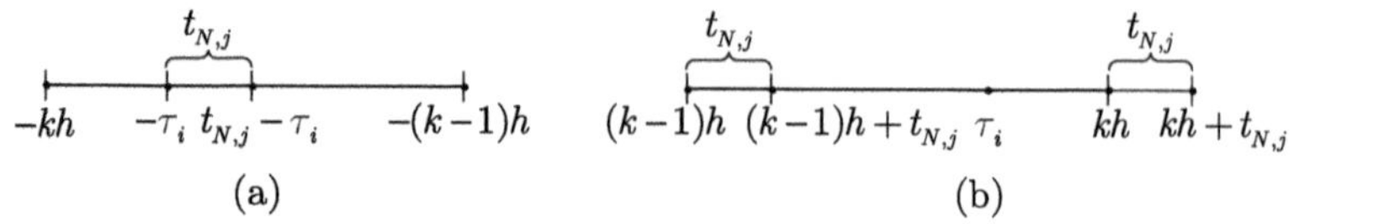

Fig. 5.2 The judgement of the k-th ($k = 1, 2, \ldots, Q-1$) sub-interval $(-kh, -(k-1)h]$ on which $t_{N,j} - \tau_i$ ($j = 1, 2, \ldots, N$; $i = 1, 2, \ldots, m$) locates

For the description, the definitions of four variables are provided below.

(1) For $t \in [0, \tau_{\max}]$,

$$i(t) = \begin{cases} i, & \tau_i \leqslant t < \tau_{i+1};\ i = 0, 1, \ldots, m-1 \\ m, & t = \tau_{\max} \end{cases}$$

is used to indicate that t lies between the $(i(t))$-th and the $(i(t)+1)$-th delay constants, i.e., $t - \tau_{i(t)} \geqslant 0 \cup t - \tau_{i(t)+1} < 0$. Therefore, $i(t_{N,j})$ indicates the maximal delay τ_i that satisfies $t_{N,j} - \tau_i \geqslant 0$ for the given $t_{N,j}$ ($j = 1, 2, \ldots, N$).

(2) The judgement of the sub-interval of $[-\tau_{\max}, 0]$ on which $t_{N,j} - \tau_i$ ($j = 1, 2, \ldots, N$; $i = 1, 2, \ldots, m$) locates can be accomplished through determining whether $t_{N,j} - \tau_i \in (-kh, -(k-1)h]$ or $\tau_i \in [(k-1)h + t_{N,j}, kh + t_{N,j})$ holds, as shown in Fig. 5.2a and b, respectively. In other words, the delay constant τ_i with $i \in [i((k-1)h + t_{N,j}) + 1, i(kh + t_{N,j})]$ makes $t_{N,j} - \tau_i$ ($j = 1, 2, \ldots, N$) fall into the k ($k = 1, 2, \ldots, Q-1$)-th sub-interval $(-kh, -(k-1)h]$.

(3) In particular, $t_{N,j} - \tau_{\max}$ ($j = 1, 2, \ldots, N$) may fall into the adjacent Q-th and $(Q-1)$-th sub-intervals, as shown in Fig. 5.3. For this reason, it is necessary to define

$$\hat{N} = \begin{cases} 0, & t_{N,j} \geq \tau_{\max} - (Q-1)h,\ \forall j = 1, 2, \ldots, N \\ \max\{j \mid t_{N,j} < \tau_{\max} - (Q-1)h\}, & t_{N,j} < \tau_{\max} - (Q-1)h \end{cases}$$

as the number of $t_{N,j} - \tau_{\max}$ ($j = 1, 2, \ldots, N$) that fall into the Q-th sub-interval, and the indicator

$$k_j := \begin{cases} Q, & j = 1, 2, \ldots, \hat{N} \\ Q-1, & j = \hat{N}+1, \hat{N}+2, \ldots, N \end{cases}.$$

In particular, all the discrete points $t_{N,j} - \tau_{\max}$ ($j = 1, 2, \ldots, N$) will fall

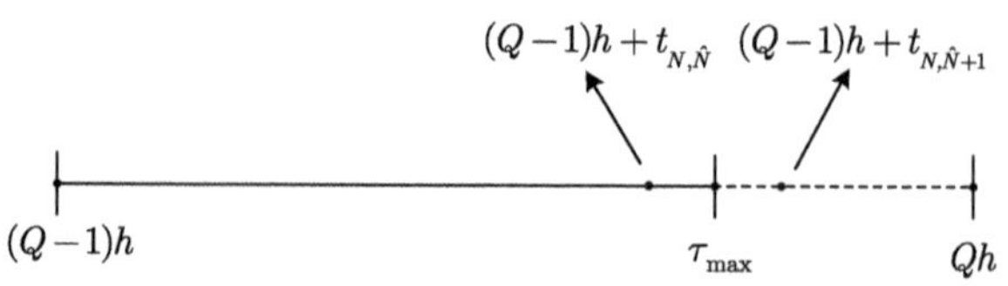

Fig. 5.3 The judgement of the Q-th or $(Q-1)$-th sub-interval on which $t_{N,j} - \tau_{\max}$ locates

into the $(Q-1)$-th sub-interval when $\tau_{\max} \approx (Q-1)h$. In this case, $\hat{N} = 0$. If $\tau_{\max} \approx Qh$, all the discrete points $t_{N,j} - \tau_{\max}$ $(j = 1,\ 2,\ \ldots,\ N)$ fall into the Q-th sub-interval and thus $\hat{N} = N$.

(4) For the convenience of presentation, define

$$t_{N,k,j} = \begin{cases} t_{N,j} + kh, & k = 0,\ 1,\ \ldots,\ k_j - 1 \\ \tau_{\max}, & k = k_j \end{cases}$$

for $j = 1,\ 2, \ldots,\ N$.

5.1.5.2 Expression of $\boldsymbol{F}_{j,1,0}$ $(j = 1,\ 2,\ \ldots,\ N)$

The submatrices $\boldsymbol{F}_{j,1,0} \in \mathbb{R}^{n\times(n_1+d_2)}$ $(j = 1,\ 2,\ \ldots,\ N)$ can be explicitly formulated as

$$\boldsymbol{F}_{j,1,0} = \left[\begin{array}{c|c} \tilde{\boldsymbol{A}}_0^{(1)} & \left[\begin{array}{c} \left[\tilde{\boldsymbol{A}}_0^{(2)}\ \boldsymbol{0}_{n\times l_2}\right] + \sum\limits_{i=1}^{i(t_{N,j})} \left[\boldsymbol{A}_i^{(2)}\ \boldsymbol{B}_i^{(2)}\right] \\ + \sum\limits_{i=i(t_{N,j})+1}^{i(t_{N,1,j})} \ell_{M,1,0}(t_{N,j} - \tau_i)\left[\boldsymbol{A}_i^{(2)}\ \boldsymbol{B}_i^{(2)}\right] \end{array}\right] \end{array}\right]. \tag{5.35}$$

The terms $\tilde{\boldsymbol{A}}_0^{(1)}$, $\tilde{\boldsymbol{A}}_0^{(2)}$ and $\sum\limits_{i=1}^{i(t_{N,j})}\left[\boldsymbol{A}_i^{(2)}\ \boldsymbol{B}_i^{(2)}\right]$ on the right-hand side of (5.35) correspond to the case where $t_{N,j} - \tau_i \in [0,\ h]$ in (5.29), and the remaining term corresponds to the case when $t_{N,j} - \tau_i \in [-h,\ 0]$.

5.1.5.3 Expression of $\boldsymbol{F}_{j,k,0}$ $(j = 1,\ 2,\ \ldots,\ N;\ k = 2,\ 3,\ \ldots,\ k_j)$

The submatrices $\boldsymbol{F}_{j,k,0} \in \mathbb{R}^{n\times d_2}$ $(j = 1,\ 2,\ \ldots,\ N;\ k = 2,\ 3,\ \ldots,\ k_j)$ can be explicitly expressed as

$$\begin{aligned} \boldsymbol{F}_{j,k,0} = &\sum_{i=i(t_{N,k-2,j})+1}^{i(t_{N,k-1,j})} \left[\boldsymbol{A}_i^{(2)}\ \boldsymbol{B}_i^{(2)}\right] \ell_{M,k-1,M}(t_{N,j} - \tau_i) + \\ &\sum_{i=i(t_{N,k-1,j})+1}^{i(t_{N,k,j})} \left[\boldsymbol{A}_i^{(2)}\ \boldsymbol{B}_i^{(2)}\right] \ell_{M,k,0}(t_{N,j} - \tau_i). \end{aligned} \tag{5.36}$$

The first term on the right-hand side of (5.36) corresponds to the case where $t_{N,j} - \tau_i$ falls into the $(k-1)$-th interval $[-(k-1)h,\ -(k-2)h]$, and the second represents the case where $t_{N,j} - \tau_i$ falls into the k-th interval $[-kh,\ -(k-1)h]$. Given that

$\theta_{M,k-1,M} = \theta_{M,k,0}$ $(k = 2,\ 3,\ \ldots,\ Q)$, the superposition of the two terms forms $\boldsymbol{F}_{j,k,0}$ $(j = 1,\ 2,\ \ldots,\ N;\ k = 2,\ 3,\ \ldots,\ k_j)$.

5.1.5.4 Expression of $\boldsymbol{F}_{j,Q,M}$ $(j = 1,\ 2,\ \ldots,\ \hat{N})$

The submatrices $\boldsymbol{F}_{j,Q,M} \in \mathbb{R}^{n\times d_2}$ $(j = 1,\ 2,\ \ldots,\ \hat{N})$ can be formulated as

$$\boldsymbol{F}_{j,Q,M} = \sum_{i=i(t_{N,Q-1,j})+1}^{m} \left[\boldsymbol{A}_i^{(2)}\ \boldsymbol{B}_i^{(2)}\right] \ell_{M,Q,M}(t_{N,j} - \tau_i).$$

It is worth noting that, $t_{N,j} - \tau_{\max}$ $(j = \hat{N}+1,\ \hat{N}+2,\ \ldots,\ N)$ falls into the $(Q-1)$-th sub-interval $[-(Q-1)h,\ -(Q-2)h]$ and thus $\ell_{M,Q,M}(t_{N,j} - \tau_{\max}) = 0$ holds. At this moment, $\boldsymbol{F}_{j,Q,M} = \boldsymbol{0}_{n\times d_2}$ $(j = \hat{N}+1,\ \hat{N}+2,\ \ldots,\ N)$ is deduced.

5.1.5.5 Expression of $\boldsymbol{F}_{j,Q,0}$ $(j = \hat{N}+1,\ \hat{N}+2,\ \ldots,\ N)$

The submatrices $\boldsymbol{F}_{j,Q,0} \in \mathbb{R}^{n\times d_2}$ $(j = \hat{N}+1,\ \hat{N}+2,\ \ldots,\ N)$ can be explicitly expressed as

$$\boldsymbol{F}_{j,Q,0} = \sum_{i=i(t_{N,Q-2,j})+1}^{m} \left[\boldsymbol{A}_i^{(2)}\ \boldsymbol{B}_i^{(2)}\right] \ell_{M,Q-1,M}(t_{N,j} - \tau_i).$$

It should be noted that, $\boldsymbol{F}_{j,Q,0}$ $(j = 1,\ 2,\ \ldots,\ \hat{N})$ is computed according to (5.36) since $t_{N,j} - \tau_i$ $(j = 1,\ 2,\ \ldots,\ \hat{N})$ falls into the Q-th sub-interval $[-\tau_{\max},\ -(Q-1)h]$.

5.1.5.6 Expression of the Remaining Submatrices

The submatrices $\boldsymbol{F}_{j,k,l} \in \mathbb{R}^{n\times d_2}$ $(j = 1,\ 2,\ \ldots,\ N;\ k = 1,\ 2,\ \ldots,\ k_j;\ l = 1,\ 2,\ \ldots,\ M-1)$ can be explicitly expressed as

$$\boldsymbol{F}_{j,k,l} = \sum_{i=i(t_{N,k-1,j})+1}^{i(t_{N,k,j})+1} \left[\boldsymbol{A}_i^{(2)}\ \boldsymbol{B}_i^{(2)}\right] \ell_{M,k,l}(t_{N,j} - \tau_i). \tag{5.37}$$

Given $t_{N,j}$ $(j = 1,\ 2,\ \ldots,\ N)$, the term on the right-hand side of (5.37) corresponds to the case where $t_{N,j} - \tau_i \in [-kh,\ -(k-1)h]$.

5.1.5.7 The Resulting $\bar{\boldsymbol{\Sigma}}_{M,N}$ and $\boldsymbol{\Sigma}_{M,N}$

It can be found from (5.35)–(5.37) that the matrix $\bar{\boldsymbol{\Sigma}}_{M,N}$ is associated with the augmented system state matrices $\tilde{\boldsymbol{A}}_0$, $\boldsymbol{A}_i^{(2)}$ and $\boldsymbol{B}_i^{(2)}$ $(i = 1,\ 2,\ \ldots,\ m)$. Thus, $\bar{\boldsymbol{\Sigma}}_{M,N}$ can be reformulated into the sum of the Kronecker products of the Lagrange interpolation coefficient matrices $\boldsymbol{L}_{M,N}^i \in \mathbb{R}^{N\times(QM+1)}$, and $\tilde{\boldsymbol{A}}_0$, $\boldsymbol{A}_i^{(2)}$ and $\boldsymbol{B}_i^{(2)}$ $(i = 1,\ 2,\ \ldots,\ m)$, i.e.,

$$\bar{\boldsymbol{\Sigma}}_{M,N} = \left[\mathbf{1}_N \otimes \tilde{\boldsymbol{A}}_0^{(1)} \ \left[\mathbf{1}_N \otimes \tilde{\boldsymbol{A}}_0^{(2)} \ \ \mathbf{0}_{Nn\times(l_2+QMd_2)}\right] + \sum_{i=1}^{m} \boldsymbol{L}_{M,N}^i \otimes \left[\boldsymbol{A}_i^{(2)} \ \ \boldsymbol{B}_i^{(2)}\right]\right] \tag{5.38}$$

where $\mathbf{1}_N = [1,\ \ldots,\ 1]^{\mathrm{T}} \in \mathbb{R}^{N\times 1}$.

Based on (5.38) and (5.30), the right-hand of (5.17) can be equivalently expressed in terms of $\boldsymbol{\Phi}$, whose coefficient matrix is

$$\boldsymbol{\Sigma}_{M,N} = \begin{bmatrix} \left[\mathbf{1}_N \otimes \tilde{\boldsymbol{A}}_0^{(1)} \ \boldsymbol{M}_1 \ \boldsymbol{N}_1 \ \boldsymbol{M}_2 \ \boldsymbol{N}_2\right] \\ \hdashline \mathbf{0}_{Nl\times(n_1+(QM+1)d_2)} \end{bmatrix} \tag{5.39}$$

with $\boldsymbol{M}_2 \in \mathbb{R}^{Nn\times((Q-1)M+1)n_2}$, $\boldsymbol{N}_2 \in \mathbb{R}^{Nn\times((Q-1)M+1)l_2}$, $\boldsymbol{M}_1 \in \mathbb{R}^{Nn\times Mn_2}$ and $\boldsymbol{N}_1 \in \mathbb{R}^{Nn\times Ml_2}$ given by

$$\begin{cases} \boldsymbol{M}_1 = \left[\mathbf{1}_N \otimes \tilde{\boldsymbol{A}}_0^{(2)} \ \ \mathbf{0}_{Nn\times(M-1)n_2}\right] + \sum\limits_{i=1}^{m} \boldsymbol{L}_{M,N,1}^i \otimes \boldsymbol{A}_i^{(2)} \\ \boldsymbol{N}_1 = \sum\limits_{i=1}^{m} \boldsymbol{L}_{M,N,1}^i \otimes \boldsymbol{B}_i^{(2)} \\ \boldsymbol{M}_2 = \sum\limits_{i=1}^{m} \boldsymbol{L}_{M,N,2}^i \otimes \boldsymbol{A}_i^{(2)} \\ \boldsymbol{N}_2 = \sum\limits_{i=1}^{m} \boldsymbol{L}_{M,N,2}^i \otimes \boldsymbol{B}_i^{(2)} \end{cases} \tag{5.40}$$

where $\boldsymbol{L}_{M,N,1}^i = \boldsymbol{L}_{M,N}^i(:,\ 1:M) \in \mathbb{R}^{N\times M}$ and $\boldsymbol{L}_{M,N,2}^i = \boldsymbol{L}_{M,N}^i(:,\ M+1:QM+1) \in \mathbb{R}^{N\times((Q-1)M+1)}$, $i = 1,\ 2,\ \ldots,\ m$.

5.1.6 Matrix $\boldsymbol{\Sigma}_N$

According to (5.9) and (5.6), $R_N^+ F V_2 P_N^+ \boldsymbol{Z}$ in the right-hand of (5.18) can be further written as

$$
\begin{aligned}
\left[R_N^+ F V_2 P_N^+ \boldsymbol{Z}\right]_j = & [\boldsymbol{A}_0 \ \ \boldsymbol{B}_0]\left(V_2 P_N^+ \begin{bmatrix} \boldsymbol{Z} \\ \boldsymbol{W} \end{bmatrix}\right)(t_{N,j}) \\
& + \sum_{i=1}^{m} \left[\boldsymbol{A}_i^{(2)} \ \ \boldsymbol{B}_i^{(2)}\right]\left(V_2 P_N^+ \begin{bmatrix} \boldsymbol{Z}^{(2)*} \\ \boldsymbol{W}^{(2)*} \end{bmatrix}\right)(t_{N,j} - \tau_i)
\end{aligned} \tag{5.41}
$$

for $j = 1, 2, \ldots, N$.

Considering that the operator V_2 defined as (5.6) contains two pieces, determining whether $t_{N,j} - \tau_i$ $(i = 0, 1, \ldots, m)$ is positive is the first step towards deriving (5.41). Specifically, if $t_{N,j} - \tau_i$ lies in the interval $[-\tau_{\max}, 0]$, $\left(V_2 P_N^+\left[\left(\boldsymbol{Z}^{(2)*}\right)^{\mathrm{T}}, \left(\boldsymbol{W}^{(2)*}\right)^{\mathrm{T}}\right]^{\mathrm{T}}\right)(t_{N,j} - \tau_i) = \boldsymbol{0}$; if $t_{N,j} - \tau_i$ locates on the interval $[0, h]$, $\left(V_2 P_N^+\left(\boldsymbol{Z}^{(2)*}\right)\right)(t_{N,j} - \tau_i)$ and $\left(V_2 P_N^+\left(\boldsymbol{W}^{(2)*}\right)\right)(t_{N,j} - \tau_i)$ are required to be estimated by (5.2) and (5.3) defined on the sub-interval. Thus, (5.41) can be written as

$$
\begin{aligned}
\left[R_N^+ F V_2 P_N^+ \boldsymbol{Z}\right]_j = & \sum_{k=1}^{N} [\boldsymbol{A}_0 \ \ \boldsymbol{B}_0] \int_0^{t_{N,j}} \ell_{N,k}^+(t) \begin{bmatrix} \boldsymbol{z}_k^* \\ \boldsymbol{w}_k^* \end{bmatrix} \mathrm{d}t \\
& + \sum_{k=1}^{N} \sum_{i=1}^{i(t_{N,j})} \left[\boldsymbol{A}_i^{(2)} \ \ \boldsymbol{B}_i^{(2)}\right] \int_0^{t_{N,j}-\tau_i} \ell_{N,k}^+(t) \begin{bmatrix} \boldsymbol{z}_k^{(2)*} \\ \boldsymbol{w}_k^{(2)*} \end{bmatrix} \mathrm{d}t
\end{aligned}
$$

Then, one can obtain

$$
\boldsymbol{\Sigma}_N = \begin{bmatrix} \begin{bmatrix} \boldsymbol{G}_{1,1} & \boldsymbol{G}_{1,2} & \cdots & \boldsymbol{G}_{1,N} & \boldsymbol{H}_{1,1} & \boldsymbol{H}_{1,2} & \cdots & \boldsymbol{H}_{1,N} \\ \boldsymbol{G}_{2,1} & \boldsymbol{G}_{2,2} & \cdots & \boldsymbol{G}_{2,N} & \boldsymbol{H}_{2,1} & \boldsymbol{H}_{2,2} & \cdots & \boldsymbol{H}_{2,N} \\ \vdots & \vdots & & \vdots & \vdots & \vdots & & \vdots \\ \boldsymbol{G}_{N,1} & \boldsymbol{G}_{N,2} & \cdots & \boldsymbol{G}_{N,N} & \boldsymbol{H}_{N,1} & \boldsymbol{H}_{N,2} & \cdots & \boldsymbol{H}_{N,N} \end{bmatrix} \\ \hdashline \boldsymbol{0}_{Nl \times Nd} \end{bmatrix}
$$

where $\boldsymbol{G}_{j,k} \in \mathbb{R}^{n \times n}$ and $\boldsymbol{H}_{j,k} \in \mathbb{R}^{n \times l}$ $(j = 1, 2, \ldots, N;\ k = 1, 2, \ldots, N)$ are given by

$$
\boldsymbol{G}_{j,k} = \left[\boldsymbol{A}_0^{(1)} \int_0^{t_{N,j}} \ell_{N,k}^+(t)\mathrm{d}t \ \ \sum_{i=0}^{i(t_{N,j})} \boldsymbol{A}_i^{(2)} \int_0^{t_{N,j}-\tau_i} \ell_{N,k}^+(t)\mathrm{d}t\right]
$$

and

$$
\boldsymbol{H}_{j,k} = \left[\boldsymbol{B}_0^{(1)} \int_0^{t_{N,j}} \ell_{N,k}^+(t)\mathrm{d}t \ \ \sum_{i=0}^{i(t_{N,j})} \boldsymbol{B}_i^{(2)} \int_0^{t_{N,j}-\tau_i} \ell_{N,k}^+(t)\mathrm{d}t\right].
$$

In fact, the matrix $\boldsymbol{\Sigma}_N$ is associated with the augmented system state matrices $\boldsymbol{A}_i$ and $\boldsymbol{B}_i$ $(i = 0, 1, \ldots, m)$. Thus, it can be further reformulated into the sum of the Kronecker products between the Lagrange interpolation coefficient matrix $\boldsymbol{L}_N^i \in \mathbb{R}^{N\times N}$, and $\boldsymbol{A}_i$ and $\boldsymbol{B}_i$ $(i = 0,\ 1,\ \ldots,\ m)$, namely

$$\boldsymbol{\Sigma}_N = \left[\begin{array}{c} \left[\sum_{i=0}^{m} \boldsymbol{L}_N^i \otimes \boldsymbol{A}_i \ \ \sum_{i=0}^{m} \boldsymbol{L}_N^i \otimes \boldsymbol{B}_i\right] \\ \hdashline \boldsymbol{0}_{Nl\times Nd} \end{array}\right]. \tag{5.42}$$

5.2 An Alternative Derivation of PS Partial Discretization Matrix

This section gives an alternative form of the partial discretization matrix $\boldsymbol{T}_{M,N}$ based on (3.24). The process can be attributed to derive the analytical expression of $\Delta\hat{\boldsymbol{x}}_h(0)$ and $\Delta\hat{\boldsymbol{x}}_h^{(2)}$ at the discrete points over the interval $[-\tau_{\max},\ 0)$ in terms of $\Delta\hat{\boldsymbol{x}}_0(0)$ and $\Delta\hat{\boldsymbol{x}}_0^{(2)}$ at the discrete points over the interval $[-\tau_{\max},\ 0)$, which is divided into the partial discretization of the time integration segment and that of the shift segment.

5.2.1 *Partial Discretization of the Time Integration Segment*

5.2.1.1 Basic Idea

The partial discretization of the *Time Integration* segment turns out to be the evaluation of $\Delta\hat{\boldsymbol{x}}_h$ at $\theta_{M,1,k}$ $(k = 0,\ 1,\ \ldots,\ M-1)$ of the subinterval $(-h,\ 0]$, i.e., $\Delta\hat{\boldsymbol{x}}_{1,1,k}$. First, the derivatives $\boldsymbol{z}_j$ and $\boldsymbol{w}_j$ $(j = 1,\ 2,\ \ldots,\ N)$ can be solved from

$$\begin{cases} \boldsymbol{z}_j = [\boldsymbol{A}_0\ \ \boldsymbol{B}_0]\,\Delta\hat{\boldsymbol{x}}_{N,j} + \sum_{i=1}^{m} \left[\,\boldsymbol{A}_i^{(2)}\ \ \boldsymbol{B}_i^{(2)}\,\right] \Delta\hat{\boldsymbol{x}}^{(2)}(t_{N,j} - \tau_i) \\ \boldsymbol{0} = \boldsymbol{C}_0\boldsymbol{z}_j + \boldsymbol{D}_0\boldsymbol{w}_j \end{cases} \tag{5.43}$$

by using the PS method, where $\Delta\hat{\boldsymbol{x}}_{N,j} = \left[\Delta\boldsymbol{x}_{N,j}^{\mathrm{T}},\ \Delta\boldsymbol{y}_{N,j}^{\mathrm{T}}\right]^{\mathrm{T}}$, $\Delta\boldsymbol{x}_{N,j} = \Delta\boldsymbol{x}(t_{N,j})$ and $\Delta\boldsymbol{y}_{N,j} = \Delta\boldsymbol{y}(t_{N,j})$. Then, one can obtain the fitting expressions of continuous functions $\boldsymbol{z}(t)$ and $\boldsymbol{w}(t)$ during $t \in [0,\ h]$ on the basis of $\boldsymbol{z}_j$ and $\boldsymbol{w}_j$ $(j = 1,\ 2,\ \ldots,\ N)$. Finally, integrating $\left[\boldsymbol{z}^{\mathrm{T}},\ \boldsymbol{w}^{\mathrm{T}}\right]^{\mathrm{T}}$ with respect to t yields the estimates of $\Delta\hat{\boldsymbol{x}}_{1,1,k}$ $(k = 0,\ 1,\ \ldots,\ M-1)$.

5.2.1.2 Explicit Expressions of z_j and w_j ($j = 1, 2, \ldots, N$)

First, the term $[A_0 \;\; B_0]\, \Delta\hat{x}_{N,j}$ of (5.43) can be further written as

$$\begin{aligned}
&[A_0 \;\; B_0]\, \Delta\hat{x}_{N,j} \\
&= [A_0 \;\; B_0] \begin{bmatrix} \varphi_{x,1,0} \\ \varphi_{y,1,0} \end{bmatrix} + \int_0^{t_{N,j}} \sum_{k=1}^{N} \ell_{N,k}^{+}(s)\, [A_0 \;\; B_0] \begin{bmatrix} z_k \\ w_k \end{bmatrix} \mathrm{d}s \\
&= \tilde{A}_0 \varphi_{x,1,0} + \int_0^{t_{N,j}} \sum_{k=1}^{N} \ell_{N,k}^{+}(s)\, [A_0 \;\; B_0] \begin{bmatrix} z_k \\ w_k \end{bmatrix} \mathrm{d}s
\end{aligned} \tag{5.44}$$

where $\varphi_{y,1,0}$ is eliminated because of its huge dimension.

Then, the first step to derive $\sum_{i=1}^{m} \left[A_i^{(2)} \;\; B_i^{(2)} \right] \Delta\hat{x}^{(2)}(t_{N,j} - \tau_i)$ of (5.43) is determining whether $t_{N,j} - \tau_i$ ($i = 1, 2, \ldots, m$) is positive.

(1) If $i = 1, 2, \ldots, i(t_{N,j})$, $t_{N,j} - \tau_i$ lies in the interval $(0, h]$. At this point, one has

$$\begin{aligned}
&\sum_{i=1}^{i(t_{N,j})} \left[A_i^{(2)} \;\; B_i^{(2)} \right] \Delta\hat{x}^{(2)}(t_{N,j} - \tau_i) \\
&= \sum_{i=1}^{i(t_{N,j})} \left(\left[A_i^{(2)} \;\; B_i^{(2)} \right] \begin{bmatrix} \varphi_{x,1,0}^{(2)} \\ \varphi_{y,1,0}^{(2)} \end{bmatrix} + \int_0^{t_{N,j} - \tau_i} \sum_{k=1}^{N} \ell_{N,k}^{+}(s) \left[A_i^{(2)} \;\; B_i^{(2)} \right] \begin{bmatrix} z_k^{(2)} \\ w_k^{(2)} \end{bmatrix} \mathrm{d}s \right).
\end{aligned} \tag{5.45}$$

(2) If $i = i(t_{N,j}) + 1, \; i(t_{N,j}) + 2, \; \ldots, \; m$, $t_{N,j} - \tau_i$ falls into the r ($r = 1, 2, \ldots, Q$)-th sub-interval of the delay interval $[-\tau_{\max}, \; 0]$. At this point, $\Delta\hat{x}^{(2)}(t_{N,j} - \tau_i)$ can be estimated using the prolongation operator P_M defined on the corresponding sub-interval. Since that the Q-th sub-interval is shorter than the ($Q - 1$)-th sub-interval, $t_{N,j} - \tau_{\max}$ ($j = 1, 2, \ldots, \hat{N}$) and $t_{N,j} - \tau_{\max}$ ($j = \hat{N} + 1, \; \hat{N} + 2, \; \ldots, \; N$) fall into the Q-th and ($Q - 1$)-th sub-intervals, respectively. Therefore, one has

$$\begin{aligned}
&\sum_{i=i(t_{N,j})+1}^{m} \left[A_i^{(2)} \;\; B_i^{(2)} \right] \Delta\hat{x}^{(2)}(t_{N,j} - \tau_i) \\
&= \sum_{i=i(t_{N,j})+1}^{i(t_{N,1,j})} \left[A_i^{(2)} \;\; B_i^{(2)} \right] \sum_{k=0}^{M} \ell_{M,1,k}(t_{N,j} - \tau_i) \begin{bmatrix} \varphi_{x,1,k}^{(2)} \\ \varphi_{y,1,k}^{(2)} \end{bmatrix} + \cdots \\
&\quad + \sum_{i=i(t_{N,Q-2,j})+1}^{i(t_{N,Q-1,j})} \left[A_i^{(2)} \;\; B_i^{(2)} \right] \sum_{k=0}^{M} \ell_{M,Q-1,k}(t_{N,j} - \tau_i) \begin{bmatrix} \varphi_{x,Q-1,k}^{(2)} \\ \varphi_{y,Q-1,k}^{(2)} \end{bmatrix} \\
&\quad + \sum_{i=i(t_{N,Q-1,j})+1}^{m} \left[A_i^{(2)} \;\; B_i^{(2)} \right] \sum_{k=0}^{M} \ell_{M,Q,k}(t_{N,j} - \tau_i) \begin{bmatrix} \varphi_{x,Q,k}^{(2)} \\ \varphi_{y,Q,k}^{(2)} \end{bmatrix}
\end{aligned} \tag{5.46}$$

for $j = 1, 2, \ldots, \hat{N}$ and

$$\begin{aligned}
&\sum_{i=i(t_{N,j})+1}^{m} \left[\boldsymbol{A}_i^{(2)} \ \boldsymbol{B}_i^{(2)} \right] \Delta\hat{\boldsymbol{x}}^{(2)}(t_{N,j}-\tau_i) \\
&= \sum_{i=i(t_{N,j})+1}^{i(t_{N,1,j})} \left[\boldsymbol{A}_i^{(2)} \ \boldsymbol{B}_i^{(2)} \right] \sum_{k=0}^{M} \ell_{M,1,k}(t_{N,j}-\tau_i) \begin{bmatrix} \boldsymbol{\varphi}_{x,1,k}^{(2)} \\ \boldsymbol{\varphi}_{y,1,k}^{(2)} \end{bmatrix} + \cdots \\
&+ \sum_{i=i(t_{N,Q-2,j})+1}^{m} \left[\boldsymbol{A}_i^{(2)} \ \boldsymbol{B}_i^{(2)} \right] \sum_{k=0}^{M} \ell_{M,Q-1,k}(t_{N,j}-\tau_i) \begin{bmatrix} \boldsymbol{\varphi}_{x,Q-1,k}^{(2)} \\ \boldsymbol{\varphi}_{y,Q-1,k}^{(2)} \end{bmatrix}
\end{aligned} \tag{5.47}$$

for $j = \hat{N}+1, \ \hat{N}+2, \ \ldots, \ N$.

Substituting (5.44) ∼ (5.47) into (5.43) and appropriately shifting the terms yield

$$\begin{aligned}
&\boldsymbol{z}_j - \int_0^{t_{N,j}} \sum_{k=1}^{N} \ell_{N,k}^{+}(s) \left[\boldsymbol{A}_0 \ \boldsymbol{B}_0 \right] \begin{bmatrix} \boldsymbol{z}_k \\ \boldsymbol{w}_k \end{bmatrix} \mathrm{d}s \\
&- \sum_{i=1}^{i(t_{N,j})} \int_0^{t_{N,j}-\tau_i} \sum_{k=1}^{N} \ell_{N,k}^{+}(s) \left[\boldsymbol{A}_i^{(2)} \ \boldsymbol{B}_i^{(2)} \right] \begin{bmatrix} \boldsymbol{z}_k^{(2)} \\ \boldsymbol{w}_k^{(2)} \end{bmatrix} \mathrm{d}s \\
=& \tilde{\boldsymbol{A}}_0 \boldsymbol{\varphi}_{x,1,0} + \sum_{i=1}^{i(t_{N,j})} \left[\boldsymbol{A}_i^{(2)} \ \boldsymbol{B}_i^{(2)} \right] \begin{bmatrix} \boldsymbol{\varphi}_{x,1,0}^{(2)} \\ \boldsymbol{\varphi}_{y,1,0}^{(2)} \end{bmatrix} \\
&+ \sum_{i=i(t_{N,j})+1}^{i(t_{N,1,j})} \left[\boldsymbol{A}_i^{(2)} \ \boldsymbol{B}_i^{(2)} \right] \sum_{k=0}^{M} \ell_{M,1,k}(t_{N,j}-\tau_i) \begin{bmatrix} \boldsymbol{\varphi}_{x,1,k}^{(2)} \\ \boldsymbol{\varphi}_{y,1,k}^{(2)} \end{bmatrix} + \cdots \\
&+ \sum_{i=i(t_{N,Q-2,j})+1}^{i(t_{N,Q-1,j})} \left[\boldsymbol{A}_i^{(2)} \ \boldsymbol{B}_i^{(2)} \right] \sum_{k=0}^{M} \ell_{M,Q-1,k}(t_{N,j}-\tau_i) \begin{bmatrix} \boldsymbol{\varphi}_{x,Q-1,k}^{(2)} \\ \boldsymbol{\varphi}_{y,Q-1,k}^{(2)} \end{bmatrix} \\
&+ \sum_{i=i(t_{N,Q-1,j})+1}^{m} \left[\boldsymbol{A}_i^{(2)} \ \boldsymbol{B}_i^{(2)} \right] \sum_{k=0}^{M} \ell_{M,Q,k}(t_{N,j}-\tau_i) \begin{bmatrix} \boldsymbol{\varphi}_{x,Q,k}^{(2)} \\ \boldsymbol{\varphi}_{y,Q,k}^{(2)} \end{bmatrix}
\end{aligned} \tag{5.48}$$

for $j = 1, \ 2, \ \ldots, \ \hat{N}$ and

$$
\begin{aligned}
& z_j - \int_0^{t_{N,j}} \sum_{k=1}^{N} \ell_{N,k}^{+}(s)\,[\boldsymbol{A}_0\ \ \boldsymbol{B}_0]\begin{bmatrix} z_k \\ \boldsymbol{w}_k \end{bmatrix} \mathrm{d}s \\
& - \sum_{i=1}^{i(t_{N,j})} \int_0^{t_{N,j}-\tau_i} \sum_{k=1}^{N} \ell_{N,k}^{+}(s)\left[\boldsymbol{A}_i^{(2)}\ \ \boldsymbol{B}_i^{(2)}\right]\begin{bmatrix} z_k^{(2)} \\ \boldsymbol{w}_k^{(2)} \end{bmatrix} \mathrm{d}s \\
= & \tilde{\boldsymbol{A}}_0 \varphi_{x,1,0} + \sum_{i=1}^{i(t_{N,j})} \left[\boldsymbol{A}_i^{(2)}\ \ \boldsymbol{B}_i^{(2)}\right]\begin{bmatrix} \varphi_{x,1,0}^{(2)} \\ \varphi_{y,1,0}^{(2)} \end{bmatrix} \\
& + \sum_{i=i(t_{N,j})+1}^{i(t_{N,1,j})} \left[\boldsymbol{A}_i^{(2)}\ \ \boldsymbol{B}_i^{(2)}\right]\sum_{k=0}^{M} \ell_{M,1,k}(t_{N,j}-\tau_i)\begin{bmatrix} \varphi_{x,1,k}^{(2)} \\ \varphi_{y,1,k}^{(2)} \end{bmatrix} + \cdots \\
& + \sum_{i=i(t_{N,Q-2,j})+1}^{m} \left[\boldsymbol{A}_i^{(2)}\ \ \boldsymbol{B}_i^{(2)}\right]\sum_{k=0}^{M} \ell_{M,Q-1,k}(t_{N,j}-\tau_i)\begin{bmatrix} \varphi_{x,Q-1,k}^{(2)} \\ \varphi_{y,Q-1,k}^{(2)} \end{bmatrix}
\end{aligned} \tag{5.49}
$$

for $j = \hat{N}+1,\ \hat{N}+2,\ \ldots,\ N$.

According to Sect. 5.1.6, the left-hand side of (5.48) and (5.49) can be rewritten in the matrix form

$$
\left[\boldsymbol{I}_{Nn} - \sum_{i=0}^{m} \boldsymbol{L}_N^i \otimes \boldsymbol{A}_i\ \ - \sum_{i=0}^{m} \boldsymbol{L}_N^i \otimes \boldsymbol{B}_i\right]\begin{bmatrix} \boldsymbol{Z} \\ \boldsymbol{W} \end{bmatrix}. \tag{5.50}
$$

According to Sect. 5.1.5, the right-hand side of (5.48) and (5.49) can be equivalently expressed as

$$
\left[\boldsymbol{1}_N \otimes \tilde{\boldsymbol{A}}_0^{(1)}\ \ \boldsymbol{M}_1\ \ \boldsymbol{N}_1\ \ \boldsymbol{M}_2\ \ \boldsymbol{N}_2\right]\boldsymbol{\Phi} \tag{5.51}
$$

where the specific expressions of $\boldsymbol{M}_2 \in \mathbb{R}^{Nn\times((Q-1)M+1)n_2}$, $\boldsymbol{N}_2 \in \mathbb{R}^{Nn\times((Q-1)M+1)l_2}$, $\boldsymbol{M}_1 \in \mathbb{R}^{Nn\times Mn_2}$ and $\boldsymbol{N}_1 \in \mathbb{R}^{Nn\times Ml_2}$ can be found in (5.40).

Combining (5.50), (5.51) and the matrix form of (5.43), i.e., $(\boldsymbol{I}_N \otimes \boldsymbol{C}_0)\boldsymbol{Z} + (\boldsymbol{I}_N \otimes \boldsymbol{D}_0)\boldsymbol{W} = \boldsymbol{0}$, one can obtain

$$
\begin{bmatrix} \boldsymbol{I}_{Nn} - \sum_{i=0}^{m} \boldsymbol{L}_N^i \otimes \boldsymbol{A}_i & - \sum_{i=0}^{m} \boldsymbol{L}_N^i \otimes \boldsymbol{B}_i \\ \boldsymbol{I}_N \otimes \boldsymbol{C}_0 & \boldsymbol{I}_N \otimes \boldsymbol{D}_0 \end{bmatrix}\begin{bmatrix} \boldsymbol{Z} \\ \boldsymbol{W} \end{bmatrix} = \begin{bmatrix} \left[\boldsymbol{1}_N \otimes \tilde{\boldsymbol{A}}_0^{(1)}\ \ \boldsymbol{M}_1\ \ \boldsymbol{N}_1\ \ \boldsymbol{M}_2\ \ \boldsymbol{N}_2\right] \\ \boldsymbol{0}_{Nl\times(n_1+(QM+1)d_2)} \end{bmatrix}\boldsymbol{\Phi}
$$

which can be further reformulated as

$$
\begin{bmatrix} \boldsymbol{Z} \\ \boldsymbol{W} \end{bmatrix} = \left(\begin{bmatrix} \boldsymbol{I}_{Nn} & \\ \boldsymbol{I}_N \otimes \boldsymbol{C}_0 & \boldsymbol{I}_N \otimes \boldsymbol{D}_0 \end{bmatrix} - \boldsymbol{\Sigma}_N\right)^{-1} \boldsymbol{\Sigma}_{M,N}\boldsymbol{\Phi} \tag{5.52}
$$

where the specific expressions of $\boldsymbol{\Sigma}_{M,N}$ and $\boldsymbol{\Sigma}_N$ are found in (5.39) and (5.42).

5.2.1.3 Evaluation of $\Delta x_{1,1,k}$ and $\Delta y_{1,1,k}$ $(k = 0,\ 1,\ \ldots,\ M-1)$

Based on (5.52), one can obtain the fitting expressions of $z(t)$ and $w(t)$ during $t \in [0,\ h]$ by utilizing the prolongation operator P_N^+. Then, one can integrate $\left[z^{\mathrm{T}},\ w^{\mathrm{T}}\right]^{\mathrm{T}}$ with the initial condition is $\left[\varphi_{x,1,0}^{\mathrm{T}},\ \varphi_{y,1,0}^{\mathrm{T}}\right]^{\mathrm{T}}$ to estimate $\Delta\hat{x}$ at $h + \theta_{M,1,k}$ $(k = 0,\ 1,\ \ldots,\ M-1)$, i.e., $\Delta\hat{x}_{1,1,k} = \left[\Delta x_{1,1,k}^{\mathrm{T}},\ \Delta y_{1,1,k}^{\mathrm{T}}\right]^{\mathrm{T}}$ $(k = 0,\ 1,\ \ldots,\ M-1)$. The process can be expressed as

$$\Delta x_{1,1,k} = \Delta x(h + \theta_{M,1,k}) = \varphi_{x,1,0} + \sum_{j=1}^{N} \int_0^{h+\theta_{M,1,k}} \ell_{N,j}^+(s) z_j \mathrm{d}s \tag{5.53}$$

and

$$\begin{aligned}\Delta y_{1,1,k} = \Delta y(h + \theta_{M,1,k}) &= \varphi_{y,1,0} + \sum_{j=1}^{N} \int_0^{h+\theta_{M,1,k}} \ell_{N,j}^+(s) w_j \mathrm{d}s \\ &= -D_0^{-1} C_0 \varphi_{x,1,0}^{(0)} + \sum_{j=1}^{N} \int_0^{h+\theta_{M,1,k}} \ell_{N,j}^+(s) w_j \mathrm{d}s.\end{aligned} \tag{5.54}$$

5.2.1.4 Partial Discretization of the Time Integration Segment

According to (5.52)–(5.54) and $\boldsymbol{\Phi}$ defined as (5.1), the estimates $\Delta x_{1,1,k}$ and $\Delta y_{1,1,k}$ $(k = 0,\ 1,\ \ldots,\ M-1)$ during $(-h,\ 0]$ are independent of $\Delta\hat{x}^{(1)}$ over the delay interval $[-\tau_{\max},\ 0)$ and $\Delta y^{(1)}$ at 0. Therefore, combining (5.53) and (5.54) and further eliminating $\Delta x_{1,1,k}^{(1)} = \Delta x^{(1)}(h + \theta_{M,1,k})$ $(k = 1,\ 2,\ \ldots,\ M-1)$ and $\Delta y_{1,1,k}^{(1)} = \Delta y^{(1)}(h + \theta_{M,1,k})$ $(k = 0,\ 1,\ \ldots,\ M-1)$, one can obtain the PS partial discretization of the *Time Integration* segment of $\mathcal{T}(h)$, i.e.,

$$\bar{\boldsymbol{\Psi}} = \left(\tilde{\boldsymbol{\Pi}}_M + \tilde{\boldsymbol{\Pi}}_{M,N}\left(\begin{bmatrix} \boldsymbol{I}_{Nn} & \\ \boldsymbol{I}_N \otimes \boldsymbol{C}_0 & \boldsymbol{I}_N \otimes \boldsymbol{D}_0 \end{bmatrix} - \boldsymbol{\Sigma}_N\right)^{-1} \boldsymbol{\Sigma}_{M,N}\right)\boldsymbol{\Phi} \tag{5.55}$$

where $\tilde{\boldsymbol{\Pi}}_M$ and $\tilde{\boldsymbol{\Pi}}_{M,N}$ can be found from (5.22) and (5.26), and $\bar{\boldsymbol{\Psi}} \in \mathbb{R}^{(n_1+Md_2)\times 1}$ is given by

$$\begin{aligned}\bar{\boldsymbol{\Psi}} = \Big[&\left(\Delta x_{1,1,0}^{(1)}\right)^{\mathrm{T}},\ \left(\Delta x_{1,1,0}^{(2)}\right)^{\mathrm{T}},\ \left(\Delta x_{1,1,1}^{(2)}\right)^{\mathrm{T}},\ \ldots,\ \left(\Delta x_{1,1,M-1}^{(2)}\right)^{\mathrm{T}}, \\ &\left(\Delta y_{1,1,0}^{(2)}\right)^{\mathrm{T}},\ \left(\Delta y_{1,1,1}^{(2)}\right)^{\mathrm{T}},\ \ldots,\ \left(\Delta y_{1,1,M-1}^{(2)}\right)^{\mathrm{T}}\Big]^{\mathrm{T}}\end{aligned}$$

with $\Delta \boldsymbol{x}_{1,1,0}^{(1)} = \Delta \boldsymbol{x}^{(1)}(h + \theta_{M,1,0}) \in \mathbb{R}^{n_1 \times 1}$, $\Delta \boldsymbol{x}_{1,1,k}^{(2)} = \Delta \boldsymbol{x}^{(2)}(h + \theta_{M,1,k}) \in \mathbb{R}^{n_2 \times 1}$ and $\Delta \boldsymbol{y}_{1,1,k}^{(2)} = \Delta \boldsymbol{y}^{(2)}(h + \theta_{M,1,k}) \in \mathbb{R}^{l_2 \times 1}$, $k = 0, 1, \ldots, M-1$.

5.2.2 *Partial Discretization of the Shift Segment*

The partial discretization of the *Shift* segment turns out to be the evaluation of $\Delta \boldsymbol{x}_h^{(2)}$ and $\Delta \boldsymbol{y}_h^{(2)}$ at $\theta_{M,i,k}$ ($i = 2, 3, \ldots, Q$; $k = 0, 1, \ldots, M$) of the subinterval $[-\tau_{\max}, -h]$, i.e., $\Delta \boldsymbol{x}_{1,i,k}^{(2)} \triangleq \Delta \boldsymbol{x}_h^{(2)}(\theta_{M,i,k}) = \Delta \boldsymbol{x}^{(2)}(h + \theta_{M,i,k})$ and $\Delta \boldsymbol{y}_{1,i,k}^{(2)} \triangleq \Delta \boldsymbol{y}_h^{(2)}(\theta_{M,i,k}) = \Delta \boldsymbol{y}^{(2)}(h + \theta_{M,i,k})$.

(1) If $i = 2, 3, \ldots, Q-1$, there exist

$$\Delta \boldsymbol{x}_{1,i,k}^{(2)} = \boldsymbol{\varphi}_{x,i-1,k}^{(2)}, \ \Delta \boldsymbol{y}_{1,i,k}^{(2)} = \boldsymbol{\varphi}_{y,i-1,k}^{(2)}, \ k = 0, 1, \ldots, M-1. \tag{5.56}$$

(2) If $i = Q$, $h + \theta_{M,i,k}$ ($k = 0, 1, \ldots, M-1$) falls into the $(Q-1)$-th sub-interval, i.e., $h + \theta_{M,i,k} \in [-(Q-1)h, -(Q-2)h]$. Considering that the Q-th sub-interval is generally shorter than the $(Q-1)$-th sub-interval, $h + \theta_{M,Q,k}$ does not coincide with $\theta_{M,Q-1,k}$ ($k = 0, 1, \ldots, M$). In this case, it is necessary to compute $\Delta \boldsymbol{x}_{1,Q,k}^{(2)}$ and $\Delta \boldsymbol{y}_{1,Q,k}^{(2)}$ ($k = 0, 1, \ldots, M$) as

$$\begin{cases} \Delta \boldsymbol{x}_{1,Q,k}^{(2)} = \sum_{j=0}^{M} \ell_{M,Q-1,j}(h + \theta_{M,Q,k}) \boldsymbol{\varphi}_{x,Q-1,j}^{(2)} \\ \Delta \boldsymbol{y}_{1,Q,k}^{(2)} = \sum_{j=0}^{M} \ell_{M,Q-1,j}(h + \theta_{M,Q,k}) \boldsymbol{\varphi}_{y,Q-1,j}^{(2)} \end{cases} \tag{5.57}$$

by using the Lagrange interpolation method.

Combining (5.56) and (5.57), one can obtain the PS partial discretization of the *Shift* segment of $\mathcal{T}(h)$, i.e.,

$$\tilde{\boldsymbol{\Psi}} = \begin{bmatrix} \boldsymbol{0}_{Mn_2 \times n_1} & \boldsymbol{I}_{Mn_2} & & & & & \\ & & & \boldsymbol{I}_{(Q-3)Mn_2} & & & \\ & & & & \boldsymbol{U}_{Mn} & & \\ & & \boldsymbol{I}_{Ml_2} & & & & \\ & & & & & \boldsymbol{I}_{(Q-3)Ml_2} & \\ & & & & & & \boldsymbol{U}_{Ml} \end{bmatrix} \boldsymbol{\Phi} \tag{5.58}$$

where the specific expressions of $\boldsymbol{U}_{Mn}$ and $\boldsymbol{U}_{Ml}$ are found in (5.22), and $\tilde{\boldsymbol{\Psi}} \in \mathbb{R}^{((Q-1)M+1)d_2 \times 1}$ is given by

$$\tilde{\boldsymbol{\Psi}} = \left[\left(\Delta \boldsymbol{x}_{1,2,0}^{(2)}\right)^{\mathrm{T}}, \left(\Delta \boldsymbol{x}_{1,2,1}^{(2)}\right)^{\mathrm{T}}, \ldots, \left(\Delta \boldsymbol{x}_{1,Q,M}^{(2)}\right)^{\mathrm{T}}, \right.$$
$$\left. \left(\Delta \boldsymbol{y}_{1,2,0}^{(2)}\right)^{\mathrm{T}}, \left(\Delta \boldsymbol{y}_{1,2,1}^{(2)}\right)^{\mathrm{T}}, \ldots, \left(\Delta \boldsymbol{y}_{1,Q,M}^{(2)}\right)^{\mathrm{T}}\right]^{\mathrm{T}}.$$

5.2.3 *Partial Discretization Matrix by PS Method*

Combining (5.55) and (5.58) yields the expression between $\boldsymbol{\Psi} = \left[\bar{\boldsymbol{\Psi}}^{\mathrm{T}}, \tilde{\boldsymbol{\Psi}}^{\mathrm{T}}\right]^{\mathrm{T}} \in \mathbb{R}^{(n_1+(QM+1)d_2)\times 1}$ and $\boldsymbol{\Phi}$, whose coefficient matrix is the PS partial solution operator discretization matrix

$$\boldsymbol{T}_{M,N} = \left[\begin{array}{c} \tilde{\boldsymbol{\Pi}}_M + \tilde{\boldsymbol{\Pi}}_{M,N}\left(\begin{bmatrix} \boldsymbol{I}_{Nn} \\ \boldsymbol{I}_N \otimes \boldsymbol{C}_0 \ \boldsymbol{I}_N \otimes \boldsymbol{D}_0 \end{bmatrix} - \boldsymbol{\Sigma}_N\right)^{-1} \boldsymbol{\Sigma}_{M,N} \\ \hdashline \begin{bmatrix} \boldsymbol{0}_{Mn_2\times n_1} & \boldsymbol{I}_{Mn_2} & & & & & \\ & & \boldsymbol{I}_{(Q-3)Mn_2} & & & & \\ & & & \boldsymbol{U}_{Mn} & & & \\ & \boldsymbol{I}_{Ml_2} & & & & & \\ & & & & \boldsymbol{I}_{(Q-3)Ml_2} & & \\ & & & & & \boldsymbol{U}_{Ml} \end{bmatrix} \end{array}\right] \tag{5.59}$$

where $\tilde{\boldsymbol{\Pi}}_M$ and $\tilde{\boldsymbol{\Pi}}_{M,N}$ are the first $n_1 + Md_2$ rows of $\boldsymbol{\Pi}_M$ and $\boldsymbol{\Pi}_{M,N}$ that have been given in Sects. 5.1.3 and 5.1.4.

It should be noted that the discretization matrices obtained through the two derivations, i.e., (5.14) and (5.59), are exactly the same apart from the different types of forms. The following content is operating on (5.59).

5.3 Eigenvalue Computation for Large-Scale Time-Delay Systems

5.3.1 *Inexact Coordinate Rotation*

To capture the eigenvalues with damping ratios less than a threshold ζ, one can rotate the coordinate by $\theta = \arcsin \zeta$ radian in counterclockwise direction. According to Sect. 3.4.3, the rotation operation can be accomplished by directly replacing the augmented system state matrices $\boldsymbol{J}_i$ ($i = 0, 1, \ldots, m$) and its submatrices in $\boldsymbol{T}_{M,N}$ with their rotated counterparts, leading to

$$
\boldsymbol{T}'_{M,N} = \left[\begin{array}{c} \tilde{\boldsymbol{\Pi}}_M + \tilde{\boldsymbol{\Pi}}_{M,N}\left(\begin{bmatrix} \boldsymbol{I}_{Nn} & \\ \boldsymbol{I}_N \otimes \boldsymbol{C}_0 & \boldsymbol{I}_N \otimes \boldsymbol{D}_0 \end{bmatrix} - \boldsymbol{\Sigma}'_N\right)^{-1} \boldsymbol{\Sigma}'_{M,N} \\ \hdashline \begin{bmatrix} \boldsymbol{0}_{Mn_2 \times n_1} & \boldsymbol{I}_{Mn_2} & & & & & \\ & & \boldsymbol{I}_{(Q-3)Mn_2} & & & & \\ & & & \boldsymbol{U}_{Mn} & & & \\ & \boldsymbol{I}_{Ml_2} & & & & & \\ & & & & \boldsymbol{I}_{(Q-3)Ml_2} & & \\ & & & & & \boldsymbol{U}_{Ml} \end{bmatrix} \end{array}\right]
$$

of dimension $(n_1 + (QM + 1)d_2)$-by-$(n_1 + (QM + 1)d_2)$, where the submatrices $\boldsymbol{\Sigma}'_{M,N} \in \mathbb{C}^{Nd \times (n_1 + (QM+1)d_2)}$ and $\boldsymbol{\Sigma}'_N \in \mathbb{C}^{Nd \times Nd}$ are given by

$$
\boldsymbol{\Sigma}'_{M,N} = \left[\begin{array}{c} \begin{bmatrix} \boldsymbol{1}_N \otimes \tilde{\boldsymbol{A}}_0^{\prime(1)} & \boldsymbol{M}'_1 & \boldsymbol{N}'_1 & \boldsymbol{M}'_2 & \boldsymbol{N}'_2 \end{bmatrix} \\ \hdashline \boldsymbol{0}_{Nl \times (n_1 + (QM+1)d_2)} \end{array}\right]
$$

and

$$
\boldsymbol{\Sigma}'_N = \left[\begin{array}{c} \begin{bmatrix} \sum_{i=0}^{m} \boldsymbol{L}^i_N \otimes \boldsymbol{A}'_i & \sum_{i=0}^{m} \boldsymbol{L}^i_N \otimes \boldsymbol{B}'_i \end{bmatrix} \\ \hdashline \boldsymbol{0}_{Nl \times Nd} \end{array}\right]
$$

with $\boldsymbol{M}'_2 \in \mathbb{C}^{Nn \times ((Q-1)M+1)n_2}$, $\boldsymbol{N}'_2 \in \mathbb{C}^{Nn \times ((Q-1)M+1)l_2}$, $\boldsymbol{M}'_1 \in \mathbb{C}^{Nn \times Mn_2}$ and $\boldsymbol{N}'_1 \in \mathbb{C}^{Nn \times Ml_2}$ described as

$$
\begin{cases} \boldsymbol{M}'_1 = \begin{bmatrix} \boldsymbol{1}_N \otimes \tilde{\boldsymbol{A}}_0^{\prime(2)} & \boldsymbol{0}_{Nn \times (M-1)n_2} \end{bmatrix} + \sum_{i=1}^{m} \boldsymbol{L}^i_{M,N,1} \otimes \boldsymbol{A}_i^{\prime(2)} \\ \boldsymbol{N}'_1 = \sum_{i=1}^{m} \boldsymbol{L}^i_{M,N,1} \otimes \boldsymbol{B}_i^{\prime(2)} \\ \boldsymbol{M}'_2 = \sum_{i=1}^{m} \boldsymbol{L}^i_{M,N,2} \otimes \boldsymbol{A}_i^{\prime(2)} \\ \boldsymbol{N}'_2 = \sum_{i=1}^{m} \boldsymbol{L}^i_{M,N,2} \otimes \boldsymbol{B}_i^{\prime(2)} \end{cases}
$$

5.3.2 Rotation-and-Multiplication Preconditioning

After the above inexact coordinate rotation operation, the dispersion of the eigenvalues μ of $\mathcal{T}(h)$ is required to be enhanced for an accelerated convergence rate of IRA, which is also called the multiplication preconditioning. Therefore, this section first introduces two implementations of the multiplication preconditioning given in Sect. 3.4.3 and then proves their consistency.

5.3.2.1 The First Implementation

The first implementation of the multiplication preconditioning can be concluded as follows:

- h remains and τ_i $(i = 1, 2, \ldots, m)$ is reduced to $1/\alpha$ of it; $\boldsymbol{A}'_i$, $\boldsymbol{B}'_i$ and $\tilde{\boldsymbol{A}}'_i$ $(i = 0, 1, \ldots, m)$ are expanded into α times, i.e., $\boldsymbol{A}''_i$, $\boldsymbol{B}''_i$ and $\tilde{\boldsymbol{A}}''_i$;
- the delay interval $[-\tau_{\max}/\alpha, 0]$ is redivided into Q' sub-intervals, where $Q' = \lceil \tau_{\max}/(\alpha h) \rceil$;
- $\boldsymbol{U}'_M$, $\boldsymbol{U}_{Mn}$ and $\boldsymbol{U}_{Ml}$ are reformed to $\boldsymbol{U}''_M \in \mathbb{R}^{(M+1)\times(M+1)}$, $\boldsymbol{U}''_{Mn} \in \mathbb{R}^{(M+1)n_2\times(2M+1)n_2}$ and $\boldsymbol{U}''_{Ml} \in \mathbb{R}^{(M+1)l_2\times(2M+1)l_2}$, respectively;
- $\boldsymbol{U}_{M,N}$, $\bar{\boldsymbol{U}}_{M,N}$ and $\tilde{\boldsymbol{U}}_{M,N}$ are reformed to $\boldsymbol{U}'_{M,N} \in \mathbb{R}^{M\times N}$, $\bar{\boldsymbol{U}}'_{M,N} = \boldsymbol{U}'_{M,N}(1, :) \in \mathbb{R}^{1\times N}$ and $\tilde{\boldsymbol{U}}'_{M,N} = \boldsymbol{U}'_{M,N}(2 : M, :) \in \mathbb{R}^{(M-1)\times N}$, respectively;
- $\boldsymbol{L}^i_{M,N}$, $\boldsymbol{L}^i_{M,N,1}$ and $\boldsymbol{L}^i_{M,N,2}$ $(i = 1, 2, \ldots, m)$ are reformed to $\tilde{\boldsymbol{L}}^i_{M,N} \in \mathbb{R}^{N\times(Q'M+1)}$, $\tilde{\boldsymbol{L}}^i_{M,N,1} = \tilde{\boldsymbol{L}}^i_{M,N}(:, 1 : M) \in \mathbb{R}^{N\times M}$ and $\tilde{\boldsymbol{L}}^i_{M,N,2} = \tilde{\boldsymbol{L}}^i_{M,N}(:, M+1 : Q'M+1) \in \mathbb{R}^{N\times((Q'-1)M+1)}$, respectively;
- $\boldsymbol{L}^i_N$ $(i = 0, 1, \ldots, m)$ are reformed to $\left(\tilde{\boldsymbol{L}}^i_N\right)' \in \mathbb{R}^{N\times N}$;
- finally, the preconditioned PS partial solution operator discretization matrix $\hat{\boldsymbol{T}}''_{M,N} \in \mathbb{C}^{(n_1+(Q'M+1)d_2)\times(n_1+(Q'M+1)d_2)}$ is built as

$$
\hat{\boldsymbol{T}}''_{M,N} = \begin{bmatrix} \tilde{\boldsymbol{\Pi}}''_M + \hat{\boldsymbol{\Pi}}''_{M,N}\left(\begin{bmatrix} & \boldsymbol{I}_{Nn} \\ \boldsymbol{I}_N \otimes \boldsymbol{C}_0 & \boldsymbol{I}_N \otimes \boldsymbol{D}_0 \end{bmatrix} - \hat{\boldsymbol{\Sigma}}''_N\right)^{-1} \hat{\boldsymbol{\Sigma}}''_{M,N} \\ \hdashline \begin{bmatrix} \boldsymbol{0}_{Mn_2\times n_1} & \boldsymbol{I}_{Mn_2} & & & & \\ & & \boldsymbol{I}_{(Q'-3)Mn_2} & & & \\ & & & \boldsymbol{U}''_{Mn} & & \\ & \boldsymbol{I}_{Ml_2} & & & & \\ & & & & \boldsymbol{I}_{(Q'-3)Ml_2} & \\ & & & & & \boldsymbol{U}''_{Ml} \end{bmatrix} \end{bmatrix}
$$

where $\tilde{\boldsymbol{\Pi}}''_M \in \mathbb{R}^{(n_1+Md_2)\times(n_1+(Q'M+1)d_2)}$, $\hat{\boldsymbol{\Pi}}''_{M,N} \in \mathbb{R}^{(n_1+Md_2)\times Nd}$, $\hat{\boldsymbol{\Sigma}}''_N \in \mathbb{C}^{Nd\times Nd}$ and $\hat{\boldsymbol{\Sigma}}''_{M,N} \in \mathbb{C}^{Nd\times(n_1+(Q'M+1)d_2)}$ are given by

$$
\tilde{\boldsymbol{\Pi}}''_M = \left[\begin{bmatrix} & \boldsymbol{I}_n & \\ & \boldsymbol{1}_{M-1} \otimes \begin{bmatrix} \boldsymbol{0}_{n_2\times n_1} & \boldsymbol{I}_{n_2} \end{bmatrix} & \\ -\boldsymbol{1}_M \otimes \begin{bmatrix} \boldsymbol{0}_{l_2\times l_1} & \boldsymbol{I}_{l_2} \end{bmatrix} \boldsymbol{D}_0^{-1}\boldsymbol{C}_0 & \end{bmatrix} \;\vdots\; \boldsymbol{0}_{(n_1+Md_2)\times(l_2+Q'Md_2)}\right] \tag{5.60}
$$

$$\hat{\mathbf{\Pi}}''_{M,N} = \begin{bmatrix} \bar{\boldsymbol{U}}'_{M,N} \otimes \boldsymbol{I}_n & \\ \tilde{\boldsymbol{U}}'_{M,N} \otimes \left[\boldsymbol{0}_{n_2 \times n_1} \ \boldsymbol{I}_{n_2} \right] & \\ & \boldsymbol{U}'_{M,N} \otimes \left[\boldsymbol{0}_{l_2 \times l_1} \ \boldsymbol{I}_{l_2} \right] \end{bmatrix} \tag{5.61}$$

$$\hat{\mathbf{\Sigma}}''_{M,N} = \begin{bmatrix} \left[\boldsymbol{1}_N \otimes \tilde{\boldsymbol{A}}_0^{''(1)} \ \hat{\boldsymbol{M}}''_1 \ \hat{\boldsymbol{N}}''_1 \ \hat{\boldsymbol{M}}''_2 \ \hat{\boldsymbol{N}}''_2 \right] \\ \boldsymbol{0}_{Nl \times (n_1 + (Q'M+1)d_2)} \end{bmatrix}$$

$$\hat{\mathbf{\Sigma}}''_{N} = \begin{bmatrix} \left[\sum\limits_{i=0}^{m} (\tilde{\boldsymbol{L}}^i_N)' \otimes \boldsymbol{A}''_i \ \sum\limits_{i=0}^{m} (\tilde{\boldsymbol{L}}^i_N)' \otimes \boldsymbol{B}''_i \right] \\ \boldsymbol{0}_{Nl \times Nd} \end{bmatrix}$$

with $\hat{\boldsymbol{M}}''_2 \in \mathbb{C}^{Nn \times ((Q'-1)M+1)n_2}$, $\hat{\boldsymbol{N}}''_2 \in \mathbb{C}^{Nn \times ((Q'-1)M+1)l_2}$, $\hat{\boldsymbol{M}}''_1 \in \mathbb{C}^{Nn \times Mn_2}$ and $\hat{\boldsymbol{N}}''_1 \in \mathbb{C}^{Nn \times Ml_2}$ described as

$$\begin{cases} \hat{\boldsymbol{M}}''_1 = \left[\boldsymbol{1}_N \otimes \tilde{\boldsymbol{A}}_0^{''(2)} \ \boldsymbol{0}_{Nn \times (M-1)n_2} \right] + \sum\limits_{i=1}^{m} \tilde{\boldsymbol{L}}^i_{M,N,1} \otimes \boldsymbol{A}_i^{''(2)} \\ \hat{\boldsymbol{N}}''_1 = \sum\limits_{i=1}^{m} \tilde{\boldsymbol{L}}^i_{M,N,1} \otimes \boldsymbol{B}_i^{''(2)} \\ \hat{\boldsymbol{M}}''_2 = \sum\limits_{i=1}^{m} \tilde{\boldsymbol{L}}^i_{M,N,2} \otimes \boldsymbol{A}_i^{''(2)} \\ \hat{\boldsymbol{N}}''_2 = \sum\limits_{i=1}^{m} \tilde{\boldsymbol{L}}^i_{M,N,2} \otimes \boldsymbol{B}_i^{''(2)} \end{cases}$$

5.3.2.2 The Second Implementation

The second implementation can be summarized as follows:

- τ_i ($i = 1, 2, \ldots, m$), $\boldsymbol{A}'_i$, $\boldsymbol{B}'_i$ and $\tilde{\boldsymbol{A}}'_i$ ($i = 0, 1, \ldots, m$) are kept; h is expanded into α times.
- the delay interval $[-\tau_{\max}, 0]$ is redivided into Q' sub-intervals, where $Q' = \lceil \tau_{\max}/(\alpha h) \rceil$;
- $\boldsymbol{U}'_M$, $\boldsymbol{U}_{Mn}$ and $\boldsymbol{U}_{Ml}$ are reformed to $\boldsymbol{U}''_M$, $\boldsymbol{U}''_{Mn}$ and $\boldsymbol{U}''_{Ml}$, respectively;
- $\boldsymbol{U}_{M,N}$, $\bar{\boldsymbol{U}}_{M,N}$ and $\tilde{\boldsymbol{U}}_{M,N}$ are reformed to $\boldsymbol{U}''_{M,N} \in \mathbb{R}^{M \times N}$, $\bar{\boldsymbol{U}}''_{M,N} = \boldsymbol{U}''_{M,N}(1, :) \in \mathbb{R}^{1 \times N}$ and $\tilde{\boldsymbol{U}}''_{M,N} = \boldsymbol{U}''_{M,N}(2 : M, :) \in \mathbb{R}^{(M-1) \times N}$, respectively;
- $\boldsymbol{L}^i_{M,N}$, $\boldsymbol{L}^i_{M,N,1}$ and $\boldsymbol{L}^i_{M,N,2}$ ($i = 1, 2, \ldots, m$) are reformed to $\tilde{\boldsymbol{L}}^i_{M,N}$, $\tilde{\boldsymbol{L}}^i_{M,N,1}$ and $\tilde{\boldsymbol{L}}^i_{M,N,2}$, respectively;
- $\boldsymbol{L}^i_N$ ($i = 0, 1, \ldots, m$) are reformed to $\tilde{\boldsymbol{L}}^i_N \in \mathbb{R}^{N \times N}$;

- finally, one can obtain $\boldsymbol{T}''_{M,N} \in \mathbb{C}^{(n_1+(Q'M+1)d_2)\times(n_1+(Q'M+1)d_2)}$ that is expressed as

$$\hat{\boldsymbol{T}}''_{M,N} = \left[\begin{array}{c} \tilde{\boldsymbol{\Pi}}''_M + \tilde{\boldsymbol{\Pi}}''_{M,N}\left(\begin{bmatrix} \boldsymbol{I}_{Nn} \\ \boldsymbol{I}_N \otimes \boldsymbol{C}_0 \ \boldsymbol{I}_N \otimes \boldsymbol{D}_0 \end{bmatrix} - \boldsymbol{\Sigma}''_N\right)^{-1} \boldsymbol{\Sigma}''_{M,N} \\ \hdashline \begin{bmatrix} \boldsymbol{0}_{Mn_2\times n_1} & \boldsymbol{I}_{Mn_2} & & & & & \\ & & \boldsymbol{I}_{(Q'-3)Mn_2} & & & & \\ & & & \boldsymbol{U}''_{Mn} & & & \\ & \boldsymbol{I}_{Ml_2} & & & & & \\ & & & & \boldsymbol{I}_{(Q'-3)Ml_2} & & \\ & & & & & \boldsymbol{U}''_{Ml} \end{bmatrix} \end{array}\right]$$

where $\tilde{\boldsymbol{\Pi}}''_{M,N} \in \mathbb{R}^{(n_1+Md_2)\times Nd}$, $\boldsymbol{\Sigma}''_N \in \mathbb{C}^{Nd\times Nd}$ and $\boldsymbol{\Sigma}''_{M,N} \in \mathbb{C}^{Nd\times(n_1+(Q'M+1)d_2)}$ are given by

$$\tilde{\boldsymbol{\Pi}}''_{M,N} = \begin{bmatrix} \bar{\boldsymbol{U}}''_{M,N} \otimes \boldsymbol{I}_n & \\ \tilde{\boldsymbol{U}}''_{M,N} \otimes \left[\boldsymbol{0}_{n_2\times n_1} \ \boldsymbol{I}_{n_2}\right] & \\ & \boldsymbol{U}''_{M,N} \otimes \left[\boldsymbol{0}_{l_2\times l_1} \ \boldsymbol{I}_{l_2}\right] \end{bmatrix}$$

$$\boldsymbol{\Sigma}''_{M,N} = \left[\begin{array}{c} \left[\boldsymbol{1}_N \otimes \tilde{\boldsymbol{A}}_0^{\prime(1)} \ \boldsymbol{M}''_1 \ \boldsymbol{N}''_1 \ \boldsymbol{M}''_2 \ \boldsymbol{N}''_2\right] \\ \hdashline \boldsymbol{0}_{Nl\times(n_1+(Q'M+1)d_2)} \end{array}\right]$$

$$\boldsymbol{\Sigma}''_N = \left[\begin{array}{c} \left[\sum_{i=0}^{m} \tilde{\boldsymbol{L}}_N^i \otimes \boldsymbol{A}'_i \ \sum_{i=0}^{m} \tilde{\boldsymbol{L}}_N^i \otimes \boldsymbol{B}'_i\right] \\ \hdashline \boldsymbol{0}_{Nd\times Nd} \end{array}\right]$$

with $\boldsymbol{M}''_2 \in \mathbb{C}^{Nn\times((Q'-1)M+1)n_2}$, $\boldsymbol{N}''_2 \in \mathbb{C}^{Nn\times((Q'-1)M+1)l_2}$, $\boldsymbol{M}''_1 \in \mathbb{C}^{Nn\times Mn_2}$ and $\boldsymbol{N}''_1 \in \mathbb{C}^{Nn\times Ml_2}$ described as

$$\begin{cases} \boldsymbol{M}''_1 = \left[\boldsymbol{1}_N \otimes \tilde{\boldsymbol{A}}_0^{\prime(2)} \ \boldsymbol{0}_{Nn\times(M-1)n_2}\right] + \sum_{i=1}^{m} \tilde{\boldsymbol{L}}^i_{M,N,1} \otimes \boldsymbol{A}_i^{\prime(2)} \\ \boldsymbol{N}''_1 = \sum_{i=1}^{m} \tilde{\boldsymbol{L}}^i_{M,N,1} \otimes \boldsymbol{B}_i^{\prime(2)} \\ \boldsymbol{M}''_2 = \sum_{i=1}^{m} \tilde{\boldsymbol{L}}^i_{M,N,2} \otimes \boldsymbol{A}_i^{\prime(2)} \\ \boldsymbol{N}''_2 = \sum_{i=1}^{m} \tilde{\boldsymbol{L}}^i_{M,N,2} \otimes \boldsymbol{B}_i^{\prime(2)} \end{cases}.$$

Table 5.1 Comparison of the discretization matrices $\hat{\boldsymbol{T}}_{M,N}$ and the preconditioned ones $\hat{\boldsymbol{T}}''_{M,N}$ and $\boldsymbol{T}''_{M,N}$

$\boldsymbol{T}_{M,N}$	$\hat{\boldsymbol{T}}''_{M,N}$	$\boldsymbol{T}''_{M,N}$
$\tilde{\boldsymbol{\Pi}}_M, \bar{\boldsymbol{\Pi}}_{M,N}, \boldsymbol{\Sigma}_N, \boldsymbol{\Sigma}_{M,N}$	$\tilde{\boldsymbol{\Pi}}''_M, \hat{\boldsymbol{\Pi}}''_{M,N}, \hat{\boldsymbol{\Sigma}}''_N, \hat{\boldsymbol{\Sigma}}''_{M,N}$	$\tilde{\boldsymbol{\Pi}}''_M, \tilde{\boldsymbol{\Pi}}''_{M,N}, \boldsymbol{\Sigma}''_N, \boldsymbol{\Sigma}''_{M,N}$
$\boldsymbol{U}_{M,N}, \bar{\boldsymbol{U}}_{M,N}, \tilde{\boldsymbol{U}}_{M,N}$	$\boldsymbol{U}'_{M,N}, \bar{\boldsymbol{U}}'_{M,N}, \tilde{\boldsymbol{U}}'_{M,N}$	$\boldsymbol{U}''_{M,N}, \bar{\boldsymbol{U}}''_{M,N}, \tilde{\boldsymbol{U}}''_{M,N}$
$\boldsymbol{M}_1, \boldsymbol{N}_1, \boldsymbol{M}_2, \boldsymbol{N}_2,$ $\boldsymbol{L}^i_{M,N}, \boldsymbol{L}^i_{M,N,1}, \boldsymbol{L}^i_{M,N,2}$	$\hat{\boldsymbol{M}}''_1, \hat{\boldsymbol{N}}''_1, \hat{\boldsymbol{M}}''_2, \hat{\boldsymbol{N}}''_2,$ $\tilde{\boldsymbol{L}}^i_{M,N}, \tilde{\boldsymbol{L}}^i_{M,N,1}, \tilde{\boldsymbol{L}}^i_{M,N,2}$	$\boldsymbol{M}''_1, \boldsymbol{N}''_1, \boldsymbol{M}''_2, \boldsymbol{N}''_2,$ $\tilde{\boldsymbol{L}}^i_{M,N}, \tilde{\boldsymbol{L}}^i_{M,N,1}, \tilde{\boldsymbol{L}}^i_{M,N,2}$
$\boldsymbol{L}^i_N$	$(\tilde{\boldsymbol{L}}^i_N)'$	$\tilde{\boldsymbol{L}}^i_N$
$\boldsymbol{U}'_M, \boldsymbol{U}_{Mn}, \boldsymbol{U}_{Ml}$	$\boldsymbol{U}''_M, \boldsymbol{U}''_{Mn}, \boldsymbol{U}''_{Ml}$	$\boldsymbol{U}''_M, \boldsymbol{U}''_{Mn}, \boldsymbol{U}''_{Ml}$

5.3.2.3 Consistency of the Two Preconditioning Implementations

To verify the consistency of the preconditioned discretization matrices $\hat{\boldsymbol{T}}''_{M,N}$ and $\boldsymbol{T}''_{M,N}$ under two implementations of rotation-and-multiplication preconditioning, the corresponding relationship between each discretization submatrix is proofed as follows. For illustration purposes, $\boldsymbol{T}_{M,N}$, $\hat{\boldsymbol{T}}''_{M,N}$ and $\boldsymbol{T}''_{M,N}$ as well as their sub-matrices are listed in Table 5.1.

The Last $((Q-1)M+1)d_2$ Rows of $\hat{\boldsymbol{T}}''_{M,N}$ and $\boldsymbol{T}''_{M,N}$ are Identical

The last $((Q-1)M+1)d_2$ rows of the discretization matrix $\boldsymbol{T}_{M,N}$ are unrelated to $\boldsymbol{J}_i$ $(i = 0, 1, \ldots, m)$. Thus, the investigation of the effect of the multiplication preconditioning turns out to be that of the effect of τ_i $(i = 1, 2, \ldots, m)$ and h against $\boldsymbol{U}'_M$ and its elements $t_{Q,k,j}$ $(k, j = 0, 1, \ldots, m)$.

Under the first implementation, the values of $\theta_{M,i,k}$ $(i = 1, 2, \ldots, Q'; k = 0, 1, \ldots, M)$ are kept and $t_{Q,k,j}$ $(k, j = 0, 1, \ldots, M)$ are evolved into

$$t_{Q',k,j} = \prod_{l=0,\ l\neq j}^{M} \frac{h + \theta_{M,Q',k} - \theta_{M,Q'-1,l}}{\theta_{M,Q'-1,j} - \theta_{M,Q'-1,l}}. \tag{5.62}$$

Under the second implementation, the values of $\theta_{M,i,k}$ $(i = 1, 2, \ldots, Q'; k = 0, 1, \ldots, M)$ are transformed to α times. Subsequently, $t_{Q,k,j}$ $(k, j = 0, 1, \ldots, M)$ are evolved into

$$t_{Q',k,j} = \prod_{l=0,\ l\neq j}^{M} \frac{\alpha h + \alpha\theta_{M,Q',k} - \alpha\theta_{M,Q'-1,l}}{\alpha\theta_{M,Q'-1,j} - \alpha\theta_{M,Q'-1,l}} = \prod_{l=0,\ l\neq j}^{M} \frac{h + \theta_{M,Q',k} - \theta_{M,Q'-1,l}}{\theta_{M,Q'-1,j} - \theta_{M,Q'-1,l}}. \tag{5.63}$$

It can be seen from (5.62) and (5.63) that the transformed values $t_{Q',k,j}$ $(Q, j = 0, 1, \ldots, M)$ under two implementations are identical, leading to the same $\boldsymbol{U}''_M$ and the last $((Q-1)M+1)d_2$ rows of $\hat{\boldsymbol{T}}''_{M,N}$ and $\boldsymbol{T}''_{M,N}$.

$\tilde{\boldsymbol{\Pi}}''_{M,N} = \alpha\hat{\boldsymbol{\Pi}}''_{M,N}$

The discretization submatrix $\tilde{\mathbf{\Pi}}_{M,N}$ has nothing to do with $\boldsymbol{J}_i$ $(i = 0,\ 1,\ \ldots,\ m)$. Thus, the investigation of the effect of the multiplication preconditioning is reduced to that of the effect of τ_i $(i = 1,\ 2,\ \ldots,\ m)$ and h against $\boldsymbol{U}_{M,N}$ and its elements $E_{k+1,j}$ $(k = 0,\ 1,\ \ldots,\ M-1;\ j = 1,\ 2,\ \ldots,\ N)$.

Under the first implementation, $\theta_{M,1,k}$ $(k = 0,\ 1,\ \ldots,\ M)$ and $t_{N,i}$ $(i = 1,\ 2,\ \ldots,\ N)$ are kept. So is $E_{k+1,j}$ $(k = 0,\ 1,\ \ldots,\ M-1;\ j = 1,\ 2,\ \ldots,\ N)$. Under the second implementation, $\theta_{M,1,k}$ $(k = 0,\ 1,\ \ldots,\ M)$ and $t_{N,i}$ $(i = 1,\ 2,\ \ldots,\ N)$ are transformed to α times. Subsequently, $E_{k+1,j}$ are evolved into

$$
\begin{aligned}
E'_{k+1,j} &= \int_0^{\alpha h+\alpha\theta_{M,1,k}} \prod_{l=1,\ l\neq j}^{N} \frac{s-\alpha t_{N,l}}{\alpha t_{N,j}-\alpha t_{N,l}} \mathrm{d}s \\
&\overset{s=\alpha s_0}{=} \alpha \int_0^{h+\theta_{M,1,k}} \prod_{l=1,\ l\neq j}^{N} \frac{s_0 - t_{N,l}}{t_{N,j}-t_{N,l}} \mathrm{d}s_0, \\
& k = 0,\ 1,\ \ldots,\ M-1;\ j = 1,\ 2,\ \ldots,\ N.
\end{aligned} \tag{5.64}
$$

Combining (5.27) and (5.64), it can be concluded that $E'_{k+1,j} = \alpha E_{k+1,j}$ and then $\boldsymbol{U}''_{M,N} = \alpha \boldsymbol{U}'_{M,N}$. Thus, under two implementations of multiplication preconditioning, one has $\tilde{\mathbf{\Pi}}''_{M,N} = \alpha \hat{\mathbf{\Pi}}''_{M,N}$.

$$
\mathbf{\Sigma}''_{M,N} = \frac{1}{\alpha}\hat{\mathbf{\Sigma}}''_{M,N}
$$

The discretization submatrix $\mathbf{\Sigma}_{M,N}$ is related to $\boldsymbol{J}_i$ $(i = 0,\ 1,\ \ldots,\ m)$ and $\boldsymbol{L}^i_{M,N}$ $(i = 0,\ 1,\ \ldots,\ m)$ as well as their submatrices. The first step is to investigate the effect of τ_i $(i = 1,\ 2,\ \ldots,\ m)$ and h against the elements of $\boldsymbol{L}^i_{M,N}$ $(i = 0,\ 1,\ \ldots,\ m)$, i.e., $\ell_{M,k,l}$ $(k = 1,\ 2,\ \ldots,\ Q;\ l = 0,\ 1,\ \ldots,\ M)$.

Under the first implementation, since that $\theta_{M,1,k}$ $(k = 0,\ 1,\ \ldots,\ M)$ and $t_{N,i}$ $(i = 1,\ 2,\ \ldots,\ N)$ remains unchanged, $\ell_{M,k,l}$ $(k = 1,\ 2,\ \ldots,\ Q;\ l = 0,\ 1,\ \ldots,\ M)$ are evolved into

$$
\begin{aligned}
\ell'_{M,k,l}(t_{N,j}-\tau_i/\alpha) &= \prod_{r=0,\ r\neq l}^{M} \frac{t_{N,j}-\tau_i/\alpha-\theta_{M,k,r}}{\theta_{M,k,l}-\theta_{M,k,r}} = \prod_{r=0,\ r\neq l}^{M} \frac{\alpha t_{N,j}-\tau_i-\alpha\theta_{M,k,r}}{\alpha\theta_{M,k,l}-\alpha\theta_{M,k,r}}, \\
& k = 1,\ 2,\ \ldots,\ Q';\ l = 0,\ 1,\ \ldots,\ M;\ j = 1,\ 2,\ \ldots,\ N.
\end{aligned} \tag{5.65}
$$

Under the second implementation, given that $\theta_{M,1,k}$ $(k = 0,\ 1,\ \ldots,\ M)$ and $t_{N,i}$ $(i = 1,\ 2,\ \ldots,\ N)$ are transformed to α times, the last item of (5.65) also holds. It can thus be concluded that the transformed $\tilde{\boldsymbol{L}}^i_{M,N}$ $(i = 0,\ 1,\ \ldots,\ m)$ are identical under two implementations.

Then, considering the system state matrices $\boldsymbol{A}_i^{\prime\prime(2)}$, $\boldsymbol{B}_i^{\prime\prime(2)}$ and $\tilde{\boldsymbol{A}}_0^{\prime\prime}$ under the first implementation are α times of $\boldsymbol{A}_i^{\prime(2)}$, $\boldsymbol{B}_i^{\prime(2)}$ $(i = 1,\ 2,\ \ldots,\ m)$ and $\tilde{\boldsymbol{A}}_0^{\prime}$ under the second implementation, one has

$$\boldsymbol{\Sigma}''_{M,N} = \frac{1}{\alpha}\hat{\boldsymbol{\Sigma}}''_{M,N}.$$

$$\boldsymbol{\Sigma}''_N = \hat{\boldsymbol{\Sigma}}''_N$$

The discretization submatrix $\boldsymbol{\Sigma}_N$ is related to $\boldsymbol{J}_i$ $(i = 0,\ 1,\ \ldots,\ m)$ and $\boldsymbol{L}_N^i$ $(i = 0,\ 1,\ \ldots,\ m)$. The first step is to investigate the effect of τ_i $(i = 1,\ 2,\ \ldots,\ m)$ and h against the elements of $\boldsymbol{L}_N^i$ $(i = 0,\ 1,\ \ldots,\ m)$, i.e., $g_{j,k}$ $(j = 1,\ 2,\ \ldots,\ N;\ k = 1,\ 2,\ \ldots,\ N)$.

For a given τ_i $(i = 0,\ 1,\ \ldots,\ m)$, $g_{j,k}$ can be expressed as

$$g_{j,k} = \int_0^{t_{N,j}-\tau_i} \ell_{N,k}^{+}\mathrm{d}t = \int_0^{t_{N,j}-\tau_i} \prod_{l=1,\ l\neq k}^{N} \frac{t - t_{N,l}}{t_{N,k} - t_{N,l}}\mathrm{d}t,$$
$$j = 1,\ 2,\ \ldots,\ N;\ k = 1,\ 2,\ \ldots,\ N.$$

Under the first implementation, $g_{j,k}$ are evolved into

$$g'_{j,k} = \int_0^{t_{N,j}-\tau_i/\alpha} \prod_{l=1,\ l\neq k}^{N} \frac{t - t_{N,l}}{t_{N,k} - t_{N,l}}\mathrm{d}t,\ j = 1,\ 2,\ \ldots,\ N;\ k = 1,\ 2,\ \ldots,\ N. \tag{5.66}$$

Under the second implementation, given that $t_{N,k}$ $(k = 1,\ 2,\ \ldots,\ N)$ are transformed to α times, $g_{j,k}$ are evolved into

$$g''_{j,k} = \int_0^{\alpha t_{N,j}-\tau_i} \prod_{l=1,\ l\neq k}^{N} \frac{t - \alpha t_{N,l}}{\alpha t_{N,k} - \alpha t_{N,l}}\mathrm{d}t \overset{t=\alpha t_0}{=} \alpha \int_0^{t_{N,j}-\tau_i/\alpha} \prod_{l=1,\ l\neq k}^{N} \frac{t_0 - t_{N,l}}{t_{N,k} - t_{N,l}}\mathrm{d}t_0,$$
$$j = 1,\ 2,\ \ldots,\ N;\ k = 1,\ 2,\ \ldots,\ N. \tag{5.67}$$

It can be concluded from (5.66) and (5.67) that $g''_{j,k} = \alpha g'_{j,k}$ and then $\tilde{\boldsymbol{L}}_N^i = \alpha(\tilde{\boldsymbol{L}}_N^i)'$.

Then, considering the system state matrices $\boldsymbol{A}_i^{\prime\prime(2)}$, $\boldsymbol{B}_i^{\prime\prime(2)}$ and $\tilde{\boldsymbol{A}}_0^{\prime\prime}$ under the first implementation are α times of $\boldsymbol{A}_i^{\prime(2)}$, $\boldsymbol{B}_i^{\prime(2)}$ $(i = 1,\ 2,\ \ldots,\ m)$ and $\tilde{\boldsymbol{A}}_0^{\prime}$ under the second implementation, one can obtain

$$\boldsymbol{\Sigma}''_N = \hat{\boldsymbol{\Sigma}}''_N.$$

$$\hat{T}''_{M,N} = T''_{M,N}$$

To sum up, $\hat{T}''_{M,N}$ and $T''_{M,N}$ are identical under two implementations of rotation-and-multiplication preconditioning.

5.3.3 *Sparse Eigenvalue Computation*

In this subsection, the IRA algorithm is utilized to successively compute eigenvalues μ'' from $\hat{T}''_{M,N}$ in a descending order of modulus. In the process, the most critical and computationally expensive operation is generating Krylov subspace. Let $q_j \in \mathbb{C}^{(n_1+(Q'M+1)d_2)\times 1}$ be the j-th Krylov vector, then the $(j+1)$-th vector $q_{j+1} \in \mathbb{C}^{(n_1+(Q'M+1)d_2)\times 1}$ can be obtained by the MVP

$$q_{j+1} = \hat{T}''_{M,N} q_j.$$

Since that the last $((Q-1)M+1)d_2$ rows of $\hat{T}''_{M,N}$ are highly sparse, only the computation of the first $n_1 + Md_2$ elements of q_{j+1} are elaborated here. The computation can be further divided into the following three steps:

$$z = \hat{\Sigma}''_{M,N} q_j \tag{5.68}$$

$$w = \left(\begin{bmatrix} I_{Nn} & \\ I_N \otimes C_0 & I_N \otimes D_0 \end{bmatrix} - \hat{\Sigma}''_N \right)^{-1} z \tag{5.69}$$

$$q_{j+1}(1 : n_1 + Md_2,\ 1) = \tilde{\Pi}''_M q_j + \hat{\Pi}''_{M,N} w \tag{5.70}$$

by introducing two ancillary vectors $w \in \mathbb{C}^{Nd\times 1}$ and $z \in \mathbb{C}^{Nd\times 1}$.

5.3.3.1 Implementation of (5.68)

The vector $q_j(n_1+1 : \text{end},\ 1)$ is firstly reshaped in column-wise direction, leading to the matrices $Q_1 \in \mathbb{C}^{n_2\times M}$, $Q_2 \in \mathbb{C}^{l_2\times M}$, $Q_3 \in \mathbb{C}^{n_2\times((Q'-1)M+1)}$ and $Q_4 \in \mathbb{C}^{l_2\times((Q'-1)M+1)}$, respectively, namely

$$\begin{cases} \text{vec}(Q_1) = q_j(n_1+1 : n_1+Mn_2,\ 1) \\ \text{vec}(Q_2) = q_j(n_1+Mn_2+1 : n_1+Md_2,\ 1) \\ \text{vec}(Q_3) = q_j\left(n_1+Md_2+1 : n_1+\left(Q'M+1\right)n_2+Ml_2,\ 1\right) \\ \text{vec}(Q_4) = q_j\left(n_1+\left(Q'M+1\right)n_2+Ml_2+1 : n_1+\left(Q'M+1\right)d_2,\ 1\right) \end{cases}.$$

By using the unique property of Kronecker product, (5.68) can be efficiently computed as

$$z = \boldsymbol{\Sigma}'_{M,N} \begin{bmatrix} \boldsymbol{q}_j(1:n_1,\ 1) \\ \mathrm{vec}(\boldsymbol{Q}_1) \\ \mathrm{vec}(\boldsymbol{Q}_2) \\ \mathrm{vec}(\boldsymbol{Q}_3) \\ \mathrm{vec}(\boldsymbol{Q}_4) \end{bmatrix} = \begin{bmatrix} \mathrm{vec}\Bigg(\tilde{\boldsymbol{A}}''_0 \boldsymbol{q}_j(1:n,\ 1)\boldsymbol{1}_N^{\mathrm{T}} \\ +\sum_{i=1}^{m} \left(\boldsymbol{A}_i''^{(2)} \boldsymbol{Q}_1 + \boldsymbol{B}_i''^{(2)} \boldsymbol{Q}_2\right)\left(\tilde{\boldsymbol{L}}^i_{M,N,1}\right)^{\mathrm{T}} \\ +\sum_{i=1}^{m} \left(\boldsymbol{A}_i''^{(2)} \boldsymbol{Q}_3 + \boldsymbol{B}_i''^{(2)} \boldsymbol{Q}_4\right)\left(\tilde{\boldsymbol{L}}^i_{M,N,2}\right)^{\mathrm{T}}\Bigg) \\ \hdashline \boldsymbol{0}_{Nl\times 1} \end{bmatrix}. \tag{5.71}$$

5.3.3.2 Implementation of (5.69)

The coefficient matrix in the right-hand side of (5.69) can be written as the block matrix

$$\boldsymbol{J}'_N \triangleq \begin{bmatrix} \boldsymbol{A}'_N & \boldsymbol{B}'_N \\ \boldsymbol{C}_N & \boldsymbol{D}_N \end{bmatrix} = \begin{bmatrix} \boldsymbol{I}_{Nn} & \\ \boldsymbol{I}_N \otimes \boldsymbol{C}_0 & \boldsymbol{I}_N \otimes \boldsymbol{D}_0 \end{bmatrix} - \hat{\boldsymbol{\Sigma}}''_N$$

of dimension Nd-by-Nd, where $\boldsymbol{A}'_N \in \mathbb{C}^{Nn\times Nn}$, $\boldsymbol{B}'_N \in \mathbb{C}^{Nn\times Nl}$, $\boldsymbol{C}_N \in \mathbb{R}^{Nl\times Nn}$ and $\boldsymbol{D}_N \in \mathbb{R}^{Nl\times Nl}$ expressed as

$$\begin{cases} \boldsymbol{A}'_N = \boldsymbol{I}_{Nn} - \sum_{i=0}^{m} (\tilde{\boldsymbol{L}}^i_N)' \otimes \boldsymbol{A}''_i,\ \boldsymbol{B}'_N = -\sum_{i=0}^{m} (\tilde{\boldsymbol{L}}^i_N)' \otimes \boldsymbol{B}''_i \\ \boldsymbol{C}_N = \boldsymbol{I}_N \otimes \boldsymbol{C}_0,\ \boldsymbol{D}_N = \boldsymbol{I}_N \otimes \boldsymbol{D}_0 \end{cases} \tag{5.72}$$

are Kronecker products associated with $\boldsymbol{A}''_i$, $\boldsymbol{B}''_i$ $(i = 0,\ 1,\ \ldots,\ m)$, $\boldsymbol{C}_0$ and $\boldsymbol{D}_0$, respectively.

Considering the high sparsity of $\boldsymbol{A}'_N$, $\boldsymbol{B}'_N$, $\boldsymbol{C}_N$ and $\boldsymbol{D}_N$, the inverse of $\boldsymbol{J}'_N$ are solved by LU decomposition and thus (5.69) can be efficiently computed.

5.3.3.3 Implementation of (5.70)

Substituting (5.60) and (5.61) into (5.70), one can obtain

$$\begin{aligned}
\boldsymbol{q}_{j+1}(1:n_1+Md_2,\ 1) &= \tilde{\boldsymbol{\Pi}}''_M\boldsymbol{q}_j + \hat{\boldsymbol{\Pi}}''_{M,N}\boldsymbol{w} \\
&= \begin{bmatrix} \boldsymbol{I}_n \\ \boldsymbol{1}_{M-1}\otimes\left[\boldsymbol{0}_{n_2\times n_1}\ \boldsymbol{I}_{n_2}\right] \\ -\boldsymbol{1}_M\otimes\left[\boldsymbol{0}_{l_2\times l_1}\ \boldsymbol{I}_{l_2}\right]\boldsymbol{D}_0^{-1}\boldsymbol{C}_0 \end{bmatrix}\boldsymbol{q}_j(1:n,\ 1) \\
&+ \begin{bmatrix} \bar{\boldsymbol{U}}'_{M,N}\otimes\boldsymbol{I}_n & \\ \tilde{\boldsymbol{U}}'_{M,N}\otimes\left[\boldsymbol{0}_{n_2\times n_1}\ \boldsymbol{I}_{n_2}\right] & \\ & \boldsymbol{U}'_{M,N}\otimes\left[\boldsymbol{0}_{l_2\times l_1}\ \boldsymbol{I}_{l_2}\right] \end{bmatrix}\boldsymbol{w}.
\end{aligned}$$

5.3.4 Eigenvalue Correction

After obtaining μ'' from $\hat{\boldsymbol{T}}''_{M,N}$, the estimates $\hat{\lambda}$ to accurate eigenvalues λ of the time-delay system can be restored from

$$\hat{\lambda} = \frac{1}{\alpha h}\mathrm{e}^{\mathrm{j}\theta}\ln\mu''. \tag{5.73}$$

Correspondingly, the first n entries of the eigenvector of $\hat{\mu}$ can be chosen as a good estimation $\hat{\boldsymbol{v}}$ for eigenvector $\boldsymbol{v}$. By taking $(\hat{\lambda},\ \hat{\boldsymbol{v}})$ as initial values, the accurate λ and $\boldsymbol{v}$ can be readily obtained by the Newton's method.

5.3.5 Characteristic Analyses

(1) **The low dimension**. Given that $n \gg d_2$, the dimension of the discretization matrix $\boldsymbol{T}_{M,N}$, i.e., $n_1 + (QM+1)d_2$, gets very close to the total number of state variables and minimally affected by Q and M.
(2) **The computational complexity**. The computational burden of the presented PSOD-PS method is dominated by that of $\boldsymbol{z} = \boldsymbol{\Sigma}_{M,N}\boldsymbol{q}_j$ as shown in (5.68) and $\boldsymbol{v} = \boldsymbol{J}_N^{-1}\boldsymbol{z}$ as shown in (5.69), which can be roughly weighted by the number of multiplication between the dense system state matrix $\tilde{\boldsymbol{A}}_0$ and the variable vectors.

- As shown in (5.71), considering that $\boldsymbol{A}_i^{(2)}$ and $\boldsymbol{B}_i^{(2)}$ $(i=1,\ 2,\ \ldots,\ m)$ are highly sparse and have only n_2 and l_2 columns respectively, the multiplication between them and $\boldsymbol{Q}_j$ $(j=1,\ 2,\ 3,\ 4)$ can be effortlessly computed. Therefore, the computational burden of solving (5.68) is almost identical to that of computing the MVP associated with $\tilde{\boldsymbol{A}}_0$.
- As shown in (5.72), the submatrices $\boldsymbol{A}_N$, $\boldsymbol{B}_N$, $\boldsymbol{C}_N$ and $\boldsymbol{D}_N$ of $\boldsymbol{J}_N$ are Kronecker products associated with the augmented system state submatrices $\boldsymbol{A}_i$, $\boldsymbol{B}_i$, $\boldsymbol{C}_0$ and $\boldsymbol{D}_0$ $(i=0,\ 1,\ \ldots,\ m)$, thus exhibit high sparsity. This fact results in that (5.69) can be efficiently computed, whose computational burden can be roughly weighted by

that of the MIVP associated with the augmented system state matrix $\boldsymbol{J}_0$ (denoted as T times that of computing the MVP associated with $\tilde{\boldsymbol{A}}_0$). That is to say, the computational burden of solving (5.69) is almost T times that of computing the MVP associated with $\tilde{\boldsymbol{A}}_0$.

To sum up, the computational burden of the PSOD-PS method is $T + 1$ times the traditional eigen-analysis of a delay-free system whose state matrix is $\tilde{\boldsymbol{A}}_0$.

References

1. Breda, D., Maset, S., & Vermiglio, R. (2005). Pseudospectral differencing methods for characteristic roots of delay differential equations. *Siam Journal on Scientific Computing, 27*(2), 482–495.
2. Breda, D., Maset, S., & Vermiglio, R. (2015). *Stability of linear delay differential equations: A numerical approach with MATLAB*. New York: Springer.

Chapter 6
Applications to Power Systems with Inclusion of Wide-Area Delays

The operation and control of modern power systems always depend on a reliable information system. Based on computer, communication and sensor technologies, the primary power system is closely and organically integrated with the information system, leading to a cyber-physical power system (CPPS) [1–3]. With the consideration of time delays introduced by the information system, the CPPS becomes a time-delay CPPS. Accordingly, there is a demand for establishing a rounded system of small-signal stability modeling, analysis and control for time-delay CPPSs [4, 5]. To this end, this chapter illustrates applications of PSD-based eigenvalue computation methods elaborated in previous sections to time-delay CPPSs.

6.1 Configuration

In the modern power system, the energy management system utilizes the information provided by the supervisory control and data acquisition (SCADA) system, to carry out online safety monitoring and remote regulation of power system equipment. Since the 1990s, due to the emergence of satellite timing system and the sustained development of power communication network and digital signal processing technology, the wide-area measurement system (WAMS) based on phasor measurement unit (PMU) has been significantly developed, providing a novel information platform for state sensing of large-scale interconnected power systems and a new means for wide-area protection and coordinated control [6].

Along with the increasing scale and in-depth construction of modern power systems, the WAMS has made great development in recent years. By the end of 2013, all provincial power dispatch and control centers in China had installed their WAMSs, with more than 2400 PMUs operating in power plants of 500 kV and above [7]. According to the U.S. Department of Energy (DOE), there were 1126 PMUs operating in the U.S. power grids by the end of 2013 [8]. PMUs in a WAMS environment,

H. Ye et al., *Spectral Discretization-Based Eigen-Analysis of Time-Delay Systems*,
https://doi.org/10.1007/978-981-96-0738-9_6

provide measurement signals which are transmitted to the phasor data concentrator (PDC) through the communication system [6, 9]. With the aid of the timing service of the global positioning system (GPS), PMUs sample the state of system components at the high speed (30–60 Hz with the highest being 120–240 Hz) in phasor form and simultaneously creates a unique timestamp. Subsequently, the sampled data is processed properly in PDC, allowing dynamic monitoring and other advanced applications of the power system [6]. Unlike the SCADA system based on the remote terminal unit with a sampling period of 2–4 s, WAMS is capable of making real-time and synchronous acquisition of the dynamic information of systems distributed geographically over thousands of kilometers. Advanced applications that can be realized by WAMS include dynamic monitoring and state estimation, parameter identification, stability monitoring and evaluation, low-frequency oscillation identification, and wide-area damping control, fault location and protection, etc.

In the past decades, a lot of researches have been conducted on the power system analysis and control based on wide-area measurement information [10–12]. In sharp contrast with local signals, wide-area measurements, such as active power on the tie-lines and relative rotor angles/speeds, show good observability of some significant interarea oscillation modes. By using wide-area measurements as remote feedback control signals, the wide-area damping controller (WADC) has a significant potential in stabilizing power system against weakly damped interarea low frequency oscillations [13–15]. The WADC in cooperation with the power system stabilizer (PSS) intended for suppressing local low-frequency oscillation can form the “local + wide-area” two-layer control [15, 16].

A salient characteristic of the power system integrated with the WAMS, however, is the heterogeneous latency in acquiring, routing, transmitting and processing wide-area signals, leading to time delays ranging from tens to several hundred milliseconds [17]. As shown in Fig. 6.1, the time delay in a wide-area damping control loop consists of four parts: measurement delay $\tau_{\rm m}$ (including acquisition delay of current/voltage transformers, phasor calculation delay and data packaging and sending delay), communication delay (including uplink delay $\tau_{\rm up}$ and downlink delay $\tau_{\rm down}$), calculation delay $\tau_{\rm PDC}$ (time delay resulting from generating the wide-area damping control signals by the PDC), and control delay $\tau_{\rm c}$ (time delay resulting from executing control signals) [18, 19]. $\tau_{\rm m}$ is generated by PMU, which consists of nearly constant phasor sampling delay, calculation delay and data packaging delay, and fluctuating data sending delay. $\tau_{\rm PDC}$ is composed of random delay caused by synchronizing the data from different PMUs, and constant calculation delay. The controller action delay $\tau_{\rm c}$ remains constant approximately at 10 ms. $\tau_{\rm up}$ and $\tau_{\rm down}$ play the leading roles in the closed-loop delay in the wide-area control loop, consisting of the serial delay, routing delay and propagation delay [19, 20].

Considering the complex communication network architecture and a large amount of data that is required to be transmitted, the actual communication time delay tend to be more than 100 ms [21]. The field measurement results show that, the communication delay of a wide area damping control loop in China Southern Power Grid is about 110 ms, and that of the Pacific DC intertie wide-area damping control loop in the North American western interconnection is up to 113 ms [12]. In

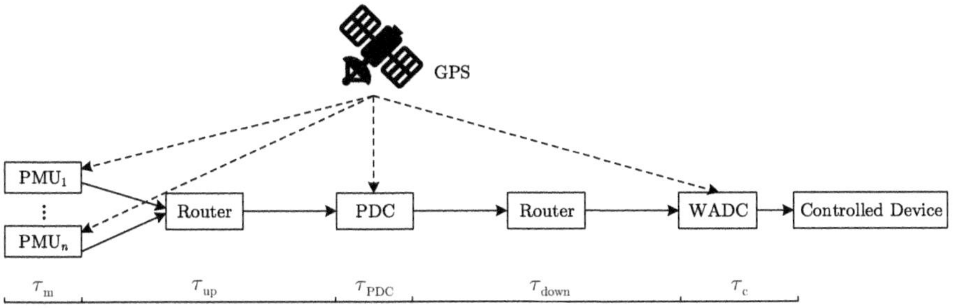

Fig. 6.1 Time delays in a wide-area damping control loop

addition, a study conducted by the Electric Power Research Institute shows that the maximum latency of wide-area measurements may reach 460 ms, under different communication architectures (unicast/multicast) and different sampling frequencies (30/60 Hz) [22].

Time delays can drastically change the performance of power systems [7, 23], inducing dangerous instability and loss of performance as well as improving stability. Therefore, it is critical to accurately evaluate the impacts of time delays on the system stability so that engineers can design WADCs adaptive to stringent time delays [7].

6.2 Dynamic Model

6.2.1 Overview

Figure 6.2 shows the framework of the dynamic model of a multi-machine power system for small-signal stability analysis and control. Differential equations describe the dynamics of synchronous generators, excitation systems, PSSs, prime movers and their speed-governing systems, induction motors, high voltage direct current (HVDC) transmission systems, flexible alternative current transmission system (FACTS) equipments such as static var compensator (SVC) and thyristor controlled series compensator (TCSC), etc. Algebraic equations include power network equations, static loads, as well as interface equations between the dynamic devices and the power network, and so on. The synchronous generator unit and other dynamic devices are independent of each other and linked together by the power network.

Therefore, the complete dynamic model of the multi-machine power system can be generally expressed as a set of ordinary DAEs

$$\begin{cases} \dot{\boldsymbol{x}} = \boldsymbol{f}(\boldsymbol{x},\ \boldsymbol{y}) \\ \boldsymbol{0} = \boldsymbol{g}(\boldsymbol{x},\ \boldsymbol{y}) \end{cases} \tag{6.1}$$

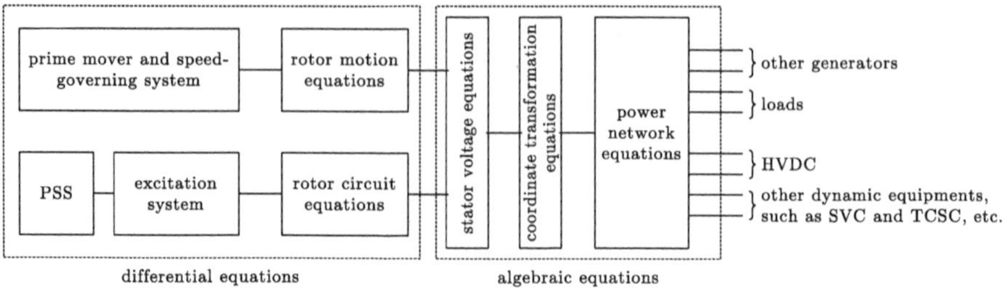

Fig. 6.2 Framework of the dynamic model of a multi-machine power system

where $\boldsymbol{x} \in \mathbb{R}^{n\times 1}$ and $\boldsymbol{y} \in \mathbb{R}^{l\times 1}$ are the vector of state and algebraic variables, respectively. $\boldsymbol{f}$ represents the set of differential equations describing the dynamics of the system, and $\boldsymbol{g}$ denotes the set of algebraic equations describing the statics of system.

Specifically, $\boldsymbol{f}$ includes:

(1) the swing equation describing the rotor motion of the synchronous generator;
(2) the differential equation describing the transient and subtransient electromagnetic processes of the rotor windings of the synchronous generator;
(3) the differential equation describing the dynamics of the excitation system associated with the synchronous generator;
(4) the differential equation describing the dynamics of the prime mover and its speed-governing system associated with the synchronous generator;
(5) the differential equation describing the dynamics of induction motor and synchronous motor;
(6) the differential equation describing the dynamics of the HVDC transmission systems;
(7) and the differential equation describing the dynamics of the FACTS equipments.

$\boldsymbol{g}$ includes:

(1) the stator voltage balance equations for synchronous generator;
(2) the transformation equations between the synchronous generator's d-q reference frame and the common x-y reference frame;
(3) the power network equations, i.e., the equations describing the relationship between the node voltage and the node injection current in the x-y reference frame;
(4) the voltage equations for the HVDC transmission line;
(5) and the equations for the static voltage characteristics of loads, etc.

6.2.2 Modeling of the Synchronous Generator Unit

This section gives the differential equations and algebraic equations for the synchronous generator unit, the most important component in the power system, including the synchronous generator, excitation system, PSS, as well as prime mover and its speed-governing system. The models of other dynamic devices can be found in relevant monographs, e.g., [24].

6.2.2.1 Synchronous Generator

In the d-q reference frame, by taking X_{ad} as the base of reactance, the mutual reactances of each winding satisfy

$$X_{af}X_D = X_{aD}X_{fD}, \quad X_{ag}X_Q = X_{aQ}X_{gQ}$$

where X_D and X_Q are the self-reactances of the damper windings D and Q, respectively. X_{af}, X_{aD}, X_{ag} and X_{aQ} are the mutual reactances between the d-axis winding and the excitation winding f, the damper windings D, g and Q, respectively. X_{fD} is the mutual reactance between the excitation winding f and the damper winding D, and X_{gQ} is the mutual reactance between the damper windings g and Q.

Let e_{q1}, e_{q2}, e_{d1} and e_{d2} be the potentials proportional to the currents of each winding on the rotor. Denote the transient potentials as e'_q, e'_d and the subtransient potentials as e''_q, e''_d, which are proportional to the flux linkage of each winding on the rotor. By substituting them into the synchronous machine model expressed in per unit, the differential equations of the synchronous generator expressed in terms of machine parameters can be obtained.

The dynamics of the round-rotor synchronous generator can be described as

$$\psi_d = -X_d i_d + e_{q1} + e_{q2} \tag{6.2}$$

$$\psi_q = -X_q i_q - e_{d1} - e_{d2} \tag{6.3}$$

$$e'_q = -\left(X_d - X'_d\right) i_d + e_{q1} + \frac{X_d - X'_d}{X_d - X''_d} e_{q2} \tag{6.4}$$

$$e''_q = -\left(X_d - X''_d\right) i_d + e_{q1} + e_{q2} \tag{6.5}$$

$$e'_d = \left(X_q - X'_q\right) i_q + e_{d1} + \frac{X_q - X'_q}{X_q - X''_q} e_{d2} \tag{6.6}$$

$$e''_d = \left(X_q - X''_q\right) i_q + e_{d1} + e_{d2} \tag{6.7}$$

$$u_d = \frac{\mathrm{d}\psi_d}{\mathrm{d}t} - \omega\psi_q - R_a i_d \tag{6.8}$$

$$u_q = \frac{\mathrm{d}\psi_q}{\mathrm{d}t} + \omega\psi_d - R_a i_q \tag{6.9}$$

$$T'_{d0}\frac{\mathrm{d}e'_q}{\mathrm{d}t} = E_{fq} - e_{q1} \tag{6.10}$$

$$T''_{d0}\frac{\mathrm{d}e''_q}{\mathrm{d}t} = -\frac{X'_d - X''_d}{X_d - X''_d}e_{q2} \tag{6.11}$$

$$T'_{q0}\frac{\mathrm{d}e'_d}{\mathrm{d}t} = -e_{d1} \tag{6.12}$$

$$T''_{q0}\frac{\mathrm{d}e''_d}{\mathrm{d}t} = -\frac{X'_q - X''_q}{X_q - X''_q}e_{d2} \tag{6.13}$$

$$\frac{\mathrm{d}\delta}{\mathrm{d}t} = \omega - 1 \tag{6.14}$$

$$\frac{\mathrm{d}\omega}{\mathrm{d}t} = \frac{1}{T_J}(-D\omega + T_\mathrm{m} - T_\mathrm{e}) = \frac{1}{T_J}\left(-D\omega + \frac{P_\mathrm{m}}{\omega} - \frac{P_\mathrm{e}}{\omega}\right). \tag{6.15}$$

Here the subtransient and transient electromagnetic processes of the excitation winding f and damper winding D on the d-axis as well as the damper windings Q and g on the q-axis are intensively considered.

In (6.2)–(6.13), all physical quantities (including time and time constants) are in per unit. X_d (X_q), X'_d (X'_q) and X''_d (X''_q) are the d-axis (q-axis) synchronous reactances, transient reactances and subtransient reactances, respectively. T'_{d0} (T'_{q0}) and T''_{d0} (T''_{q0}) are the d-axis (q-axis) open-circuit transient and subtransient time constants, respectively. ψ_d (ψ_q), i_d (i_q) and u_d (u_q) are the d-axis (q-axis) flux linkage, current and voltage of the stator, respectively. R_a is the resistance of the stator winding and E_{fq} is the no-load steady-state stator voltage of the synchronous generator. Let U_f be the excitation voltage of the generator, then $E_{fq} = \frac{X_{af}}{R_f}U_f$. In the per unit system based on the concept of unit excitation voltage/unit stator voltage, X_{af} and R_f are equal. Consequently, $E_{fq} = U_f$.

In rotor motion equations (6.14) and (6.15), δ is the electrical angle between the q-axis of the synchronous generator and the reference axis x of the system rotating at synchronous speed. ω is the electrical angular velocity of the synchronous generator. T_J is the inertia time constant of the synchronous generator. T_m (P_m) and T_e (P_e) are mechanical torque (power) inputs from the prime mover and the electromagnetic torque (power) outputs from the generator, respectively. D is the damping coefficient.

In practical applications, the abovementioned sixth-order model is usually simplified according to different accuracy requirements.

(1) For salient pole synchronous generators, there is no damper winding g. In this case, $e_{d1} = 0$ and $X'_q = X_q$. Thus, (6.6) and (6.12) are removed, leading to the fifth-order model of the salient pole synchronous generator.
(2) When the damper windings on the d and q-axes are neglected, there remains the excitation winding f on the d-axis. Let $e_{q2} = e_{d1} = e_{d2} = 0$, then (6.5)–(6.7) and (6.11)–(6.13) are automatically removed. Accordingly, a third-order synchronous generator model can be obtained, where the dynamics of transient potential e'_q and the motion of the rotor are described.
(3) The damper windings on the d and q-axes are neglected. In addition, it is assumed that the transient potential e'_q is unchanged and fixed to be a constant, so as to approximate the effect of the excitation regulator. To achieve this, in (6.2)–(6.13), let $e_{q2} = e_{d1} = e_{d2} = 0$. As a result, (6.5)–(6.7) and (6.11)–(6.13) can be removed and the right-hand side of (6.10) is equal to 0. Finally, a second-order synchronous generator model can be obtained.
(4) It is assumed that the fictitious potential e' representing the voltage behind X'_d is fixed to be a constant. Let $X_q = X'_d$, then (6.4)–(6.7) and (6.10)–(6.13) can be removed, leading to the classical second-order synchronous generator model.
(5) If the electromagnetic transient processes of the armature windings are neglected, i.e., $\dfrac{\mathrm{d}\psi_d}{\mathrm{d}t} = \dfrac{\mathrm{d}\psi_q}{\mathrm{d}t} = 0$. Then, (6.8) and (6.9) become

$$u_d = -\omega\psi_q - R_a i_d$$

$$u_q = \omega\psi_d - R_a i_q.$$

(6) If the electromagnetic transient processes of the stator is neglected, then (6.8) and (6.9) become

$$u_d = -\psi_q - R_a i_d$$

$$u_q = \psi_d - R_a i_q.$$

(7) If the small change in rotor speed is neglected, then $\omega = 1$ and the mechanical torque and electrical power in per unit are identical, i.e., $P_\mathrm{m} = T_\mathrm{m}$ and $P_\mathrm{e} = T_\mathrm{e}$.

Based on the above analyses, the sixth-order mathematical model of the synchronous generator includes rotor motion equations

$$\begin{cases} \dfrac{\mathrm{d}\delta}{\mathrm{d}t} = \omega - 1 \\ \dfrac{\mathrm{d}\omega}{\mathrm{d}t} = \dfrac{1}{T_J}\left\{P_\mathrm{m} - \left[E''_q I_q + E''_d I_d - (X''_d - X''_q) I_d I_q\right] - D\omega\right\} \end{cases} \tag{6.16}$$

the rotor electromagnetic transient equations

$$\begin{cases} \dfrac{\mathrm{d}E'_q}{\mathrm{d}t} = \dfrac{1}{T'_{d0}}\left[E_{fq} - k_d E'_q + (k_d - 1)E''_q\right] \\ \dfrac{\mathrm{d}E''_q}{\mathrm{d}t} = \dfrac{1}{T''_{d0}}\left[E'_q - E''_q - (X'_d - X''_d)I_d\right] \\ \dfrac{\mathrm{d}E'_d}{\mathrm{d}t} = \dfrac{1}{T'_{q0}}\left[-k_q E'_d + (k_q - 1)E''_d\right] \\ \dfrac{\mathrm{d}E''_d}{\mathrm{d}t} = \dfrac{1}{T''_{q0}}\left[E'_d - E''_d + (X'_q - X''_q)I_q\right] \end{cases} \tag{6.17}$$

and stator voltage balance equations

$$\begin{cases} U_d = E''_d - R_a I_d + X''_q I_q \\ U_q = E''_q - X''_d I_d - R_a I_q \end{cases}$$

where $k_d = \dfrac{X_d - X''_d}{X'_d - X''_d}$ and $k_q = \dfrac{X_q - X''_q}{X'_q - X''_q}$. Note that, the lowercase letters u, i and e are replaced by the corresponding uppercase letters U, I and E to represent the steady-state values.

6.2.2.2 Excitation System

There are several types of excitation systems for the synchronous generator. Here, the IEEE type DC1A excitation system [24] is taken as an example. As illustrated in Fig. 6.3, differential equations that describe its dynamic behaviors are

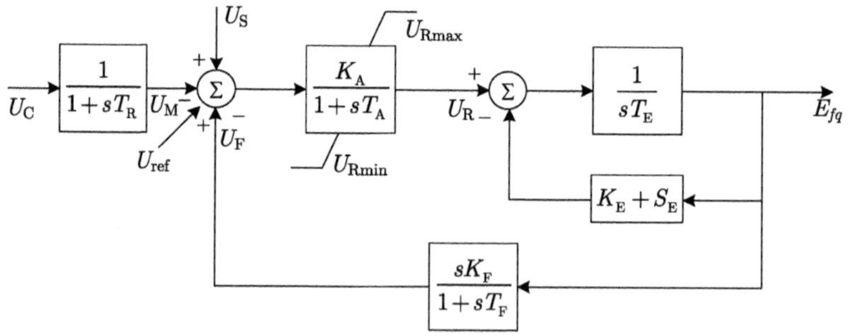

Fig. 6.3 Block diagram of IEEE type DC1A excitation system

$$\begin{cases} \dfrac{\mathrm{d}E_{fq}}{\mathrm{d}t} = \dfrac{1}{T_{\mathrm{E}}}\left[U_{\mathrm{R}} - (K_{\mathrm{E}} + S_{\mathrm{E}})E_{fq}\right] \\ \dfrac{\mathrm{d}U_{\mathrm{R}}}{\mathrm{d}t} = \dfrac{1}{T_{\mathrm{A}}}\left[K_{\mathrm{A}}(U_{\mathrm{ref}} - U_{\mathrm{F}} + U_{\mathrm{S}} - U_{\mathrm{M}}) - U_{\mathrm{R}}\right] \\ \dfrac{\mathrm{d}U_{\mathrm{F}}}{\mathrm{d}t} = \dfrac{1}{T_{\mathrm{F}}}\left(K_{\mathrm{F}}\dfrac{\mathrm{d}E_{fq}}{\mathrm{d}t} - U_{\mathrm{F}}\right) \\ \dfrac{\mathrm{d}U_{\mathrm{M}}}{\mathrm{d}t} = \dfrac{1}{T_{\mathrm{R}}}(U_{\mathrm{C}} - U_{\mathrm{M}}) \end{cases} \tag{6.18}$$

where T_{R} is the time constant of the transducer. U_{C} is the output of the load compensator. U_{M} is the output of the transducer. U_{S} is the output of PSS and used as the supplementary control signal of excitation system. U_{ref} is the voltage reference. K_{A} and T_{A} are the gain and time constant of the amplifier, respectively. U_{R} is the output of the amplifier. U_{Rmax} and U_{Rmin} are the non-windup limits of U_{R}, respectively. U_{F} is the output of the soft negative feedback element of the excitation voltage. K_{F} and T_{F} are the gain and time constant of the soft negative feedback element, respectively. K_{E} and T_{E} are the gain and time constant of the exciter, respectively. S_{E} is the saturation coefficient of the exciter. Let c_{E} and n_{E} be saturation parameters and let X_{C} be the reactance of the load compensator, then S_{E} and U_{C} can be represented as

$$S_{\mathrm{E}} = c_{\mathrm{E}}E_{fq}^{n_{\mathrm{E}}-1}$$

$$U_{\mathrm{C}} = \sqrt{(U_d - X_{\mathrm{C}}I_q)^2 + (U_q + X_{\mathrm{C}}I_d)^2}.$$

6.2.2.3 PSS

PSS has been widely used as an auxiliary regulator of the excitation system so as to improve power system dynamic performance. In principle, its output is added to the excitation system and thus the generator can generate an electromagnetic torque component in phase with the rotor speed deviation. Accordingly, the system damping is improved and low frequency oscillation can be effectively damped out.

Figure 6.4 shows the block diagram of a typical PSS, whose differential equations can be expressed as

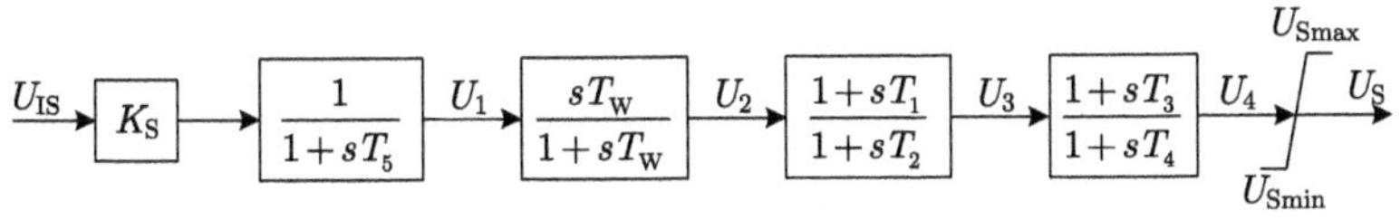

Fig. 6.4 Block diagram of the typical PSS

$$\begin{cases} \dfrac{dU_1}{dt} = \dfrac{1}{T_5}(K_S U_{IS} - U_1) \\ \dfrac{d(U_1 - U_2)}{dt} = \dfrac{1}{T_W} U_2 \\ \dfrac{d(T_1 U_2 - T_2 U_3)}{dt} = U_3 - U_2 \\ \dfrac{d(T_3 U_3 - T_4 U_4)}{dt} = U_4 - U_3 \end{cases} \quad (6.19)$$

and output denoted by U_S is limited to

$$U_S = \begin{cases} U_{Smax}, & U_4 \geqslant U_{Smax} \\ U_4, & U_{Smin} < U_4 < U_{Smax} \\ U_{Smin}, & U_4 \leqslant U_{Smin} \end{cases}$$

where U_{IS} is the input of PSS, usually chosen as rotor speed, terminal voltage, electromagnetic power of the synchronous generator or a combination of them. K_S is the gain of the PSS. T_5 is the time constant of the transducer. T_W is the time constant of the washout block. T_1, T_2, T_3 and T_4 are the time constants of the two lead-lag blocks. U_{Smax} and U_{Smin} are the upper and lower limits of U_S, respectively.

6.2.2.4 Prime Mover and Speed-Governing System

The block diagram of hydraulic turbine and its speed-governing system is shown in Fig. 6.5. The corresponding differential and algebraic equations consist of

$$\eta = K_\delta(\omega_{ref} - \omega)$$

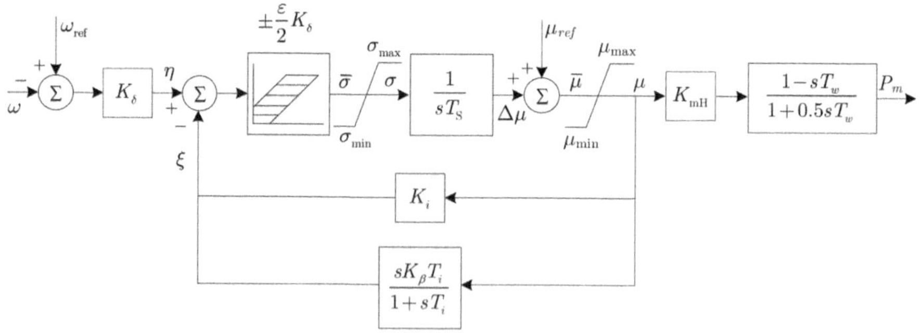

Fig. 6.5 Block diagram of hydraulic turbine and its speed-governing system

for the entrifugal pendulum mechanism (e.g., flyball),

$$\bar{\sigma} = \begin{cases} 0, & -\frac{\varepsilon K_\delta}{2} < \eta - \xi < \frac{\varepsilon K_\delta}{2} \\ \eta - \xi - \frac{\varepsilon K_\delta}{2}, & \eta - \xi \geqslant \frac{\varepsilon K_\delta}{2} \\ \eta - \xi + \frac{\varepsilon K_\delta}{2}, & \eta - \xi \leqslant -\frac{\varepsilon K_\delta}{2} \end{cases}$$

for the distributing valve failure zone,

$$\sigma = \begin{cases} \bar{\sigma}, & \sigma_{\min} < \bar{\sigma} < \sigma_{\max} \\ \sigma_{\max}, & \bar{\sigma} \geqslant \sigma_{\max} \\ \sigma_{\min}, & \bar{\sigma} \leqslant \sigma_{\min} \end{cases}$$

for the distributing valve travel limit,

$$\frac{\mathrm{d}\bar{\mu}}{\mathrm{d}t} = \frac{\sigma}{T_{\mathrm{S}}} \tag{6.20}$$

for the servomotor,

$$\mu = \begin{cases} \bar{\mu}, & \mu_{\min} < \bar{\mu} < \mu_{\max} \\ \mu_{\max}, & \bar{\mu} \geqslant \mu_{\max} \\ \mu_{\min}, & \bar{\mu} \leqslant \mu_{\min} \end{cases}$$

for the gate position limit,

$$\frac{\mathrm{d}\left[\xi - (K_\beta + K_\mathrm{i})\mu\right]}{\mathrm{d}t} = \frac{1}{T_\mathrm{i}}(K_\mathrm{i}\mu - \xi) \tag{6.21}$$

for the feedback link, and

$$\frac{\mathrm{d}(P_\mathrm{m} + 2K_{\mathrm{mH}}\mu)}{\mathrm{d}t} = \frac{2}{T_\mathrm{w}}(K_{\mathrm{mH}}\mu - P_\mathrm{m}) \tag{6.22}$$

for the hydraulic turbine, where ω_{ref} is the reference speed. η is the relative displacement of the flyballs sleeve. k_δ is the coefficient of proportionality. σ is the displacement of the distributing valve piston. $\sigma_{\max}$ and $\sigma_{\min}$ are the upper and lower limits of σ, respectively. T_S is the time constant of the servomotor. μ is the displacement of the servomotor piston. $\bar{\mu}$ is the reference of the servomotor piston displacement. $\mu_{\max}$ and $\mu_{\min}$ are the upper and lower limits of μ, respectively. T_i is the reset time. K_β and K_i are the temporary droop and the permanent droop, respectively. ξ is the displacement of the negative feedback to μ. K_{mH} is the ratio of the rated power of generators to the system reference capacity. T_w is the time constant of the equivalent effect of the water hammer. P_m is the mechanical power output of the prime mover.

6.2.3 Linearized Differential Equations

6.2.3.1 Synchronous Generator

Based on (6.16) and (6.17), the linearized differential equations of the synchronous generator can be expressed as

$$\begin{cases} \dfrac{\mathrm{d}\Delta\delta}{\mathrm{d}t} = \Delta\omega \\ \dfrac{\mathrm{d}\Delta\omega}{\mathrm{d}t} = \dfrac{1}{T_J}\{\Delta P_\mathrm{m} - D\Delta\omega - I_{q(0)}\Delta E_q'' - I_{d(0)}\Delta E_d'' \\ \qquad - [E_{d(0)}'' - (X_d'' - X_q'')\, I_{q(0)}]\Delta I_d \\ \qquad - [E_{q(0)}'' - (X_d'' - X_q'')\, I_{d(0)}]\Delta I_q\} \\ \dfrac{\mathrm{d}\Delta E_q'}{\mathrm{d}t} = \dfrac{1}{T_{d0}'}\left[\Delta E_{fq} - k_d\Delta E_q' + (k_d - 1)\Delta E_q''\right] \\ \dfrac{\mathrm{d}\Delta E_q''}{\mathrm{d}t} = \dfrac{1}{T_{d0}''}\left[\Delta E_q' - \Delta E_q'' - (X_d' - X_d'')\,\Delta I_d\right] \\ \dfrac{\mathrm{d}\Delta E_d'}{\mathrm{d}t} = \dfrac{1}{T_{q0}'}\left[-k_q\Delta E_d' + (k_q - 1)\Delta E_d''\right] \\ \dfrac{\mathrm{d}\Delta E_d''}{\mathrm{d}t} = \dfrac{1}{T_{q0}''}\left[\Delta E_d' - \Delta E_d'' + (X_q' - X_q'')\,\Delta I_q\right] \end{cases} \tag{6.23}$$

where the variables with subscript "(0)" indicate their steady-state values.

6.2.3.2 Excitation System

Linearizing (6.18) yields the linearized differential equations associated with the excitation system shown in Fig. 6.3, i.e.,

$$\begin{cases} \dfrac{\mathrm{d}\Delta E_{fq}}{\mathrm{d}t} = -\dfrac{1}{T_\mathrm{E}}\left(K_\mathrm{E} + n_\mathrm{E} c_\mathrm{E} E_{fq(0)}^{n_\mathrm{E}-1}\right)\Delta E_{fq} + \dfrac{1}{T_\mathrm{E}}\Delta U_\mathrm{R} \\ \dfrac{\mathrm{d}\Delta U_\mathrm{R}}{\mathrm{d}t} = -\dfrac{1}{T_\mathrm{A}}\Delta U_\mathrm{R} - \dfrac{K_\mathrm{A}}{T_\mathrm{A}}\Delta U_\mathrm{F} - \dfrac{K_\mathrm{A}}{T_\mathrm{A}}\Delta U_\mathrm{M} + \dfrac{K_\mathrm{A}}{T_\mathrm{A}}\Delta U_\mathrm{S} \\ \dfrac{\mathrm{d}\Delta U_\mathrm{F}}{\mathrm{d}t} = -\dfrac{K_\mathrm{F}}{T_\mathrm{E} T_\mathrm{F}}\left(K_\mathrm{E} + n_\mathrm{E} c_\mathrm{E} E_{fq(0)}^{n_\mathrm{E}-1}\right)\Delta E_{fq} + \dfrac{K_\mathrm{F}}{T_\mathrm{E} T_\mathrm{F}}\Delta U_\mathrm{R} - \dfrac{1}{T_\mathrm{F}}\Delta U_\mathrm{F} \\ \dfrac{\mathrm{d}\Delta U_\mathrm{M}}{\mathrm{d}t} = -\dfrac{1}{T_\mathrm{R}}\Delta U_\mathrm{M} + \dfrac{K_{\mathrm{C}q} X_\mathrm{C}}{T_\mathrm{R}}\Delta I_d - \dfrac{K_{\mathrm{C}d} X_\mathrm{C}}{T_\mathrm{R}}\Delta I_q + \dfrac{K_{\mathrm{C}d}}{T_\mathrm{R}}\Delta U_d + \dfrac{K_{\mathrm{C}q}}{T_\mathrm{R}}\Delta U_q \end{cases} \tag{6.24}$$

where

$$\begin{cases} K_{\mathrm{C}d} = (U_{d(0)} - X_{\mathrm{C}} I_{q(0)})/U_{\mathrm{C}(0)} \\ K_{\mathrm{C}q} = (U_{q(0)} + X_{\mathrm{C}} I_{d(0)})/U_{\mathrm{C}(0)} \\ U_{\mathrm{C}(0)} = \sqrt{(U_{d(0)} - X_{\mathrm{C}} I_{q(0)})^2 + (U_{q(0)} + X_{\mathrm{C}} I_{d(0)})^2} \end{cases}. \tag{6.25}$$

6.2.3.3 PSS

Without loss of generality, the input signal of the PSS here is chosen as the deviation of the rotor speed, and the output limiting link in Fig. 6.4 is neglected. According to (6.19), the linearized differential equations of the PSS can be derived and expressed as

$$\begin{cases} \dfrac{\mathrm{d}\Delta U_1}{\mathrm{d}t} = \dfrac{K_{\mathrm{S}}}{T_5}\Delta\omega - \dfrac{1}{T_5}\Delta U_1 \\ \dfrac{\mathrm{d}\Delta U_2}{\mathrm{d}t} = \dfrac{K_{\mathrm{S}}}{T_5}\Delta\omega - \dfrac{1}{T_5}\Delta U_1 - \dfrac{1}{T_{\mathrm{W}}}\Delta U_2 \\ \dfrac{\mathrm{d}\Delta U_3}{\mathrm{d}t} = \dfrac{K_{\mathrm{S}} T_1}{T_2 T_5}\Delta\omega - \dfrac{T_1}{T_2 T_5}\Delta U_1 - \dfrac{T_1 - T_{\mathrm{W}}}{T_2 T_{\mathrm{W}}}\Delta U_2 - \dfrac{1}{T_2}\Delta U_3 \\ \dfrac{\mathrm{d}\Delta U_{\mathrm{S}}}{\mathrm{d}t} = \dfrac{K_{\mathrm{S}} T_1 T_3}{T_2 T_4 T_5}\Delta\omega - \dfrac{T_1 T_3}{T_2 T_4 T_5}\Delta U_1 - \dfrac{T_3(T_1 - T_{\mathrm{W}})}{T_2 T_4 T_{\mathrm{W}}}\Delta U_2 \\ \qquad - \dfrac{T_3 - T_2}{T_2 T_4}\Delta U_3 - \dfrac{1}{T_4}\Delta U_{\mathrm{S}} \end{cases}. \tag{6.26}$$

6.2.3.4 Prime Mover and Its Speed-Governing System

Neglecting the measurement failure zone and the travel and opening limiting links of the distributing valve and then linearizing (6.20), (6.21) and (6.22), one can obtain

$$\begin{cases} \dfrac{\mathrm{d}\Delta\mu}{\mathrm{d}t} = -\dfrac{K_\delta}{T_{\mathrm{S}}}\Delta\omega - \dfrac{1}{T_{\mathrm{S}}}\Delta\xi \\ \dfrac{\mathrm{d}\Delta\xi}{\mathrm{d}t} = -\dfrac{K_\delta(K_{\mathrm{i}} + K_\beta)}{T_{\mathrm{S}}}\Delta\omega + \dfrac{K_{\mathrm{i}}}{T_{\mathrm{i}}}\Delta\mu - \left(\dfrac{1}{T_{\mathrm{i}}} + \dfrac{K_{\mathrm{i}} + K_\beta}{T_{\mathrm{S}}}\right)\Delta\xi \\ \dfrac{\mathrm{d}\Delta P_{\mathrm{m}}}{\mathrm{d}t} = \dfrac{2K_{\mathrm{mH}}K_\delta}{T_{\mathrm{S}}}\Delta\omega + \dfrac{2K_{\mathrm{mH}}}{T_{\mathrm{w}}}\Delta\mu + \dfrac{2K_{\mathrm{mH}}}{T_{\mathrm{S}}}\Delta\xi - \dfrac{2}{T_{\mathrm{w}}}\Delta P_{\mathrm{m}} \end{cases} \tag{6.27}$$

for the hydraulic turbine and its speed-governing system shown in Fig. 6.5.

6.2.3.5 Linearized Differential Equations

By combining (6.23)–(6.27), the linearized differential equations of a synchronous generator unit can be arranged in a matrix form

$$\frac{\mathrm{d}\Delta \boldsymbol{x}_{\mathrm{g}}}{\mathrm{d}t} = \bar{\boldsymbol{A}}_{\mathrm{g}}\Delta \boldsymbol{x}_{\mathrm{g}} + \bar{\boldsymbol{B}}_{I\mathrm{g}}\Delta \boldsymbol{I}_{dq\mathrm{g}} + \bar{\boldsymbol{B}}_{U\mathrm{g}}\Delta \boldsymbol{U}_{dq\mathrm{g}} \tag{6.28}$$

where $\bar{\boldsymbol{A}}_{\mathrm{g}} \in \mathbb{R}^{17\times 17}$, $\bar{\boldsymbol{B}}_{I\mathrm{g}} \in \mathbb{R}^{17\times 2}$, $\bar{\boldsymbol{B}}_{U\mathrm{g}} \in \mathbb{R}^{17\times 2}$, and

$$\begin{cases} \Delta \boldsymbol{x}_{\mathrm{g}} = \big[\Delta\delta,\ \Delta\omega,\ \Delta E'_q,\ \Delta E''_q,\ \Delta E'_d,\ \Delta E''_d,\ \Delta E_{fq},\ \Delta U_{\mathrm{R}}, \\ \qquad\qquad \Delta U_{\mathrm{F}},\ \Delta U_{\mathrm{M}},\ \Delta U_1,\ \Delta U_2,\ \Delta U_3,\ \Delta U_{\mathrm{S}},\ \Delta\mu,\ \Delta\xi,\ \Delta P_{\mathrm{m}}\big]^{\mathrm{T}} \in \mathbb{R}^{17\times 1} \\ \Delta \boldsymbol{U}_{dq\mathrm{g}} = [\Delta U_d,\ \Delta U_q]^{\mathrm{T}} \in \mathbb{R}^{2\times 1} \\ \Delta \boldsymbol{I}_{dq\mathrm{g}} = [\Delta I_d,\ \Delta I_q]^{\mathrm{T}} \in \mathbb{R}^{2\times 1} \end{cases}.$$

Then, the linearized differential equations of a multi-machine power system can be established as

$$\frac{\mathrm{d}\Delta \boldsymbol{x}_{\mathrm{G}}}{\mathrm{d}t} = \bar{\boldsymbol{A}}_{\mathrm{G}}\Delta \boldsymbol{x}_{\mathrm{G}} + \bar{\boldsymbol{B}}_{I\mathrm{G}}\Delta \boldsymbol{I}_{dq\mathrm{G}} + \bar{\boldsymbol{B}}_{U\mathrm{G}}\Delta \boldsymbol{U}_{dq\mathrm{G}} \tag{6.29}$$

where

$$\begin{cases} \bar{\boldsymbol{A}}_{\mathrm{G}} = \mathrm{diag}\left\{\bar{\boldsymbol{A}}_{\mathrm{g}1},\ \bar{\boldsymbol{A}}_{\mathrm{g}2},\ \ldots\right\} \\ \bar{\boldsymbol{B}}_{I\mathrm{G}} = \mathrm{diag}\left\{\bar{\boldsymbol{B}}_{I\mathrm{g}1},\ \bar{\boldsymbol{B}}_{I\mathrm{g}2},\ \ldots\right\} \\ \bar{\boldsymbol{B}}_{U\mathrm{G}} = \mathrm{diag}\left\{\bar{\boldsymbol{B}}_{U\mathrm{g}1},\ \bar{\boldsymbol{B}}_{U\mathrm{g}2},\ \ldots\right\} \\ \Delta \boldsymbol{x}_{\mathrm{G}} = \left[\boldsymbol{x}_{\mathrm{g}1}^{\mathrm{T}},\ \boldsymbol{x}_{\mathrm{g}2}^{\mathrm{T}},\ \ldots\right]^{\mathrm{T}} \\ \Delta \boldsymbol{I}_{dq\mathrm{G}} = \left[\Delta \boldsymbol{I}_{dq\mathrm{g}1}^{\mathrm{T}},\ \Delta \boldsymbol{I}_{dq\mathrm{g}2}^{\mathrm{T}},\ \ldots\right]^{\mathrm{T}} \\ \Delta \boldsymbol{U}_{dq\mathrm{G}} = \left[\Delta \boldsymbol{U}_{dq\mathrm{g}1}^{\mathrm{T}},\ \Delta \boldsymbol{U}_{dq\mathrm{g}2}^{\mathrm{T}},\ \ldots\right]^{\mathrm{T}} \end{cases}.$$

6.2.4 Linearized Algebraic Equations

6.2.4.1 Stator Voltage Equations

The linearized stator voltage equations of a synchronous generator are

$$\begin{cases} \Delta U_d = \Delta E''_d - R_a \Delta I_d + X''_q \Delta I_q \\ \Delta U_q = \Delta E''_q - X''_d \Delta I_d - R_a \Delta I_q \end{cases}. \tag{6.30}$$

Let $\bar{\boldsymbol{P}}_{\mathrm{g}} \in \mathbb{R}^{2\times 17}$ and $\bar{\boldsymbol{Z}}_{\mathrm{g}} \in \mathbb{R}^{2\times 2}$ be

$$\bar{\boldsymbol{P}}_{\mathrm{g}} = \begin{bmatrix} 0 & 0 & 0 & 0 & 0 & 1 & 0 & 0 & 0 & 0 & 0 & 0 & 0 & 0 & 0 & 0 & 0 \\ 0 & 0 & 0 & 1 & 0 & 0 & 0 & 0 & 0 & 0 & 0 & 0 & 0 & 0 & 0 & 0 & 0 \end{bmatrix},\quad \bar{\boldsymbol{Z}}_{\mathrm{g}} = \begin{bmatrix} -R_a & X''_q \\ -X''_d & -R_a \end{bmatrix} \tag{6.31}$$

then (6.30) can be compactly rewritten as

$$\Delta \boldsymbol{U}_{dq\mathrm{g}} = \bar{\boldsymbol{P}}_{\mathrm{g}} \Delta \boldsymbol{x}_{\mathrm{g}} + \bar{\boldsymbol{Z}}_{\mathrm{g}} \Delta \boldsymbol{I}_{dq\mathrm{g}}. \tag{6.32}$$

Based on (6.32), the linearized stator voltage equations of a multi-machine power system is formulated as

$$\Delta \boldsymbol{U}_{dq\mathrm{G}} = \bar{\boldsymbol{P}}_{\mathrm{G}} \Delta \boldsymbol{x}_{\mathrm{G}} + \bar{\boldsymbol{Z}}_{\mathrm{G}} \Delta \boldsymbol{I}_{dq\mathrm{G}} \tag{6.33}$$

where

$$\begin{cases} \bar{\boldsymbol{P}}_{\mathrm{G}} = \mathrm{diag}\left\{\bar{\boldsymbol{P}}_{\mathrm{g}1},\ \bar{\boldsymbol{P}}_{\mathrm{g}2},\ \ldots\right\} \\ \bar{\boldsymbol{Z}}_{\mathrm{G}} = \mathrm{diag}\left\{\bar{\boldsymbol{Z}}_{\mathrm{g}1},\ \bar{\boldsymbol{Z}}_{\mathrm{g}2},\ \ldots\right\} \end{cases}.$$

6.2.4.2 *dq-xy* Reference Transformation Equations

The $\Delta \boldsymbol{U}_{dq\mathrm{g}}$ and $\Delta \boldsymbol{I}_{dq\mathrm{g}}$ in (6.28) and (6.32) are the voltage and current deviation vectors in each generator's own d-q reference frame which rotates with the rotor. To interconnect them with the network, all voltages and currents must be expressed in the common x-y reference frame according to

$$\begin{bmatrix} I_d \\ I_q \end{bmatrix} = \begin{bmatrix} \sin\delta & -\cos\delta \\ \cos\delta & \sin\delta \end{bmatrix} \begin{bmatrix} I_x \\ I_y \end{bmatrix} \tag{6.34}$$

and

$$\begin{bmatrix} U_d \\ U_q \end{bmatrix} = \begin{bmatrix} \sin\delta & -\cos\delta \\ \cos\delta & \sin\delta \end{bmatrix} \begin{bmatrix} U_x \\ U_y \end{bmatrix}. \tag{6.35}$$

Linearizing (6.34) and (6.35) yields

$$\begin{bmatrix} \Delta I_d \\ \Delta I_q \end{bmatrix} = \begin{bmatrix} \sin\delta_{(0)} & -\cos\delta_{(0)} \\ \cos\delta_{(0)} & \sin\delta_{(0)} \end{bmatrix} \begin{bmatrix} \Delta I_x \\ \Delta I_y \end{bmatrix} + \begin{bmatrix} I_{q(0)} \\ -I_{d(0)} \end{bmatrix} \Delta\delta \tag{6.36}$$

and

$$\begin{bmatrix} \Delta U_d \\ \Delta U_q \end{bmatrix} = \begin{bmatrix} \sin\delta_{(0)} & -\cos\delta_{(0)} \\ \cos\delta_{(0)} & \sin\delta_{(0)} \end{bmatrix} \begin{bmatrix} \Delta U_x \\ \Delta U_y \end{bmatrix} + \begin{bmatrix} U_{q(0)} \\ -U_{d(0)} \end{bmatrix} \Delta\delta. \tag{6.37}$$

Let $\boldsymbol{I}_{\mathrm{g}} \in \mathbb{R}^{2\times1}$, $\boldsymbol{T}_{\mathrm{g}(0)} \in \mathbb{R}^{2\times2}$, $\boldsymbol{R}_{\mathrm{Ig}} \in \mathbb{R}^{2\times17}$, $\boldsymbol{U}_{\mathrm{g}} \in \mathbb{R}^{2\times1}$ and $\boldsymbol{R}_{\mathrm{Ug}} \in \mathbb{R}^{2\times17}$ be

$$\begin{cases} \boldsymbol{I}_{\mathrm{g}} = \begin{bmatrix} \Delta I_x \\ \Delta I_y \end{bmatrix}, & \boldsymbol{T}_{\mathrm{g}(0)} = \begin{bmatrix} \sin\delta_{(0)} & -\cos\delta_{(0)} \\ \cos\delta_{(0)} & \sin\delta_{(0)} \end{bmatrix}, \quad \boldsymbol{R}_{I\mathrm{g}} = \begin{bmatrix} I_{q(0)} & 0 \cdots 0 \\ -I_{d(0)} & 0 \cdots 0 \end{bmatrix} \\ \boldsymbol{U}_{\mathrm{g}} = \begin{bmatrix} \Delta U_x \\ \Delta U_y \end{bmatrix}, & \boldsymbol{R}_{U\mathrm{g}} = \begin{bmatrix} U_{q(0)} & 0 \cdots 0 \\ -U_{d(0)} & 0 \cdots 0 \end{bmatrix} \end{cases} \tag{6.38}$$

then (6.36) and (6.37) can be reduced to

$$\Delta \boldsymbol{I}_{dq\text{g}} = \boldsymbol{T}_{\text{g}(0)} \Delta \boldsymbol{I}_{\text{g}} + \boldsymbol{R}_{I\text{g}} \Delta \boldsymbol{x}_{\text{g}}$$

and

$$\Delta \boldsymbol{U}_{dq\text{g}} = \boldsymbol{T}_{\text{g}(0)} \Delta \boldsymbol{U}_{\text{g}} + \boldsymbol{R}_{U\text{g}} \Delta \boldsymbol{x}_{\text{g}}.$$

The reference transformation equations for a multi-machine power system can be expressed as

$$\begin{cases} \Delta \boldsymbol{I}_{dq\text{G}} = \boldsymbol{T}_{\text{G}(0)} \Delta \boldsymbol{I}_{\text{G}} + \boldsymbol{R}_{I\text{G}} \Delta \boldsymbol{x}_{\text{G}} \\ \Delta \boldsymbol{U}_{dq\text{G}} = \boldsymbol{T}_{\text{G}(0)} \Delta \boldsymbol{U}_{\text{G}} + \boldsymbol{R}_{U\text{G}} \Delta \boldsymbol{x}_{\text{G}} \end{cases} \tag{6.39}$$

where

$$\begin{cases} \boldsymbol{I}_{\text{G}} = \left[\boldsymbol{I}_{\text{g}1}^{\text{T}},\ \boldsymbol{I}_{\text{g}2}^{\text{T}},\ \ldots \right]^{\text{T}} \\ \boldsymbol{U}_{\text{G}} = \left[\boldsymbol{U}_{\text{g}1}^{\text{T}},\ \boldsymbol{U}_{\text{g}2}^{\text{T}},\ \ldots \right]^{\text{T}} \\ \boldsymbol{T}_{\text{G}(0)} = \text{diag}\left\{ \boldsymbol{T}_{\text{g}1(0)},\ \boldsymbol{T}_{\text{g}2(0)},\ \ldots \right\} \\ \boldsymbol{R}_{I\text{G}} = \text{diag}\left\{ \boldsymbol{R}_{I\text{g}1},\ \boldsymbol{R}_{I\text{g}2},\ \ldots \right\} \\ \boldsymbol{R}_{U\text{G}} = \text{diag}\left\{ \boldsymbol{R}_{U\text{g}1},\ \boldsymbol{R}_{U\text{g}2},\ \ldots \right\} \end{cases}.$$

6.2.4.3 Power Network and Static Load Equations

The power network connects all dynamic devices in the system. For a power network with k nodes, by neglecting the electromagnetic transient processes, the relationship between the deviations of node voltage and node current injection can be expressed by the augmented admittance matrix, i.e.,

$$\begin{bmatrix} \Delta \boldsymbol{I}_1 \\ \vdots \\ \Delta \boldsymbol{I}_i \\ \vdots \\ \Delta \boldsymbol{I}_k \end{bmatrix} = \begin{bmatrix} \boldsymbol{Y}_{11} & \cdots & \boldsymbol{Y}_{1i} & \cdots & \boldsymbol{Y}_{1k} \\ \vdots & & \vdots & & \vdots \\ \boldsymbol{Y}_{i1} & \cdots & \boldsymbol{Y}_{ii} & \cdots & \boldsymbol{Y}_{ik} \\ \vdots & & \vdots & & \vdots \\ \boldsymbol{Y}_{k1} & \cdots & \boldsymbol{Y}_{ki} & \cdots & \boldsymbol{Y}_{kk} \end{bmatrix} \begin{bmatrix} \Delta \boldsymbol{U}_1 \\ \vdots \\ \Delta \boldsymbol{U}_i \\ \vdots \\ \Delta \boldsymbol{U}_k \end{bmatrix} \tag{6.40}$$

with $\Delta \boldsymbol{I}_i \in \mathbb{R}^{2\times 1}$, $\Delta \boldsymbol{U}_i \in \mathbb{R}^{2\times 1}$ and $\boldsymbol{Y}_{ij} \in \mathbb{R}^{2\times 2}$ $(i,\ j = 1,\ 2,\ \vdots,\ k)$ given by

$$\Delta \boldsymbol{I}_i = \begin{bmatrix} \Delta I_{xi} \\ \Delta I_{yi} \end{bmatrix},\quad \Delta \boldsymbol{U}_i = \begin{bmatrix} \Delta U_{xi} \\ \Delta U_{yi} \end{bmatrix},\quad \boldsymbol{Y}_{ij} = \begin{bmatrix} G_{ij} & -B_{ij} \\ B_{ij} & G_{ij} \end{bmatrix}$$

where G_{ij} and B_{ij} are the conductance and susceptance of line i-j, respectively.

In the load model, the small-signal relationship between the voltage and injection current of the load nodes can be written as

$$\Delta \boldsymbol{I}_{\ell} = \boldsymbol{Y}_{\ell} \Delta \boldsymbol{U}_{\ell} \tag{6.41}$$

with $\Delta \boldsymbol{I}_{\ell} \in \mathbb{R}^{2\times1}$, $\boldsymbol{Y}_{\ell} \in \mathbb{R}^{2\times2}$ and $\Delta \boldsymbol{U}_{\ell} \in \mathbb{R}^{2\times1}$ described as

$$\Delta \boldsymbol{I}_{\ell} = \begin{bmatrix} \Delta I_x \\ \Delta I_y \end{bmatrix}, \quad \boldsymbol{Y}_{\ell} = \begin{bmatrix} G_{xx} & B_{xy} \\ -B_{yx} & G_{yy} \end{bmatrix}, \quad \Delta \boldsymbol{U}_{\ell} = \begin{bmatrix} \Delta U_x \\ \Delta U_y \end{bmatrix}$$

where the specific expressions of G_{xx}, B_{xy}, B_{yx} and G_{yy} depend on the adopted static voltage characteristic.

By substituting (6.41), the deviation of current injecting to the load node $\Delta \boldsymbol{I}_{\ell}$ in (6.40) can be eliminated. Without loss of generality, it is assumed that the load is connected to node i. Therefore, (6.40) becomes

$$\begin{bmatrix} \Delta \boldsymbol{I}_1 \\ \vdots \\ \boldsymbol{0} \\ \vdots \\ \Delta \boldsymbol{I}_k \end{bmatrix} = \begin{bmatrix} \boldsymbol{Y}_{11} & \cdots & \boldsymbol{Y}_{1i} & \cdots & \boldsymbol{Y}_{1k} \\ \vdots & & \vdots & & \vdots \\ \boldsymbol{Y}_{i1} & \cdots & \boldsymbol{Y}_{ii} - \boldsymbol{Y}_{\ell i} & \cdots & \boldsymbol{Y}_{ik} \\ \vdots & & \vdots & & \vdots \\ \boldsymbol{Y}_{k1} & \cdots & \boldsymbol{Y}_{ki} & \cdots & \boldsymbol{Y}_{kk} \end{bmatrix} \begin{bmatrix} \Delta \boldsymbol{U}_1 \\ \vdots \\ \Delta \boldsymbol{U}_i \\ \vdots \\ \Delta \boldsymbol{U}_k \end{bmatrix}.$$

Subsequently, the current deviation of connection nodes can be directly set to be zero due to no current injection, leading to

$$\begin{bmatrix} \Delta \boldsymbol{I}_{\mathrm{G}} \\ \boldsymbol{0} \end{bmatrix} = \begin{bmatrix} \boldsymbol{Y}_{\mathrm{GG}} & \boldsymbol{Y}_{\mathrm{GL}} \\ \boldsymbol{Y}_{\mathrm{LG}} & \boldsymbol{Y}_{\mathrm{LL}} \end{bmatrix} \begin{bmatrix} \Delta \boldsymbol{U}_{\mathrm{G}} \\ \Delta \boldsymbol{U}_{\mathrm{L}} \end{bmatrix} \tag{6.42}$$

where $\Delta \boldsymbol{U}_{\mathrm{L}}$ is the voltage deviation vector of load nodes and connection nodes.

6.2.5 *Linearized DAEs*

For the ease of representation, it is assumed that power system have no dynamic devices other than synchronous generator units. In this case, by combining the aforementioned linearized differential equations (6.29) and linearized algebraic equations (6.33), (6.39) and (6.42), one can obtain the linearized DAE model of a multi-machine system, the general form of which can be expressed as

$$\begin{bmatrix} \Delta \dot{\boldsymbol{x}} \\ \boldsymbol{0} \end{bmatrix} = \begin{bmatrix} \boldsymbol{A} & \boldsymbol{B} \\ \boldsymbol{C} & \boldsymbol{D} \end{bmatrix} \begin{bmatrix} \Delta \boldsymbol{x} \\ \Delta \boldsymbol{y} \end{bmatrix} \tag{6.43}$$

where $\boldsymbol{A} \in \mathbb{R}^{n\times n}$, $\boldsymbol{B} \in \mathbb{R}^{n\times l}$, $\boldsymbol{C} \in \mathbb{R}^{l\times n}$ and $\boldsymbol{D} \in \mathbb{R}^{l\times l}$ are Jacobian matrices. In the remainder of this subsection, two methods on formulating (6.43) are presented.

To facilitate the analysis of the dimension of augmented state matrix, it is assumed that there are k nodes and q synchronous generator units in the system.

6.2.5.1 Method 1: Algebraic Equations of Dynamic Devices Are Kept

In this method, (6.43) is directly formulated by combining (6.29), (6.33), (6.39) and (6.42), where

$$\begin{cases} \Delta \boldsymbol{x} \triangleq \Delta \boldsymbol{x}_{\mathrm{G}} \in \mathbb{R}^{17q\times 1} \\ \Delta \boldsymbol{y} \triangleq \left[\Delta \boldsymbol{U}_{dq\mathrm{G}}^{\mathrm{T}},\ \Delta \boldsymbol{I}_{dq\mathrm{G}}^{\mathrm{T}},\ \Delta \boldsymbol{I}_{\mathrm{G}}^{\mathrm{T}},\ \Delta \boldsymbol{U}_{\mathrm{G}}^{\mathrm{T}},\ \Delta \boldsymbol{U}_{\mathrm{L}}^{\mathrm{T}}\right]^{\mathrm{T}} \in \mathbb{R}^{(6q+2k)\times 1} \\ \boldsymbol{A} \triangleq \bar{\boldsymbol{A}}_{\mathrm{G}} \in \mathbb{R}^{17q\times 17q} \\ \boldsymbol{B} \triangleq \left[\bar{\boldsymbol{B}}_{U\mathrm{G}},\ \bar{\boldsymbol{B}}_{I\mathrm{G}},\ \boldsymbol{0},\ \boldsymbol{0},\ \boldsymbol{0}\right] \in \mathbb{R}^{17q\times(6q+2k)} \\ \boldsymbol{C} \triangleq \left[\bar{\boldsymbol{P}}_{\mathrm{G}}^{\mathrm{T}},\ \boldsymbol{R}_{I\mathrm{G}}^{\mathrm{T}},\ \boldsymbol{R}_{U\mathrm{G}}^{\mathrm{T}},\ \boldsymbol{0},\ \boldsymbol{0}\right]^{\mathrm{T}} \in \mathbb{R}^{(6q+2k)\times 17q} \\ \boldsymbol{D} \triangleq \begin{bmatrix} -\boldsymbol{I} & \bar{\boldsymbol{Z}}_{\mathrm{G}} & \boldsymbol{0} & \boldsymbol{0} & \boldsymbol{0} \\ \boldsymbol{0} & -\boldsymbol{I} & \boldsymbol{T}_{\mathrm{G}(0)} & \boldsymbol{0} & \boldsymbol{0} \\ -\boldsymbol{I} & \boldsymbol{0} & \boldsymbol{0} & \boldsymbol{T}_{\mathrm{G}(0)} & \boldsymbol{0} \\ \boldsymbol{0} & \boldsymbol{0} & -\boldsymbol{I} & \boldsymbol{Y}_{\mathrm{GG}} & \boldsymbol{Y}_{\mathrm{GL}} \\ \boldsymbol{0} & \boldsymbol{0} & \boldsymbol{0} & \boldsymbol{Y}_{\mathrm{LG}} & \boldsymbol{Y}_{\mathrm{LL}} \end{bmatrix} \in \mathbb{R}^{(6q+2k)\times(6q+2k)} \end{cases}.$$

The linearized differential equations are composed of those of all generator units and the total number is $n = 17q$. While, the number of linearized algebraic equations is $l = 6q + 2k$, which consist 6 algebraic equations for each generator unit and 2 voltage equations for each node (network equations).

6.2.5.2 Method 2: Algebraic Equations of Dynamic Devices Are Eliminated

In this method, the linearized algebraic equations for each generator unit are eliminated. Specifically, (6.39) are substituted into (6.33) to eliminate $\Delta \boldsymbol{I}_{dq\mathrm{G}}$ and $\Delta \boldsymbol{U}_{dq\mathrm{G}}$. By taking $\boldsymbol{T}_{\mathrm{g}(0)}^{-1} = \boldsymbol{T}_{\mathrm{g}(0)}^{\mathrm{T}}$ into account, one can obtain

$$\Delta \boldsymbol{I}_{\mathrm{G}} = \boldsymbol{C}_{\mathrm{G}} \Delta \boldsymbol{x}_{\mathrm{G}} + \boldsymbol{D}_{\mathrm{G}} \Delta \boldsymbol{U}_{\mathrm{G}} \tag{6.44}$$

where

$$\begin{cases} \boldsymbol{C}_{\mathrm{G}} = \mathrm{diag}\left\{\boldsymbol{C}_{\mathrm{g}1},\ \boldsymbol{C}_{\mathrm{g}2},\ \ldots\right\},\quad \boldsymbol{D}_{\mathrm{G}} = \mathrm{diag}\left\{\boldsymbol{D}_{\mathrm{g}1},\ \boldsymbol{D}_{\mathrm{g}2},\ \ldots\right\} \\ \boldsymbol{C}_{\mathrm{g}} = \boldsymbol{T}_{\mathrm{g}(0)}^{\mathrm{T}}\left[\bar{\boldsymbol{Z}}_{\mathrm{g}}^{-1}\left(\boldsymbol{R}_{U\mathrm{g}} - \bar{\boldsymbol{P}}_{\mathrm{g}}\right) - \boldsymbol{R}_{I\mathrm{g}}\right],\quad \boldsymbol{D}_{\mathrm{g}} = \boldsymbol{T}_{\mathrm{g}(0)}^{\mathrm{T}}\bar{\boldsymbol{Z}}_{\mathrm{g}}^{-1}\boldsymbol{T}_{\mathrm{g}(0)} \end{cases}. \tag{6.45}$$

By substituting (6.31) and (6.38) into (6.45), the expression of $\Delta \boldsymbol{I}_{\mathrm{g}}$, an entry of $\Delta \boldsymbol{I}_{\mathrm{G}}$, can be obtained as follows:

$$
\begin{aligned}
\Delta \boldsymbol{I}_{\mathrm{g}} &= \boldsymbol{C}_{\mathrm{g}} \Delta \boldsymbol{x}_{\mathrm{g}} + \boldsymbol{D}_{\mathrm{g}} \Delta \boldsymbol{U}_{\mathrm{g}} \\
&= \begin{bmatrix} a_\delta \\ b_\delta \end{bmatrix} \Delta\delta + \begin{bmatrix} g_d & b_q \\ b_d & g_q \end{bmatrix} \begin{bmatrix} \Delta E''_d \\ \Delta E''_q \end{bmatrix} + \begin{bmatrix} g_{xx} & b_{xy} \\ b_{yx} & g_{yy} \end{bmatrix} \Delta \boldsymbol{U}_{\mathrm{g}}
\end{aligned} \tag{6.46}
$$

where

$$
\begin{cases}
a_\delta = \dfrac{X''_d \cos^2\delta_{(0)} + X''_q \sin^2\delta_{(0)}}{R_a^2 + X''_d X''_q} U_{x(0)} \\
\quad + \dfrac{-R_a + (X''_d - X''_q)\sin\delta_{(0)}\cos\delta_{(0)}}{R_a^2 + X''_d X''_q} U_{y(0)} - I_{y(0)} \\
b_\delta = \dfrac{R_a + (X''_d - X''_q)\sin\delta_{(0)}\cos\delta_{(0)}}{R_a^2 + X''_d X''_q} U_{x(0)} \\
\quad + \dfrac{X''_d \sin^2\delta_{(0)} + X''_q \cos^2\delta_{(0)}}{R_a^2 + X''_d X''_q} U_{y(0)} + I_{x(0)} \\
g_d = \dfrac{R_a \sin\delta_{(0)} - X''_d \cos\delta_{(0)}}{R_a^2 + X''_d X''_q}, \quad b_q = \dfrac{R_a \cos\delta_{(0)} + X''_q \sin\delta_{(0)}}{R_a^2 + X''_d X''_q} \\
b_d = \dfrac{-R_a \cos\delta_{(0)} - X''_d \sin\delta_{(0)}}{R_a^2 + X''_d X''_q}, \quad g_q = \dfrac{R_a \sin\delta_{(0)} - X''_q \cos\delta_{(0)}}{R_a^2 + X''_d X''_q} \\
g_{xx} = \dfrac{-R_a + (X''_d - X''_q)\sin\delta_{(0)}\cos\delta_{(0)}}{R_a^2 + X''_d X''_q}, \quad b_{xy} = -\dfrac{X''_d \cos^2\delta_{(0)} + X''_q \sin^2\delta_{(0)}}{R_a^2 + X''_d X''_q} \\
b_{yx} = \dfrac{X''_d \sin^2\delta_{(0)} + X''_q \cos^2\delta_{(0)}}{R_a^2 + X''_d X''_q}, \quad g_{yy} = -\dfrac{R_a + (X''_d - X''_q)\sin\delta_{(0)}\cos\delta_{(0)}}{R_a^2 + X''_d X''_q}
\end{cases}.
$$

In (6.29), $\Delta \boldsymbol{I}_{dq\mathrm{G}}$ and $\Delta \boldsymbol{U}_{dq\mathrm{G}}$ are eliminated by using (6.39), respectively. Then, the introduced $\Delta \boldsymbol{I}_{\mathrm{G}}$ is further eliminated by utilizing (6.44). As a result, (6.29) becomes

$$
\frac{\mathrm{d}\Delta \boldsymbol{x}_{\mathrm{G}}}{\mathrm{d}t} = \boldsymbol{A}_{\mathrm{G}} \Delta \boldsymbol{x}_{\mathrm{G}} + \boldsymbol{B}_{\mathrm{G}} \Delta \boldsymbol{U}_{\mathrm{G}} \tag{6.47}
$$

where

$$
\begin{cases}
\boldsymbol{A}_{\mathrm{G}} = \mathrm{diag}\left\{\boldsymbol{A}_{\mathrm{g}1}, \ \boldsymbol{A}_{\mathrm{g}2}, \ \ldots\right\}, \ \ \boldsymbol{B}_{\mathrm{G}} = \mathrm{diag}\left\{\boldsymbol{B}_{\mathrm{g}1}, \ \boldsymbol{B}_{\mathrm{g}2}, \ \ldots\right\} \\
\boldsymbol{A}_{\mathrm{g}} = \bar{\boldsymbol{A}}_{\mathrm{g}} + \bar{\boldsymbol{B}}_{I\mathrm{g}} \bar{\boldsymbol{Z}}_{\mathrm{g}}^{-1} \left(\boldsymbol{R}_{U\mathrm{g}} - \bar{\boldsymbol{P}}_{\mathrm{g}}\right) + \bar{\boldsymbol{B}}_{U\mathrm{g}} \boldsymbol{R}_{U\mathrm{g}} \\
\boldsymbol{B}_{\mathrm{g}} = \left(\bar{\boldsymbol{B}}_{I\mathrm{g}} \bar{\boldsymbol{Z}}_{\mathrm{g}}^{-1} + \bar{\boldsymbol{B}}_{U\mathrm{g}}\right) \boldsymbol{T}_{\mathrm{g}(0)}.
\end{cases}.
$$

Followed by substituting (6.44) into the power network Eq. (6.42) to eliminate $\Delta \boldsymbol{I}_{\mathrm{G}}$, one can obtain

$$
\boldsymbol{0} = \begin{bmatrix} -\boldsymbol{C}_{\mathrm{G}} \\ \boldsymbol{0} \end{bmatrix} \Delta \boldsymbol{x}_{\mathrm{G}} + \begin{bmatrix} \boldsymbol{Y}_{\mathrm{GG}} - \boldsymbol{D}_{\mathrm{G}} & \boldsymbol{Y}_{\mathrm{GL}} \\ \boldsymbol{Y}_{\mathrm{LG}} & \boldsymbol{Y}_{\mathrm{LL}} \end{bmatrix} \begin{bmatrix} \Delta \boldsymbol{U}_{\mathrm{G}} \\ \Delta \boldsymbol{U}_{\mathrm{L}} \end{bmatrix}. \tag{6.48}
$$

Finally, the combination of (6.47) and (6.48) yields (6.43), where

$$\begin{cases} \Delta \boldsymbol{x} \triangleq \Delta \boldsymbol{x}_{\mathrm{G}} \in \mathbb{R}^{17q\times 1} \\ \Delta \boldsymbol{y} \triangleq \left[\Delta \boldsymbol{U}_{\mathrm{G}}^{\mathrm{T}},\ \Delta \boldsymbol{U}_{\mathrm{L}}^{\mathrm{T}}\right]^{\mathrm{T}} \in \mathbb{R}^{2k\times 1} \\ \boldsymbol{A} \triangleq \boldsymbol{A}_{\mathrm{G}} \in \mathbb{R}^{17q\times 17q} \\ \boldsymbol{B} \triangleq [\boldsymbol{B}_{\mathrm{G}},\ \boldsymbol{0}] \in \mathbb{R}^{17q\times 2k} \\ \boldsymbol{C} \triangleq \begin{bmatrix} -\boldsymbol{C}_{\mathrm{G}} \\ \boldsymbol{0} \end{bmatrix} \in \mathbb{R}^{2k\times 17q} \\ \boldsymbol{D} \triangleq \begin{bmatrix} \boldsymbol{Y}_{\mathrm{GG}} - \boldsymbol{D}_{\mathrm{G}} & \boldsymbol{Y}_{\mathrm{GL}} \\ \boldsymbol{Y}_{\mathrm{LG}} & \boldsymbol{Y}_{\mathrm{LL}} \end{bmatrix} \in \mathbb{R}^{2k\times 2k} \end{cases}.$$

Note that, different from the first method, the linearized differential equations generated by the second method are the modified network equations and the total number is $l = 2k$.

6.2.5.3 Comparison

The aforementioned two methods generate identical matrices $\boldsymbol{A}$ in the linearized DAEs given by (6.43). However, in the first method, the number of linearized algebraic equations for the whole system is $6q$ more than that in the second method. Compared with the latter method, the first method has the following merits: on the one hand, the matrices $\boldsymbol{C}$ and $\boldsymbol{D}$ in the linearized algebraic equations are sparser; on the other hand, excessive equation derivations can be avoided when the input signals of the dynamic devices change. For example, if the input signal of the PSS in Fig. 6.4 is changed to the output power P_{e} of the synchronous generator, an algebraic variable ΔP_{e} can be added to $\Delta \boldsymbol{y}$ in the existing system linearized model, thus adding an algebraic equation to calculate ΔP_{e}. Then replacing the elements of the PSS state variable row and the ΔP_{e} column in the matrix $\boldsymbol{B}$ by those of the PSS state variable row and the $\Delta\omega$ column in the matrix $\bar{\boldsymbol{A}}_{\mathrm{g}}$, one can obtain a new linearized model of the system.

6.2.5.4 Sparsity of the Augmented System State Matrix

The augmented system state matrix $\boldsymbol{A}$ is a block diagonal matrix, with each diagonal sub-block corresponding to a dynamic component. The matrices $\boldsymbol{B}$, $\boldsymbol{C}$ and $\boldsymbol{D}$ are block sparse matrices. Particularly, the matrix $\boldsymbol{D}$ in the second method has the same sparse structure as the admittance matrix of the system, whose non-diagonal 2×2 sub-blocks can be obtained by augmenting the real and imaginary parts of the non-zero elements of the admittance matrix [25]. It should be noted that the sparsity of the augmented system state matrices $\boldsymbol{A}$, $\boldsymbol{B}$, $\boldsymbol{C}$ and $\boldsymbol{D}$ becomes more prominent as the system size (which is related to the number of state variables n and number of nodes k) increases, as shown in Fig. 6.6.

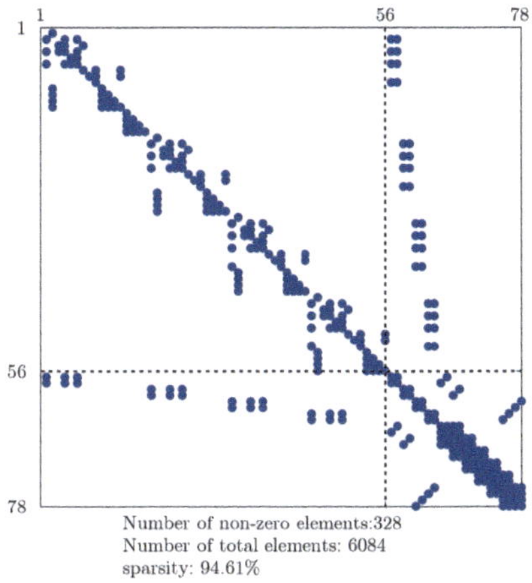

(a) the 2-area 4-generator test system

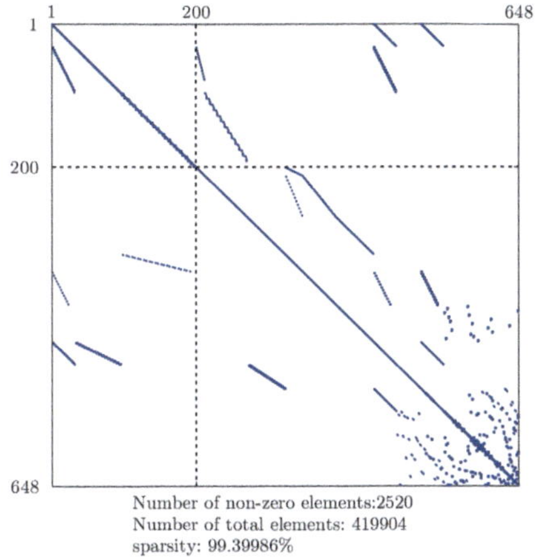

(b) the 16-generator 68-bus test system

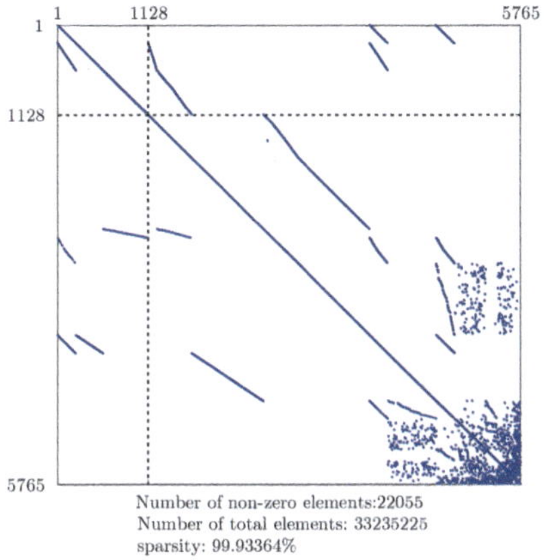

(c) a 516-bus real-life transmission system

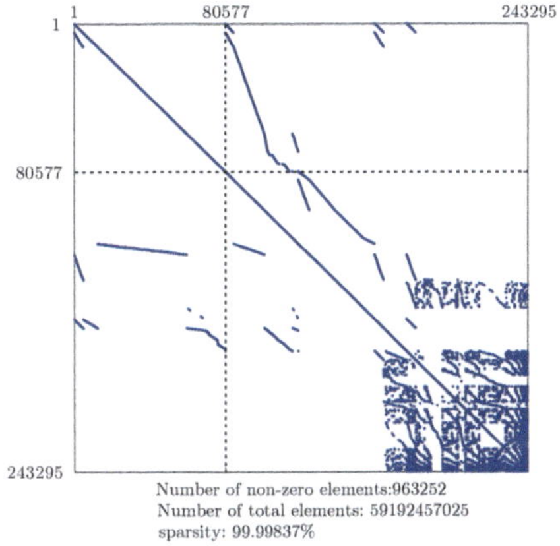

(d) a 33028-bus real-life transmission system

Fig. 6.6 Sparse structure of augmented system state matrices

6.3 Small-Signal Stability Analysis Modeling

This section elaborates how to establish the index-1 Hessenberg DDAEs and corresponding DDEs for time-delay power systems, with an emphasis on the unique property of merging all delays in a control loop as one comprehensive delay.

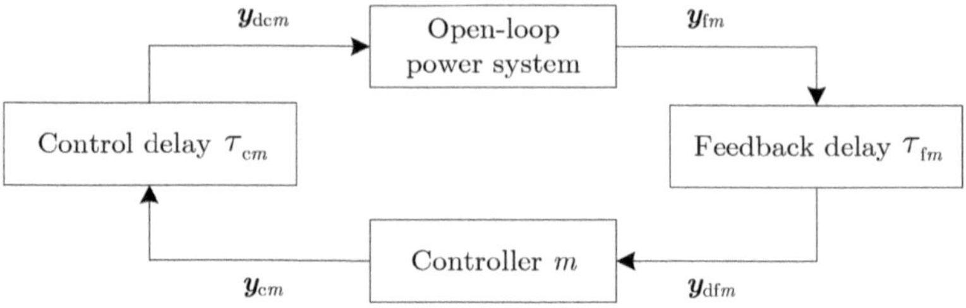

Fig. 6.7 Schematic of the time-delay power system

6.3.1 DDAEs and DDEs

Without loss of generality, it is assumed that there exist $m-1$ delays in the open-loop power system. In turn, the power system containing the m-th controller is called the closed-loop power system. As shown in Fig. 6.7, the output of the system is denoted as $\boldsymbol{y}_{\mathrm{f}m}$. $\boldsymbol{y}_{\mathrm{df}m}$ is the delayed feedback signal and used as the input of the controller. $\boldsymbol{y}_{cm}$ is the controller's output. $\boldsymbol{y}_{\mathrm{dc}m}$ is the delayed control signal and utilized as the input of the open-loop system.

6.3.1.1 The Open-Loop Power System

The DDAEs of the open-loop power system are obtained by plugging an argument $\boldsymbol{y}_{\mathrm{dc}m}$ into (2.23) and adding the corresponding algebraic equations at time $t-\tau_{\mathrm{c}m}$, that is

$$\begin{cases} \Delta\dot{\boldsymbol{x}} = \boldsymbol{A}_0\Delta\boldsymbol{x} + \boldsymbol{B}_0\Delta\boldsymbol{y} + \sum\limits_{i=1}^{m-1}\left(\boldsymbol{A}_i\Delta\boldsymbol{x}_{\mathrm{d}i} + \boldsymbol{B}_i\Delta\boldsymbol{y}_{\mathrm{d}i}\right) + \boldsymbol{E}_m\Delta\boldsymbol{y}_{\mathrm{dc}m} \\ \boldsymbol{0} = \boldsymbol{C}_0\Delta\boldsymbol{x} + \boldsymbol{D}_0\Delta\boldsymbol{y} \\ \boldsymbol{0} = \boldsymbol{C}_i\Delta\boldsymbol{x}_{\mathrm{d}i} + \boldsymbol{D}_i\Delta\boldsymbol{y}_{\mathrm{d}i},\quad i = 1,\ 2,\ \ldots,\ m-1 \\ \boldsymbol{0} = \boldsymbol{C}_{\mathrm{dc}m}\Delta\boldsymbol{x}(t-\tau_{\mathrm{c}m}) + \boldsymbol{D}_{\mathrm{dc}m}\Delta\boldsymbol{y}(t-\tau_{\mathrm{c}m}) \\ \Delta\boldsymbol{y}_{\mathrm{f}m} = \boldsymbol{K}_{1m}\Delta\boldsymbol{x} + \boldsymbol{K}_{2m}\Delta\boldsymbol{y} \\ \Delta\boldsymbol{y}_{\mathrm{df}m} = \boldsymbol{K}_{1m}\Delta\boldsymbol{x}(t-\tau_{\mathrm{f}m}) + \boldsymbol{K}_{2m}\Delta\boldsymbol{y}(t-\tau_{\mathrm{f}m}) \end{cases} \tag{6.49}$$

where $\boldsymbol{C}_{\mathrm{dc}m} = \left.\dfrac{\partial \boldsymbol{g}}{\partial \boldsymbol{x}(t-\tau_{\mathrm{c}m})}\right|_{(\bar{\boldsymbol{x}},\bar{\boldsymbol{y}})} \in \mathbb{R}^{l\times n}$ and $\boldsymbol{D}_{\mathrm{dc}m} = \left.\dfrac{\partial \boldsymbol{g}}{\partial \boldsymbol{y}(t-\tau_{\mathrm{c}m})}\right|_{(\bar{\boldsymbol{x}},\bar{\boldsymbol{y}})} \in \mathbb{R}^{l\times l}$ are Jacobian matrices, $\boldsymbol{E}_m = \left.\dfrac{\partial \boldsymbol{f}}{\partial \boldsymbol{y}_{\mathrm{dc}m}}\right|_{(\bar{\boldsymbol{x}},\bar{\boldsymbol{y}})} \in \mathbb{R}^{n\times l_c}$ is the input matrix, and $\boldsymbol{K}_{1m} = \left.\dfrac{\partial \boldsymbol{g}_{\mathrm{f}m}}{\partial \boldsymbol{x}}\right|_{(\bar{\boldsymbol{x}},\bar{\boldsymbol{y}})} \in \mathbb{R}^{f_c\times n}$ and $\boldsymbol{K}_{2m} = \left.\dfrac{\partial \boldsymbol{g}_{\mathrm{f}m}}{\partial \boldsymbol{y}}\right|_{(\bar{\boldsymbol{x}},\bar{\boldsymbol{y}})} \in \mathbb{R}^{f_c\times l}$ are the state and algebraic output matrices, respectively.

The non-zero elements in $\boldsymbol{E}_m$ characterize the connection between the controller and the controlled dynamic devices(excitation regulators, HVDC transmission systems, FACTS equipments, etc.). The rows where the non-zero elements of $\boldsymbol{E}_m$ are located correspond to the output variable of the amplifying link in the controller. The values of the non-zero elements are the ratio of the gain to the time constant of the amplifying link.

6.3.1.2 Feedback Controller

The dynamics of the m-th feedback controller and its output can be represented as

$$\begin{cases} \Delta\dot{\boldsymbol{x}}_{cm} = \boldsymbol{A}_{cm}\Delta\boldsymbol{x}_{cm} + \boldsymbol{B}_{cm}\Delta\boldsymbol{y}_{\mathrm{df}m} \\ \Delta\boldsymbol{y}_{cm} = \boldsymbol{C}_{cm}\Delta\boldsymbol{x}_{cm} + \boldsymbol{D}_{cm}\Delta\boldsymbol{y}_{\mathrm{df}m} \\ \Delta\boldsymbol{y}_{\mathrm{dc}m} = \boldsymbol{C}_{cm}\Delta\boldsymbol{x}_{cm}(t-\tau_{cm}) + \boldsymbol{D}_{cm}\Delta\boldsymbol{y}_{\mathrm{df}m}(t-\tau_{cm}) \end{cases} \tag{6.50}$$

where $\boldsymbol{x}_{cm} \in \mathbb{R}^{n_c\times 1}$ is the state variable vector of the m-th controller, $\boldsymbol{A}_{cm} = \left.\frac{\partial \boldsymbol{f}_{cm}}{\partial \boldsymbol{x}_{cm}}\right|_{(\bar{\boldsymbol{x}}_{cm},\bar{\boldsymbol{y}}_{\mathrm{df}m})}$, $\boldsymbol{B}_{cm} = \left.\frac{\partial \boldsymbol{f}_{cm}}{\partial \boldsymbol{y}_{\mathrm{df}m}}\right|_{(\bar{\boldsymbol{x}}_{cm},\bar{\boldsymbol{y}}_{\mathrm{df}m})}$, $\boldsymbol{C}_{cm} = \left.\frac{\partial \boldsymbol{g}_{cm}}{\partial \boldsymbol{x}_{cm}}\right|_{(\bar{\boldsymbol{x}}_{cm},\bar{\boldsymbol{y}}_{\mathrm{df}m})}$ and $\boldsymbol{D}_{cm} = \left.\frac{\partial \boldsymbol{g}_{cm}}{\partial \boldsymbol{y}_{\mathrm{df}m}}\right|_{(\bar{\boldsymbol{x}}_{cm},\bar{\boldsymbol{y}}_{\mathrm{df}m})}$ are the state matrix, input matrix, output matrix and feed-forward matrix with appropriate dimensions of the controller, respectively.

6.3.1.3 The Closed-Loop Power System

Combining (6.49) and (6.50) yields the DDAEs associated with the closed-loop power system, i.e.,

$$\begin{cases} \begin{aligned} \Delta\dot{\boldsymbol{x}}' = {} & \boldsymbol{A}'_0\Delta\boldsymbol{x}' + \boldsymbol{B}'_0\Delta\boldsymbol{y} + \sum_{i=1}^{m-1}(\boldsymbol{A}'_i\Delta\boldsymbol{x}'_{\mathrm{d}i} + \boldsymbol{B}'_i\Delta\boldsymbol{y}_{\mathrm{d}i}) + \boldsymbol{A}'_{\mathrm{df}m}\Delta\boldsymbol{x}'(t-\tau_{\mathrm{f}m}) \\ & + \boldsymbol{B}'_{\mathrm{df}m}\Delta\boldsymbol{y}(t-\tau_{\mathrm{f}m}) + \boldsymbol{A}'_{\mathrm{dc}m}\Delta\boldsymbol{x}'(t-\tau_{cm}) \\ & + \boldsymbol{A}'_{\mathrm{dfc}m}\Delta\boldsymbol{x}'(t-\tau_{\mathrm{f}m}-\tau_{cm}) + \boldsymbol{B}'_{\mathrm{dfc}m}\Delta\boldsymbol{y}(t-\tau_{\mathrm{f}m}-\tau_{cm}) \end{aligned} \\ \boldsymbol{0} = \boldsymbol{C}'_0\Delta\boldsymbol{x}' + \boldsymbol{D}'_0\Delta\boldsymbol{y} \\ \boldsymbol{0} = \boldsymbol{C}'_i\Delta\boldsymbol{x}'_{\mathrm{d}i} + \boldsymbol{D}'_i\Delta\boldsymbol{y}_{\mathrm{d}i},\ i = 1,\ 2,\ \ldots,\ m-1 \\ \boldsymbol{0} = \boldsymbol{C}'_{\mathrm{df}m}\Delta\boldsymbol{x}'(t-\tau_{\mathrm{f}m}) + \boldsymbol{D}'_{\mathrm{df}m}\Delta\boldsymbol{y}(t-\tau_{\mathrm{f}m}) \\ \boldsymbol{0} = \boldsymbol{C}'_{\mathrm{dc}m}\Delta\boldsymbol{x}'(t-\tau_{cm}) + \boldsymbol{D}'_{\mathrm{dc}m}\Delta\boldsymbol{y}(t-\tau_{cm}) \\ \boldsymbol{0} = \boldsymbol{C}'_{\mathrm{dfc}m}\Delta\boldsymbol{x}'(t-\tau_{\mathrm{f}m}-\tau_{cm}) + \boldsymbol{D}'_{\mathrm{dfc}m}\Delta\boldsymbol{y}(t-\tau_{\mathrm{f}m}-\tau_{cm}) \end{cases} \tag{6.51}$$

where $\boldsymbol{x}' = \left[\boldsymbol{x}^{\mathrm{T}},\ \boldsymbol{x}_{cm}^{\mathrm{T}}\right]^{\mathrm{T}} \in \mathbb{R}^{(n+n_{\mathrm{c}})\times 1}$, and coefficient matrices are given by

$$\begin{cases} \boldsymbol{A}'_0 = \left.\dfrac{\partial \boldsymbol{f}'}{\partial \boldsymbol{x}'}\right|_{(\bar{\boldsymbol{x}}',\bar{\boldsymbol{y}})} = \begin{bmatrix} \boldsymbol{A}_0 & \boldsymbol{0} \\ \boldsymbol{0} & \boldsymbol{A}_{\mathrm{c}m} \end{bmatrix}, \quad \boldsymbol{B}'_0 = \left.\dfrac{\partial \boldsymbol{f}'}{\partial \boldsymbol{y}}\right|_{(\bar{\boldsymbol{x}}',\bar{\boldsymbol{y}})} = \begin{bmatrix} \boldsymbol{B}_0 \\ \boldsymbol{0} \end{bmatrix} \\ \boldsymbol{A}'_i = \left.\dfrac{\partial \boldsymbol{f}'}{\partial \boldsymbol{x}'_{\mathrm{d}i}}\right|_{(\bar{\boldsymbol{x}}',\bar{\boldsymbol{y}})} = \begin{bmatrix} \boldsymbol{A}_i & \boldsymbol{0} \\ \boldsymbol{0} & \boldsymbol{0} \end{bmatrix} \\ \boldsymbol{B}'_i = \left.\dfrac{\partial \boldsymbol{f}'}{\partial \boldsymbol{y}_{\mathrm{d}i}}\right|_{(\bar{\boldsymbol{x}}',\bar{\boldsymbol{y}})} = \begin{bmatrix} \boldsymbol{B}_i \\ \boldsymbol{0} \end{bmatrix}, \ i = 1,\ 2,\ \ldots,\ m-1 \\ \boldsymbol{A}'_{\mathrm{df}m} = \left.\dfrac{\partial \boldsymbol{f}'}{\partial \boldsymbol{x}'(t-\tau_{\mathrm{f}m})}\right|_{(\bar{\boldsymbol{x}}',\bar{\boldsymbol{y}})} = \begin{bmatrix} \boldsymbol{0} & \boldsymbol{0} \\ \boldsymbol{B}_{\mathrm{c}m}\boldsymbol{K}_{1m} & \boldsymbol{0} \end{bmatrix} \\ \boldsymbol{B}'_{\mathrm{df}m} = \left.\dfrac{\partial \boldsymbol{f}'}{\partial \boldsymbol{y}(t-\tau_{\mathrm{f}m})}\right|_{(\bar{\boldsymbol{x}}',\bar{\boldsymbol{y}})} = \begin{bmatrix} \boldsymbol{0} \\ \boldsymbol{B}_{\mathrm{c}m}\boldsymbol{K}_{2m} \end{bmatrix} \\ \boldsymbol{A}'_{\mathrm{dc}m} = \left.\dfrac{\partial \boldsymbol{f}'}{\partial \boldsymbol{x}'(t-\tau_{\mathrm{c}m})}\right|_{(\bar{\boldsymbol{x}}',\bar{\boldsymbol{y}})} = \begin{bmatrix} \boldsymbol{0} & \boldsymbol{E}_m\boldsymbol{C}_{\mathrm{c}m} \\ \boldsymbol{0} & \boldsymbol{0} \end{bmatrix} \\ \boldsymbol{A}'_{\mathrm{dfc}m} = \left.\dfrac{\partial \boldsymbol{f}'}{\partial \boldsymbol{x}'(t-\tau_{\mathrm{f}m}-\tau_{\mathrm{c}m})}\right|_{(\bar{\boldsymbol{x}}',\bar{\boldsymbol{y}})} = \begin{bmatrix} \boldsymbol{E}_m\boldsymbol{D}_{\mathrm{c}m}\boldsymbol{K}_{1m} & \boldsymbol{0} \\ \boldsymbol{0} & \boldsymbol{0} \end{bmatrix} \\ \boldsymbol{B}'_{\mathrm{dfc}m} = \left.\dfrac{\partial \boldsymbol{f}'}{\partial \boldsymbol{y}(t-\tau_{\mathrm{f}m}-\tau_{\mathrm{c}m})}\right|_{(\bar{\boldsymbol{x}}',\bar{\boldsymbol{y}})} = \begin{bmatrix} \boldsymbol{E}_m\boldsymbol{D}_{\mathrm{c}m}\boldsymbol{K}_{2m} \\ \boldsymbol{0} \end{bmatrix} \\ \boldsymbol{C}'_i = \boldsymbol{C}'_{\mathrm{df}m} = \boldsymbol{C}'_{\mathrm{dc}m} = \boldsymbol{C}'_{\mathrm{dfc}m} = [\boldsymbol{C}_0 \ \ \boldsymbol{0}] \\ \boldsymbol{D}'_i = \boldsymbol{D}'_{\mathrm{df}m} = \boldsymbol{D}'_{\mathrm{dc}m} = \boldsymbol{D}'_{\mathrm{dfc}m} = \boldsymbol{D}_0,\ i = 0,\ 1,\ \ldots,\ m-1 \end{cases}$$

with $\boldsymbol{C}'_0 = \left.\dfrac{\partial \boldsymbol{g}}{\partial \boldsymbol{x}'}\right|_{(\bar{\boldsymbol{x}}',\bar{\boldsymbol{y}})}$, $\boldsymbol{D}'_0 = \left.\dfrac{\partial \boldsymbol{g}}{\partial \boldsymbol{y}}\right|_{(\bar{\boldsymbol{x}}',\bar{\boldsymbol{y}})}$; $\boldsymbol{C}'_i = \left.\dfrac{\partial \boldsymbol{g}}{\partial \boldsymbol{x}'_{\mathrm{d}i}}\right|_{(\bar{\boldsymbol{x}}',\bar{\boldsymbol{y}})}$, $\boldsymbol{D}'_i = \left.\dfrac{\partial \boldsymbol{g}}{\partial \boldsymbol{y}_{\mathrm{d}i}}\right|_{(\bar{\boldsymbol{x}}',\bar{\boldsymbol{y}})}$, $i = 1,\ 2,\ \ldots,\ m-1$; $\boldsymbol{C}'_{\mathrm{df}m} = \left.\dfrac{\partial \boldsymbol{g}}{\partial \boldsymbol{x}'(t-\tau_{\mathrm{f}m})}\right|_{(\bar{\boldsymbol{x}}',\bar{\boldsymbol{y}})}$, $\boldsymbol{D}'_{\mathrm{df}m} = \left.\dfrac{\partial \boldsymbol{g}}{\partial \boldsymbol{y}(t-\tau_{\mathrm{f}m})}\right|_{(\bar{\boldsymbol{x}}',\bar{\boldsymbol{y}})}$, $\boldsymbol{C}'_{\mathrm{dc}m} = \left.\dfrac{\partial \boldsymbol{g}}{\partial \boldsymbol{x}'(t-\tau_{\mathrm{c}m})}\right|_{(\bar{\boldsymbol{x}}',\bar{\boldsymbol{y}})}$, $\boldsymbol{D}'_{\mathrm{dc}m} = \left.\dfrac{\partial \boldsymbol{g}}{\partial \boldsymbol{y}(t-\tau_{\mathrm{c}m})}\right|_{(\bar{\boldsymbol{x}}',\bar{\boldsymbol{y}})}$, $\boldsymbol{C}'_{\mathrm{dfc}m} = \left.\dfrac{\partial \boldsymbol{g}}{\partial \boldsymbol{x}'(t-\tau_{\mathrm{f}m}-\tau_{\mathrm{c}m})}\right|_{(\bar{\boldsymbol{x}}',\bar{\boldsymbol{y}})}$ and $\boldsymbol{D}'_{\mathrm{dfc}m} = \left.\dfrac{\partial \boldsymbol{g}}{\partial \boldsymbol{y}(t-\tau_{\mathrm{f}m}-\tau_{\mathrm{c}m})}\right|_{(\bar{\boldsymbol{x}}',\bar{\boldsymbol{y}})}$.

Let $\Delta\hat{\boldsymbol{x}} = [(\Delta\boldsymbol{x}')^{\mathrm{T}},\ \Delta\boldsymbol{y}^{\mathrm{T}}]^{\mathrm{T}} \in \mathbb{R}^{(n+n_{\mathrm{c}}+l)\times 1}$ be the augmented state variable of the system, then (6.51) can be represented as

$$\begin{aligned} \boldsymbol{E}\Delta\dot{\hat{\boldsymbol{x}}} = {} & \boldsymbol{J}_0\Delta\hat{\boldsymbol{x}} + \sum_{i=1}^{m-1}\boldsymbol{J}_i\Delta\hat{\boldsymbol{x}}_{\mathrm{d}i} + \boldsymbol{J}_{\mathrm{df}m}\Delta\hat{\boldsymbol{x}}(t-\tau_{\mathrm{f}m}) + \boldsymbol{J}_{\mathrm{dc}m}\Delta\hat{\boldsymbol{x}}(t-\tau_{\mathrm{c}m}) \\ & + \boldsymbol{J}_{\mathrm{dfc}m}\Delta\hat{\boldsymbol{x}}(t-\tau_{\mathrm{fc}m}) \end{aligned} \tag{6.52}$$

where

$$\begin{cases} \boldsymbol{E} = \begin{bmatrix} \boldsymbol{I}_{n+n_c} & \boldsymbol{0} \\ \boldsymbol{0} & \boldsymbol{I}_l \end{bmatrix}, \ \boldsymbol{J}_0 = \begin{bmatrix} \boldsymbol{A}'_0 & \boldsymbol{B}'_0 \\ \boldsymbol{C}'_0 & \boldsymbol{D}_0 \end{bmatrix} \\ \boldsymbol{J}_i = \begin{bmatrix} \boldsymbol{A}'_i & \boldsymbol{B}'_i \\ \boldsymbol{0} & \boldsymbol{0} \end{bmatrix}, \ i = 1, 2, \ldots, m-1 \\ \boldsymbol{J}_{\mathrm{df}m} = \begin{bmatrix} \boldsymbol{A}'_{\mathrm{df}m} & \boldsymbol{B}'_{\mathrm{df}m} \\ \boldsymbol{0} & \boldsymbol{0} \end{bmatrix}, \ \boldsymbol{J}_{\mathrm{dc}m} = \begin{bmatrix} \boldsymbol{A}'_{\mathrm{dc}m} & \boldsymbol{B}'_{\mathrm{dc}m} \\ \boldsymbol{0} & \boldsymbol{0} \end{bmatrix} \\ \boldsymbol{J}_{\mathrm{dfc}m} = \begin{bmatrix} \boldsymbol{A}'_{\mathrm{dfc}m} & \boldsymbol{B}'_{\mathrm{dfc}m} \\ \boldsymbol{0} & \boldsymbol{0} \end{bmatrix} \end{cases}.$$

By eliminating the algebraic variables in (6.51) and (6.52), DDAEs can be transformed into the following DDEs:

$$\begin{aligned} \Delta\dot{\boldsymbol{x}}' = & \tilde{\boldsymbol{A}}_0 \Delta\boldsymbol{x}' + \sum_{i=1}^{m-1} \tilde{\boldsymbol{A}}_i \Delta\boldsymbol{x}'_{\mathrm{d}i} + \tilde{\boldsymbol{A}}_{\mathrm{df}m} \Delta\boldsymbol{x}'(t - \tau_{\mathrm{f}m}) \\ & + \tilde{\boldsymbol{A}}_{\mathrm{dc}m} \Delta\boldsymbol{x}'(t - \tau_{\mathrm{c}m}) + \tilde{\boldsymbol{A}}_{\mathrm{dfc}m} \Delta\boldsymbol{x}'(t - \tau_{\mathrm{f}m} - \tau_{\mathrm{c}m}) \end{aligned} \quad (6.53)$$

where

$$\begin{cases} \boldsymbol{K}_0 = -\left(\boldsymbol{D}'_0\right)^{-1} \boldsymbol{C}'_0 = \left[-\boldsymbol{D}_0^{-1}\boldsymbol{C}_0 \ \ \boldsymbol{0}\right] \\ \tilde{\boldsymbol{A}}_0 = \boldsymbol{A}'_0 + \boldsymbol{B}'_0 \boldsymbol{K}_0 = \begin{bmatrix} \boldsymbol{A}_0 - \boldsymbol{B}_0 \boldsymbol{D}_0^{-1} \boldsymbol{C}_0 & \boldsymbol{0} \\ \boldsymbol{0} & \boldsymbol{A}_{\mathrm{c}m} \end{bmatrix} \\ \tilde{\boldsymbol{A}}_i = \boldsymbol{A}'_i + \boldsymbol{B}'_i \boldsymbol{K}_0 = \begin{bmatrix} \boldsymbol{A}_i - \boldsymbol{B}_i \boldsymbol{D}_0^{-1} \boldsymbol{C}_0 & \boldsymbol{0} \\ \boldsymbol{0} & \boldsymbol{0} \end{bmatrix}, \ i = 1, 2, \ldots, m-1 \\ \tilde{\boldsymbol{A}}_{\mathrm{df}m} = \boldsymbol{A}'_{\mathrm{df}m} + \boldsymbol{B}'_{\mathrm{df}m} \boldsymbol{K}_0 = \begin{bmatrix} \boldsymbol{0} & \boldsymbol{0} \\ \boldsymbol{B}_{\mathrm{c}m} \left(\boldsymbol{K}_{1m} - \boldsymbol{K}_{2m} \boldsymbol{D}_0^{-1} \boldsymbol{C}_0\right) & \boldsymbol{0} \end{bmatrix} \\ \tilde{\boldsymbol{A}}_{\mathrm{dc}m} = \boldsymbol{A}'_{\mathrm{dc}m} + \boldsymbol{B}'_{\mathrm{dc}m} \boldsymbol{K}_0 = \begin{bmatrix} \boldsymbol{0} & \boldsymbol{E}_m \boldsymbol{C}_{\mathrm{c}m} \\ \boldsymbol{0} & \boldsymbol{0} \end{bmatrix} \\ \tilde{\boldsymbol{A}}_{\mathrm{dfc}m} = \boldsymbol{A}'_{\mathrm{dfc}m} + \boldsymbol{B}'_{\mathrm{dfc}m} \boldsymbol{K}_0 = \begin{bmatrix} \boldsymbol{E}_m \boldsymbol{D}_{\mathrm{c}m} \left(\boldsymbol{K}_{1m} - \boldsymbol{K}_{2m} \boldsymbol{D}_0^{-1} \boldsymbol{C}_0\right) & \boldsymbol{0} \\ \boldsymbol{0} & \boldsymbol{0} \end{bmatrix} \end{cases}. \quad (6.54)$$

6.3.2 Delay Merging Property

The stability analysis model of the closed-loop power system has the delay merging property of the control loop described in Theorem 6.1, which facilitates modeling and formulating the system.

Theorem 6.1 *The feedback delay $\tau_{\mathrm{f}m}$ and control delay $\tau_{\mathrm{c}m}$ in the m-th control loop can be merged into an integrated delay τ_m, i.e., $\tau_m = \tau_{\mathrm{f}m} + \tau_{\mathrm{c}m}$.*

Proof The characteristic polynomial of (6.53) is

$$
\begin{aligned}
&|\Delta(\lambda)| \\
&= \left| \lambda \boldsymbol{I}_{n+n_c} - \tilde{\boldsymbol{A}}_0 - \sum_{i=1}^{m-1} \tilde{\boldsymbol{A}}_i \mathrm{e}^{-\lambda\tau_i} - \tilde{\boldsymbol{A}}_{\mathrm{df}m} \mathrm{e}^{-\lambda\tau_{\mathrm{f}m}} - \tilde{\boldsymbol{A}}_{\mathrm{dc}m} \mathrm{e}^{-\lambda\tau_{\mathrm{c}m}} - \tilde{\boldsymbol{A}}_{\mathrm{dfc}m} \mathrm{e}^{-\lambda(\tau_{\mathrm{f}m}+\tau_{\mathrm{c}m})} \right| \\
&= \left[\begin{array}{c:c} \begin{array}{c} \lambda \boldsymbol{I}_n - \left(\boldsymbol{A}_0 - \boldsymbol{B}_0 \boldsymbol{D}_0^{-1} \boldsymbol{C}_0\right) - \sum_{i=1}^{m-1} \left(\boldsymbol{A}_i - \boldsymbol{B}_i \boldsymbol{D}_0^{-1} \boldsymbol{C}_0\right) \mathrm{e}^{-\lambda\tau_i} \\ -\boldsymbol{E}_m \boldsymbol{D}_{\mathrm{c}m} \left(\boldsymbol{K}_{1m} - \boldsymbol{K}_{2m} \boldsymbol{D}_0^{-1} \boldsymbol{C}_0\right) \mathrm{e}^{-\lambda(\tau_{\mathrm{f}m}+\tau_{\mathrm{c}m})} \end{array} & -\boldsymbol{E}_m \boldsymbol{C}_{\mathrm{c}m} \mathrm{e}^{-\lambda\tau_{\mathrm{c}m}} \\ \hdashline -\boldsymbol{B}_{\mathrm{c}m} \left(\boldsymbol{K}_{1m} - \boldsymbol{K}_{2m} \boldsymbol{D}_0^{-1} \boldsymbol{C}_0\right) \mathrm{e}^{-\lambda\tau_{\mathrm{f}m}} & \lambda \boldsymbol{I}_{n_c} - \boldsymbol{A}_{\mathrm{c}m} \end{array} \right].
\end{aligned}
\tag{6.55}
$$

By applying the Schur's determinant formula [26] for the partitioned block matrix, (6.55) can be recast into

$$
\begin{aligned}
&|\Delta(\lambda)| \\
&= \left| \lambda \boldsymbol{I}_{n_c} - \boldsymbol{A}_{\mathrm{c}m} \right| \times \\
&\left| \begin{array}{c} \lambda \boldsymbol{I}_n - \left(\boldsymbol{A}_0 - \boldsymbol{B}_0 \boldsymbol{D}_0^{-1} \boldsymbol{C}_0\right) - \sum_{i=1}^{m-1} \left(\boldsymbol{A}_i - \boldsymbol{B}_i \boldsymbol{D}_0^{-1} \boldsymbol{C}_0\right) \mathrm{e}^{-\lambda\tau_i} \\ -\boldsymbol{E}_m \left(\boldsymbol{C}_{\mathrm{c}m} \left(\lambda \boldsymbol{I}_{n_c} - \boldsymbol{A}_{\mathrm{c}m}\right)^{-1} \boldsymbol{B}_{\mathrm{c}m} + \boldsymbol{D}_{\mathrm{c}m}\right) \left(\boldsymbol{K}_{1m} - \boldsymbol{K}_{2m} \boldsymbol{D}_0^{-1} \boldsymbol{C}_0\right) \mathrm{e}^{-\lambda(\tau_{\mathrm{f}m}+\tau_{\mathrm{c}m})} \end{array} \right|.
\end{aligned}
\tag{6.56}
$$

The entry $\mathrm{e}^{-\lambda(\tau_{\mathrm{f}m}+\tau_{\mathrm{c}m})}$ in the second determinant on the right-hand side of (6.56) explicitly shows the impact of time delays emerging in the m-th control loop on system's characteristic equation. In other words, for any feedback delay $\tau_{\mathrm{f}m} \leqslant 0$ and control delay $\tau_{\mathrm{c}m} \leqslant 0$, if $\tau_m = \tau_{\mathrm{f}m} + \tau_{\mathrm{c}m}$ is fixed, the characteristic equation of the closed-loop power system will keep unchanged. Hence, the proof of Theorem 6.1 is therefore complete. □

In the following, Theorems 6.2 and 6.3 below reveal the properties of eigenvalue sensitivities with respect to feedback delay and control delay in the m-th control loop, which strongly corroborate Theorem 6.1.

Theorem 6.2 *The sensitivities of the eigenvalue λ with respect to the feedback delay $\tau_{\mathrm{f}m}$ and the control delay $\tau_{\mathrm{c}m}$ in the same control loop are equal, i.e.,*

$$
\frac{\partial \lambda}{\partial \tau_{\mathrm{f}m}} = \frac{\partial \lambda}{\partial \tau_{\mathrm{c}m}}. \tag{6.57}
$$

Proof The characteristic equations of the system are

$$
\left(\tilde{\boldsymbol{A}}_0 + \sum_{i=1}^{m-1} \tilde{\boldsymbol{A}}_i \mathrm{e}^{-\lambda\tau_i} + \tilde{\boldsymbol{A}}_{\mathrm{df}m} \mathrm{e}^{-\lambda\tau_{\mathrm{f}m}} + \tilde{\boldsymbol{A}}_{\mathrm{dc}m} \mathrm{e}^{-\lambda\tau_{\mathrm{c}m}} + \tilde{\boldsymbol{A}}_{\mathrm{dfc}m} \mathrm{e}^{-\lambda\tau_m} \right) \boldsymbol{v} = \lambda \boldsymbol{v} \tag{6.58}
$$

and

$$\boldsymbol{u}^{\mathrm{T}}\left(\tilde{\boldsymbol{A}}_0+\sum_{i=1}^{m-1}\tilde{\boldsymbol{A}}_i\mathrm{e}^{-\lambda\tau_i}+\tilde{\boldsymbol{A}}_{\mathrm{df}m}\mathrm{e}^{-\lambda\tau_{\mathrm{f}m}}+\tilde{\boldsymbol{A}}_{\mathrm{dc}m}\mathrm{e}^{-\lambda\tau_{\mathrm{c}m}}+\tilde{\boldsymbol{A}}_{\mathrm{dfc}m}\mathrm{e}^{-\lambda\tau_m}\right)=\lambda\boldsymbol{u}^{\mathrm{T}}. \tag{6.59}$$

By taking the derivative of (6.58) to $\tau_{\mathrm{f}m}$ and $\tau_{\mathrm{c}m}$, respectively, and then pre-multiplying it with $\boldsymbol{u}^{\mathrm{T}}$, one can obtain

$$\frac{\partial\lambda}{\partial\tau_{\mathrm{f}m}}=\frac{N_1}{D},\ \frac{\partial\lambda}{\partial\tau_{\mathrm{c}m}}=\frac{N_2}{D} \tag{6.60}$$

where

$$N_1=-\lambda\boldsymbol{u}^{\mathrm{T}}\left(\tilde{\boldsymbol{A}}_{\mathrm{df}m}\mathrm{e}^{-\lambda\tau_{\mathrm{f}m}}+\tilde{\boldsymbol{A}}_{\mathrm{dfc}m}\mathrm{e}^{-\lambda\tau_m}\right)\boldsymbol{v} \tag{6.61}$$

$$N_2=-\lambda\boldsymbol{u}^{\mathrm{T}}\left(\tilde{\boldsymbol{A}}_{\mathrm{dc}m}\mathrm{e}^{-\lambda\tau_{\mathrm{c}m}}+\tilde{\boldsymbol{A}}_{\mathrm{dfc}m}\mathrm{e}^{-\lambda\tau_m}\right)\boldsymbol{v} \tag{6.62}$$

$$\begin{aligned}D=\boldsymbol{u}^{\mathrm{T}}\boldsymbol{v}+\boldsymbol{u}^{\mathrm{T}}\Bigg(&\sum_{i=1}^{m-1}\tau_i\tilde{\boldsymbol{A}}_i\mathrm{e}^{-\lambda\tau_i}+\tau_{\mathrm{f}m}\tilde{\boldsymbol{A}}_{\mathrm{df}m}\mathrm{e}^{-\lambda\tau_{\mathrm{f}m}}+\tau_{\mathrm{c}m}\tilde{\boldsymbol{A}}_{\mathrm{dc}m}\mathrm{e}^{-\lambda\tau_{\mathrm{c}m}}\\&+\tau_m\tilde{\boldsymbol{A}}_{\mathrm{dfc}m}\mathrm{e}^{-\lambda\tau_m}\Bigg)\boldsymbol{v}.\end{aligned} \tag{6.63}$$

The proof of (6.57) is thus reduced to that of the equivalence between the numerators N_1 and N_2. First, $\boldsymbol{u}$ and $\boldsymbol{v}$ are partitioned into

$$\boldsymbol{u}=\left[\boldsymbol{u}_1^{\mathrm{T}}\ \boldsymbol{u}_2^{\mathrm{T}}\right]^{\mathrm{T}},\ \boldsymbol{v}=\left[\boldsymbol{v}_1^{\mathrm{T}}\ \boldsymbol{v}_2^{\mathrm{T}}\right]^{\mathrm{T}} \tag{6.64}$$

with $\boldsymbol{u}_1, \boldsymbol{v}_1 \in \mathbb{C}^{(n-n_{\mathrm{c}})\times 1}$, and $\boldsymbol{u}_2, \boldsymbol{v}_2 \in \mathbb{C}^{n_{\mathrm{c}}\times 1}$. Then, inserting (6.64) into (6.58) and (6.59) yields

$$\boldsymbol{B}_{\mathrm{c}m}\left(\boldsymbol{K}_{1m}-\boldsymbol{K}_{2m}\boldsymbol{D}_0^{-1}\boldsymbol{C}_0\right)\boldsymbol{v}_1\mathrm{e}^{-\lambda\tau_{\mathrm{f}m}}+\boldsymbol{A}_{\mathrm{c}m}\boldsymbol{v}_2=\lambda\boldsymbol{v}_2 \tag{6.65}$$

and

$$\boldsymbol{u}_1^{\mathrm{T}}\boldsymbol{E}_m\boldsymbol{C}_{\mathrm{c}m}\mathrm{e}^{-\lambda\tau_{\mathrm{c}m}}+\boldsymbol{u}_2^{\mathrm{T}}\boldsymbol{A}_{\mathrm{c}m}=\lambda\boldsymbol{u}_2^{\mathrm{T}}. \tag{6.66}$$

Next, by substituting (6.64) into (6.61) and (6.62) and then by taking (6.65) and (6.66) into account, one can obtain

$$\begin{aligned}
N_1 =& -\lambda\left[\boldsymbol{u}_1^{\mathrm{T}}\ \boldsymbol{u}_2^{\mathrm{T}}\right]\begin{bmatrix}\mathbf{0} & \mathbf{0}\\ \boldsymbol{B}_{cm}\left(\boldsymbol{K}_{1m}-\boldsymbol{K}_{2m}\boldsymbol{D}_0^{-1}\boldsymbol{C}_0\right)\mathrm{e}^{-\lambda\tau_{\mathrm{f}m}} & \mathbf{0}\end{bmatrix}\begin{bmatrix}\boldsymbol{v}_1\\ \boldsymbol{v}_2\end{bmatrix}\\
&-\lambda\left[\boldsymbol{u}_1^{\mathrm{T}}\ \boldsymbol{u}_2^{\mathrm{T}}\right]\begin{bmatrix}\boldsymbol{E}_m\boldsymbol{D}_{cm}\left(\boldsymbol{K}_{1m}-\boldsymbol{K}_{2m}\boldsymbol{D}_0^{-1}\boldsymbol{C}_0\right)\mathrm{e}^{-\lambda\tau_m} & \mathbf{0}\\ \mathbf{0} & \mathbf{0}\end{bmatrix}\begin{bmatrix}\boldsymbol{v}_1\\ \boldsymbol{v}_2\end{bmatrix}\\
=& -\lambda\boldsymbol{u}_2^{\mathrm{T}}\boldsymbol{B}_{cm}\left(\boldsymbol{K}_{1m}-\boldsymbol{K}_{2m}\boldsymbol{D}_0^{-1}\boldsymbol{C}_0\right)\mathrm{e}^{-\lambda\tau_{\mathrm{f}m}}\boldsymbol{v}_1\\
&-\lambda\boldsymbol{u}_1^{\mathrm{T}}\boldsymbol{E}_m\boldsymbol{D}_{cm}\left(\boldsymbol{K}_{1m}-\boldsymbol{K}_{2m}\boldsymbol{D}_0^{-1}\boldsymbol{C}_0\right)\mathrm{e}^{-\lambda\tau_m}\boldsymbol{v}_1\\
=& -\lambda\boldsymbol{u}_2^{\mathrm{T}}(\lambda\boldsymbol{I}_{n_{\mathrm{c}}}-\boldsymbol{A}_{cm})\boldsymbol{v}_2-\lambda\boldsymbol{u}_1^{\mathrm{T}}\boldsymbol{E}_m\boldsymbol{D}_{cm}\left(\boldsymbol{K}_{1m}-\boldsymbol{K}_{2m}\boldsymbol{D}_0^{-1}\boldsymbol{C}_0\right)\mathrm{e}^{-\lambda\tau_m}\boldsymbol{v}_1
\end{aligned}$$

and

$$\begin{aligned}
N_2 =& -\lambda\left[\boldsymbol{u}_1^{\mathrm{T}}\ \boldsymbol{u}_2^{\mathrm{T}}\right]\begin{bmatrix}\mathbf{0} & \boldsymbol{E}_m\boldsymbol{C}_{cm}\mathrm{e}^{-\lambda\tau_{cm}}\\ \mathbf{0} & \mathbf{0}\end{bmatrix}\begin{bmatrix}\boldsymbol{v}_1\\ \boldsymbol{v}_2\end{bmatrix}\\
&-\lambda\left[\boldsymbol{u}_1^{\mathrm{T}}\ \boldsymbol{u}_2^{\mathrm{T}}\right]\begin{bmatrix}\boldsymbol{E}_m\boldsymbol{D}_{cm}\left(\boldsymbol{K}_{1m}-\boldsymbol{K}_{2m}\boldsymbol{D}_0^{-1}\boldsymbol{C}_0\right)\mathrm{e}^{-\lambda\tau_m} & \mathbf{0}\\ \mathbf{0} & \mathbf{0}\end{bmatrix}\begin{bmatrix}\boldsymbol{v}_1\\ \boldsymbol{v}_2\end{bmatrix}\\
=& -\lambda\boldsymbol{u}_1^{\mathrm{T}}\boldsymbol{E}_m\boldsymbol{C}_{cm}\mathrm{e}^{-\lambda\tau_{cm}}\boldsymbol{v}_2-\lambda\boldsymbol{u}_1^{\mathrm{T}}\boldsymbol{E}_m\boldsymbol{D}_{cm}\left(\boldsymbol{K}_{1m}-\boldsymbol{K}_{2m}\boldsymbol{D}_0^{-1}\boldsymbol{C}_0\right)\mathrm{e}^{-\lambda\tau_m}\boldsymbol{v}_1\\
=& -\lambda\boldsymbol{u}_2^{\mathrm{T}}(\lambda\boldsymbol{I}_{n_{\mathrm{c}}}-\boldsymbol{A}_{cm})\boldsymbol{v}_2-\lambda\boldsymbol{u}_1^{\mathrm{T}}\boldsymbol{E}_m\boldsymbol{D}_{cm}\left(\boldsymbol{K}_{1m}-\boldsymbol{K}_{2m}\boldsymbol{D}_0^{-1}\boldsymbol{C}_0\right)\mathrm{e}^{-\lambda\tau_m}\boldsymbol{v}_1.
\end{aligned}$$

Obviously, N_1 and N_2 are equal. This completes the proof of Theorem 6.2. □

By substituting (6.64) into (6.63) and taking (6.65) and (6.66) into account, D can be further rewritten as

$$\begin{aligned}
D =& \left[\boldsymbol{u}_1^{\mathrm{T}}\ \boldsymbol{u}_2^{\mathrm{T}}\right]\begin{bmatrix}\boldsymbol{v}_1\\ \boldsymbol{v}_2\end{bmatrix}+\left[\boldsymbol{u}_1^{\mathrm{T}}\ \boldsymbol{u}_2^{\mathrm{T}}\right]\sum_{i=1}^{m-1}\begin{bmatrix}\left(\boldsymbol{A}_i-\boldsymbol{B}_i\boldsymbol{D}_0^{-1}\boldsymbol{C}_0\right)\tau_i\mathrm{e}^{-\lambda\tau_i} & \mathbf{0}\\ \mathbf{0} & \mathbf{0}\end{bmatrix}\begin{bmatrix}\boldsymbol{v}_1\\ \boldsymbol{v}_2\end{bmatrix}\\
&+\left[\boldsymbol{u}_1^{\mathrm{T}}\ \boldsymbol{u}_2^{\mathrm{T}}\right]\begin{bmatrix}\mathbf{0} & \mathbf{0}\\ \boldsymbol{B}_{cm}\left(\boldsymbol{K}_{1m}-\boldsymbol{K}_{2m}\boldsymbol{D}_0^{-1}\boldsymbol{C}_0\right)\tau_{\mathrm{f}m}\mathrm{e}^{-\lambda\tau_{\mathrm{f}m}} & \mathbf{0}\end{bmatrix}\begin{bmatrix}\boldsymbol{v}_1\\ \boldsymbol{v}_2\end{bmatrix}\\
&+\left[\boldsymbol{u}_1^{\mathrm{T}}\ \boldsymbol{u}_2^{\mathrm{T}}\right]\begin{bmatrix}\mathbf{0} & \boldsymbol{E}_m\boldsymbol{C}_{cm}\tau_{cm}\mathrm{e}^{-\lambda\tau_{cm}}\\ \mathbf{0} & \mathbf{0}\end{bmatrix}\begin{bmatrix}\boldsymbol{v}_1\\ \boldsymbol{v}_2\end{bmatrix}\\
&+\left[\boldsymbol{u}_1^{\mathrm{T}}\ \boldsymbol{u}_2^{\mathrm{T}}\right]\\
&\times\begin{bmatrix}\boldsymbol{E}_m\boldsymbol{D}_{cm}\left(\boldsymbol{K}_{1m}-\boldsymbol{K}_{2m}\boldsymbol{D}_0^{-1}\boldsymbol{C}_0\right)\tau_m\mathrm{e}^{-\lambda\tau_m} & \mathbf{0}\\ \mathbf{0} & \mathbf{0}\end{bmatrix}\begin{bmatrix}\boldsymbol{v}_1\\ \boldsymbol{v}_2\end{bmatrix}\\
=&\boldsymbol{u}_1^{\mathrm{T}}\boldsymbol{v}_1+\boldsymbol{u}_2^{\mathrm{T}}\boldsymbol{v}_2+\boldsymbol{u}_1^{\mathrm{T}}\sum_{i=1}^{m-1}\left(\boldsymbol{A}_i-\boldsymbol{B}_i\boldsymbol{D}_0^{-1}\boldsymbol{C}_0\right)\tau_i\mathrm{e}^{-\lambda\tau_i}\boldsymbol{v}_1\\
&+\boldsymbol{u}_2^{\mathrm{T}}\boldsymbol{B}_{cm}\left(\boldsymbol{K}_{1m}-\boldsymbol{K}_{2m}\boldsymbol{D}_0^{-1}\boldsymbol{C}_0\right)\tau_{\mathrm{f}m}\mathrm{e}^{-\lambda\tau_{\mathrm{f}m}}\boldsymbol{v}_1+\boldsymbol{u}_1^{\mathrm{T}}\boldsymbol{E}_m\boldsymbol{C}_{cm}\tau_{cm}\mathrm{e}^{-\lambda\tau_{cm}}\boldsymbol{v}_2\\
&+\boldsymbol{u}_1^{\mathrm{T}}\boldsymbol{E}_m\boldsymbol{D}_{cm}\left(\boldsymbol{K}_{1m}-\boldsymbol{K}_{2m}\boldsymbol{D}_0^{-1}\boldsymbol{C}_0\right)\tau_m\mathrm{e}^{-\lambda\tau_m}\boldsymbol{v}_1\\
=&\boldsymbol{u}_1^{\mathrm{T}}\boldsymbol{v}_1+\boldsymbol{u}_2^{\mathrm{T}}\boldsymbol{v}_2+\boldsymbol{u}_1^{\mathrm{T}}\sum_{i=1}^{m-1}\left(\boldsymbol{A}_i-\boldsymbol{B}_i\boldsymbol{D}_0^{-1}\boldsymbol{C}_0\right)\tau_i\mathrm{e}^{-\lambda\tau_i}\boldsymbol{v}_1\\
&+\boldsymbol{u}_2^{\mathrm{T}}(\lambda\boldsymbol{I}_{n_{\mathrm{c}}}-\boldsymbol{A}_{cm})\tau_m\boldsymbol{v}_2+\boldsymbol{u}_1^{\mathrm{T}}\boldsymbol{E}_m\boldsymbol{D}_{cm}\left(\boldsymbol{K}_{1m}-\boldsymbol{K}_{2m}\boldsymbol{D}_0^{-1}\boldsymbol{C}_0\right)\tau_m\mathrm{e}^{-\lambda\tau_m}\boldsymbol{v}_1.
\end{aligned}$$

Theorem 6.3 *Let the additive delay* $\tau_m = \tau_{\mathrm{f}m} + \tau_{\mathrm{c}m}$. *For any* $\tau'_{\mathrm{f}m} \geqslant 0$ *and* $\tau'_{\mathrm{c}m} \geqslant 0$, *if* $\tau'_{\mathrm{f}m} + \tau'_{\mathrm{c}m} = \tau_m$ *is satisfied, the sensitivities of* λ *with respect to* $\tau'_{\mathrm{f}m}$, $\tau'_{\mathrm{c}m}$ *and* τ_m *in the* m*-th control loop are equal, i.e.,*

$$\frac{\partial\lambda}{\partial\tau'_{\mathrm{f}m}} = \frac{\partial\lambda}{\partial\tau'_{\mathrm{c}m}} = \frac{\partial\lambda}{\partial\tau_m}. \tag{6.67}$$

Proof According to the delay merging theorem, the eigenvalue λ is kept for different $\tau'_{\mathrm{f}m}$ and $\tau'_{\mathrm{c}m} = \tau_m - \tau'_{\mathrm{f}m}$. Supposed that the corresponding left and right eigenvector be $\boldsymbol{u}'$ and $\boldsymbol{v}'$, then characteristic equations (6.58) and (6.59) become

$$(\boldsymbol{u}')^{\mathrm{T}}\left(\tilde{\boldsymbol{A}}_0 + \sum_{i=1}^{m-1}\tilde{\boldsymbol{A}}_i \mathrm{e}^{-\lambda\tau_i} + \tilde{\boldsymbol{A}}_{\mathrm{df}m}\mathrm{e}^{-\lambda\tau'_{\mathrm{f}m}} + \tilde{\boldsymbol{A}}_{\mathrm{dc}m}\mathrm{e}^{-\lambda\tau'_{\mathrm{c}m}} + \tilde{\boldsymbol{A}}_{\mathrm{dfc}m}\mathrm{e}^{-\lambda\tau_m}\right) = \lambda(\boldsymbol{u}')^{\mathrm{T}} \tag{6.68}$$

and

$$\left(\tilde{\boldsymbol{A}}_0 + \sum_{i=1}^{m-1}\tilde{\boldsymbol{A}}_i \mathrm{e}^{-\lambda\tau_i} + \tilde{\boldsymbol{A}}_{\mathrm{df}m}\mathrm{e}^{-\lambda\tau'_{\mathrm{f}m}} + \tilde{\boldsymbol{A}}_{\mathrm{dc}m}\mathrm{e}^{-\lambda\tau'_{\mathrm{c}m}} + \tilde{\boldsymbol{A}}_{\mathrm{dfc}m}\mathrm{e}^{-\lambda\tau_m}\right)\boldsymbol{v}' = \lambda\boldsymbol{v}'. \tag{6.69}$$

The sensitivity of λ with respect to $\tau'_{\mathrm{c}m}$ is similar to (6.60) and expressed as

$$\frac{\partial\lambda}{\partial\tau'_{\mathrm{c}m}} = \frac{N'_2}{D'}$$

where

$$N'_2 = -\lambda(\boldsymbol{u}')^{\mathrm{T}}\left(\tilde{\boldsymbol{A}}_{\mathrm{dc}m}\mathrm{e}^{-\lambda\tau'_{\mathrm{c}m}} + \tilde{\boldsymbol{A}}_{\mathrm{dfc}m}\mathrm{e}^{-\lambda\tau_m}\right)\boldsymbol{v}' \tag{6.70}$$

and

$$D' = (\boldsymbol{u}')^{\mathrm{T}}\boldsymbol{v}' + (\boldsymbol{u}')^{\mathrm{T}}\left(\sum_{i=1}^{m-1}\tau_i\tilde{\boldsymbol{A}}_i\mathrm{e}^{-\lambda\tau_i} + \tau'_{\mathrm{f}m}\tilde{\boldsymbol{A}}_{\mathrm{df}m}\mathrm{e}^{-\lambda\tau'_{\mathrm{f}m}} + \tau'_{\mathrm{c}m}\tilde{\boldsymbol{A}}_{\mathrm{dc}m}\mathrm{e}^{-\lambda\tau'_{\mathrm{c}m}} + \tau_m\tilde{\boldsymbol{A}}_{\mathrm{dfc}m}\mathrm{e}^{-\lambda\tau_m}\right)\boldsymbol{v}'. \tag{6.71}$$

First, $\boldsymbol{u}'$ and $\boldsymbol{v}'$ are partitioned into

$$\boldsymbol{u}' = \left[(\boldsymbol{u}'_1)^{\mathrm{T}}\ (\boldsymbol{u}'_2)^{\mathrm{T}}\right]^{\mathrm{T}},\ \boldsymbol{v}' = \left[(\boldsymbol{v}'_1)^{\mathrm{T}}\ (\boldsymbol{v}'_2)^{\mathrm{T}}\right]^{\mathrm{T}} \tag{6.72}$$

where $\boldsymbol{u}'_1, \boldsymbol{v}'_1 \in \mathbb{C}^{(n-n_{\mathrm{c}})\times 1}$, and $\boldsymbol{u}'_2, \boldsymbol{v}'_2 \in \mathbb{C}^{n_{\mathrm{c}}\times 1}$. Accordingly, characteristic equations (6.58) and (6.59) in the case of $\tau_{\mathrm{f}m}$ and $\tau_{\mathrm{c}m}$, and (6.68) and (6.69) in the case of $\tau'_{\mathrm{f}m}$ and $\tau'_{\mathrm{c}m}$, can be reformulated as

$$
\begin{cases}
\begin{bmatrix} \boldsymbol{A}_{11} & \boldsymbol{A}_{12} \\ \boldsymbol{A}_{21} & \boldsymbol{A}_{22} \end{bmatrix} \begin{bmatrix} \boldsymbol{v}_1 \\ \boldsymbol{v}_2 \end{bmatrix} = \lambda \begin{bmatrix} \boldsymbol{v}_1 \\ \boldsymbol{v}_2 \end{bmatrix} \\
\begin{bmatrix} \boldsymbol{u}_1^{\mathrm{T}} & \boldsymbol{u}_2^{\mathrm{T}} \end{bmatrix} \begin{bmatrix} \boldsymbol{A}_{11} & \boldsymbol{A}_{12} \\ \boldsymbol{A}_{21} & \boldsymbol{A}_{22} \end{bmatrix} = \lambda \begin{bmatrix} \boldsymbol{u}_1^{\mathrm{T}} & \boldsymbol{u}_2^{\mathrm{T}} \end{bmatrix}
\end{cases}
\tag{6.73}
$$

and

$$
\begin{cases}
\begin{bmatrix} \boldsymbol{A}_{11} & k\boldsymbol{A}_{12} \\ \frac{1}{k}\boldsymbol{A}_{21} & \boldsymbol{A}_{22} \end{bmatrix} \begin{bmatrix} \boldsymbol{v}_1' \\ \boldsymbol{v}_2' \end{bmatrix} = \lambda \begin{bmatrix} \boldsymbol{v}_1' \\ \boldsymbol{v}_2' \end{bmatrix} \\
\begin{bmatrix} (\boldsymbol{u}_1')^{\mathrm{T}} & (\boldsymbol{u}_2')^{\mathrm{T}} \end{bmatrix} \begin{bmatrix} \boldsymbol{A}_{11} & k\boldsymbol{A}_{12} \\ \frac{1}{k}\boldsymbol{A}_{21} & \boldsymbol{A}_{22} \end{bmatrix} = \lambda \begin{bmatrix} (\boldsymbol{u}_1')^{\mathrm{T}} & (\boldsymbol{u}_2')^{\mathrm{T}} \end{bmatrix}
\end{cases}
\tag{6.74}
$$

respectively, where

$$
\begin{cases}
\boldsymbol{A}_{11} = \left(\boldsymbol{A}_0 + \sum_{i=1}^{m-1} \boldsymbol{A}_i \mathrm{e}^{-\lambda\tau_i} \right) - \left(\boldsymbol{B}_0 + \sum_{i=1}^{m-1} \boldsymbol{B}_i \mathrm{e}^{-\lambda\tau_i} \right) \boldsymbol{D}_0^{-1} \boldsymbol{C}_0 \\
\qquad + \boldsymbol{E}_m \boldsymbol{D}_{\mathrm{c}m} \left(\boldsymbol{K}_{1m} - \boldsymbol{K}_{2m} \boldsymbol{D}_0^{-1} \boldsymbol{C}_0 \right) \mathrm{e}^{-\lambda\tau_m} \\
\boldsymbol{A}_{12} = \boldsymbol{E}_m \boldsymbol{C}_{\mathrm{c}m} \mathrm{e}^{-\lambda\tau_{\mathrm{c}m}} \\
\boldsymbol{A}_{21} = \boldsymbol{B}_{\mathrm{c}m} \left(\boldsymbol{K}_{1m} - \boldsymbol{K}_{2m} \boldsymbol{D}_0^{-1} \boldsymbol{C}_0 \right) \mathrm{e}^{-\lambda\tau_{\mathrm{f}m}} \\
\boldsymbol{A}_{22} = \boldsymbol{A}_{\mathrm{c}m} \\
k = \mathrm{e}^{-\lambda(\tau_{\mathrm{c}m}' - \tau_{\mathrm{c}m})} = \mathrm{e}^{-\lambda(\tau_{\mathrm{f}m} - \tau_{\mathrm{f}m}')}
\end{cases}
.
$$

□

Lemma 6.1 *The relationships between the sub-eigenvectors* $\boldsymbol{u}_1'$, $\boldsymbol{v}_1'$, $\boldsymbol{u}_2'$, $\boldsymbol{v}_2'$ *and* $\boldsymbol{u}_1$, $\boldsymbol{v}_1$, $\boldsymbol{u}_2$, $\boldsymbol{v}_2$ *can be expressed as*

$$
\begin{cases}
\boldsymbol{v}_1' = \dfrac{1}{C_1} \boldsymbol{v}_1 \\
\boldsymbol{v}_2' = \dfrac{1}{C_1 k} \boldsymbol{v}_2
\end{cases}
\tag{6.75}
$$

and

$$
\begin{cases}
(\boldsymbol{u}_1')^{\mathrm{T}} = \dfrac{1}{C_2 k} \boldsymbol{u}_1^{\mathrm{T}} \\
(\boldsymbol{u}_2')^{\mathrm{T}} = \dfrac{1}{C_2} \boldsymbol{u}_2^{\mathrm{T}}
\end{cases}
\tag{6.76}
$$

where C_1 *and* C_2 *are arbitrary non-zero constants that satisfy* $C_1 C_2 = \dfrac{1}{k}$.

Proof Let

$$
\boldsymbol{A} = \begin{bmatrix} \boldsymbol{A}_{11} & \boldsymbol{A}_{12} \\ \boldsymbol{A}_{21} & \boldsymbol{A}_{22} \end{bmatrix}, \quad \boldsymbol{A}' = \begin{bmatrix} \boldsymbol{A}_{11} & k\boldsymbol{A}_{12} \\ \frac{1}{k}\boldsymbol{A}_{21} & \boldsymbol{A}_{22} \end{bmatrix}.
$$

Since $\boldsymbol{A}$ and $\boldsymbol{A}'$ have identical eigenvalues, (6.74) can be reformulated as

$$\begin{cases} \begin{bmatrix} \boldsymbol{A}_{11} & \boldsymbol{A}_{12} \\ \boldsymbol{A}_{21} & \boldsymbol{A}_{22} \end{bmatrix} \begin{bmatrix} \boldsymbol{v}_1' \\ k\boldsymbol{v}_2' \end{bmatrix} = \lambda \begin{bmatrix} \boldsymbol{v}_1' \\ k\boldsymbol{v}_2' \end{bmatrix} \\ \begin{bmatrix} k(\boldsymbol{u}_1')^{\mathrm{T}} & (\boldsymbol{u}_2')^{\mathrm{T}} \end{bmatrix} \begin{bmatrix} \boldsymbol{A}_{11} & \boldsymbol{A}_{12} \\ \boldsymbol{A}_{21} & \boldsymbol{A}_{22} \end{bmatrix} = \lambda \begin{bmatrix} k(\boldsymbol{u}_1')^{\mathrm{T}} & (\boldsymbol{u}_2')^{\mathrm{T}} \end{bmatrix} \end{cases}. \tag{6.77}$$

It can be found by comparing (6.73) and (6.77) that both $\begin{bmatrix} \boldsymbol{u}_1^{\mathrm{T}} & \boldsymbol{u}_2^{\mathrm{T}} \end{bmatrix}^{\mathrm{T}}$ and $\begin{bmatrix} k(\boldsymbol{u}_1')^{\mathrm{T}} & (\boldsymbol{u}_2')^{\mathrm{T}} \end{bmatrix}^{\mathrm{T}}$ are the left eigenvectors of $\boldsymbol{A}$, and both $\begin{bmatrix} \boldsymbol{v}_1^{\mathrm{T}} & \boldsymbol{v}_2^{\mathrm{T}} \end{bmatrix}^{\mathrm{T}}$ and $\begin{bmatrix} (\boldsymbol{v}_1')^{\mathrm{T}} & k(\boldsymbol{v}_2')^{\mathrm{T}} \end{bmatrix}^{\mathrm{T}}$ are the right eigenvectors of $\boldsymbol{A}$. By equating the two left (right) eigenvectors, (6.75) and (6.76) can be easily derived.

Subsequently, $(\boldsymbol{u}')^{\mathrm{T}}\boldsymbol{v}'$ can be computed as

$$\begin{aligned} (\boldsymbol{u}')^{\mathrm{T}}\boldsymbol{v}' =& (\boldsymbol{u}_1')^{\mathrm{T}}\boldsymbol{v}_1' + (\boldsymbol{u}_2')^{\mathrm{T}}\boldsymbol{v}_2' \\ =& \frac{1}{C_1C_2k}\boldsymbol{u}_1^{\mathrm{T}}\boldsymbol{v}_1 + \frac{1}{C_1C_2k}\boldsymbol{u}_2^{\mathrm{T}}\boldsymbol{v}_2 = \frac{1}{C_1C_2k}\boldsymbol{u}^{\mathrm{T}}\boldsymbol{v}. \end{aligned}$$

If $(\boldsymbol{u}')^{\mathrm{T}}\boldsymbol{v}' = \boldsymbol{u}^{\mathrm{T}}\boldsymbol{v}$ is expected, then one has $C_1C_2 = \frac{1}{k}$. The proof of Lemma 6.1 is thus completed. □

Then, inserting (6.72) into (6.68) and (6.69) yields

$$\boldsymbol{B}_{\mathrm{c}m}\left(\boldsymbol{K}_{1m} - \boldsymbol{K}_{2m}\boldsymbol{D}_0^{-1}\boldsymbol{C}_0\right)\boldsymbol{v}_1'\mathrm{e}^{-\lambda\tau_{\mathrm{f}m}'} + \boldsymbol{A}_{\mathrm{c}m}\boldsymbol{v}_2' = \lambda\boldsymbol{v}_2' \tag{6.78}$$

and

$$(\boldsymbol{u}_1')^{\mathrm{T}}\boldsymbol{E}_m\boldsymbol{C}_{\mathrm{c}m}\mathrm{e}^{-\lambda\tau_{\mathrm{c}m}'} + (\boldsymbol{u}_2')^{\mathrm{T}}\boldsymbol{A}_{\mathrm{c}m} = \lambda(\boldsymbol{u}_2')^{\mathrm{T}}. \tag{6.79}$$

Next, by substituting (6.75) and (6.76) into (6.70) and (6.71), respectively, and by taking (6.78) and (6.79) into account, one can obtain

$$\begin{aligned} N_2' =& -\lambda(\boldsymbol{u}_2')^{\mathrm{T}}(\lambda\boldsymbol{I}_{n_{\mathrm{c}}} - \boldsymbol{A}_{\mathrm{c}m})\boldsymbol{v}_2' \\ & - \lambda(\boldsymbol{u}_1')^{\mathrm{T}}\boldsymbol{E}_m\boldsymbol{D}_{\mathrm{c}m}\left(\boldsymbol{K}_{1m} - \boldsymbol{K}_{2m}\boldsymbol{D}_0^{-1}\boldsymbol{C}_0\right)\mathrm{e}^{-\lambda\tau_m}\boldsymbol{v}_1' \\ =& \frac{1}{C_1C_2k}\left[-\lambda\boldsymbol{u}_2^{\mathrm{T}}(\lambda\boldsymbol{I}_{n_{\mathrm{c}}} - \boldsymbol{A}_{\mathrm{c}m})\boldsymbol{v}_2\right. \\ & \left. -\lambda\boldsymbol{u}_1^{\mathrm{T}}\boldsymbol{E}_m\boldsymbol{D}_{\mathrm{c}m}\left(\boldsymbol{K}_{1m} - \boldsymbol{K}_{2m}\boldsymbol{D}_0^{-1}\boldsymbol{C}_0\right)\mathrm{e}^{-\lambda\tau_m}\boldsymbol{v}_1\right] \\ =& -\lambda\boldsymbol{u}_2^{\mathrm{T}}(\lambda\boldsymbol{I}_{n_{\mathrm{c}}} - \boldsymbol{A}_{\mathrm{c}m})\boldsymbol{v}_2 - \lambda\boldsymbol{u}_1^{\mathrm{T}}\boldsymbol{E}_m\boldsymbol{D}_{\mathrm{c}m}\left(\boldsymbol{K}_{1m} - \boldsymbol{K}_{2m}\boldsymbol{D}_0^{-1}\boldsymbol{C}_0\right)\mathrm{e}^{-\lambda\tau_m}\boldsymbol{v}_1 \end{aligned}$$

and

$$\begin{aligned}
D' &= (\boldsymbol{u}_1')^{\mathrm{T}}\boldsymbol{v}_1' + (\boldsymbol{u}_2')^{\mathrm{T}}\boldsymbol{v}_2' + (\boldsymbol{u}_1')^{\mathrm{T}}\sum_{i=1}^{m-1}\left(\boldsymbol{A}_i - \boldsymbol{B}_i\boldsymbol{D}_0^{-1}\boldsymbol{C}_0\right)\tau_i \mathrm{e}^{-\lambda\tau_i}\boldsymbol{v}_1' \\
&\quad + (\boldsymbol{u}_2')^{\mathrm{T}}(\lambda\boldsymbol{I}_{n_c} - \boldsymbol{A}_{cm})\tau_m\boldsymbol{v}_2' \\
&\quad + (\boldsymbol{u}_1')^{\mathrm{T}}\boldsymbol{E}_m\boldsymbol{D}_{cm}\left(\boldsymbol{K}_{1m} - \boldsymbol{K}_{2m}\boldsymbol{D}_0^{-1}\boldsymbol{C}_0\right)\tau_m \mathrm{e}^{-\lambda\tau_m}\boldsymbol{v}_1' \\
&= \frac{1}{C_1C_2k}\left[\boldsymbol{u}_1^{\mathrm{T}}\boldsymbol{v}_1 + \boldsymbol{u}_2^{\mathrm{T}}\boldsymbol{v}_2 + \boldsymbol{u}_1^{\mathrm{T}}\sum_{i=1}^{m-1}\left(\boldsymbol{A}_i - \boldsymbol{B}_i\boldsymbol{D}_0^{-1}\boldsymbol{C}_0\right)\tau_i \mathrm{e}^{-\lambda\tau_i}\boldsymbol{v}_1\right. \\
&\quad + \boldsymbol{u}_2^{\mathrm{T}}(\lambda\boldsymbol{I}_{n_c} - \boldsymbol{A}_{cm})\tau_m\boldsymbol{v}_2 \\
&\quad \left. + \boldsymbol{u}_1^{\mathrm{T}}\boldsymbol{E}_m\boldsymbol{D}_{cm}\left(\boldsymbol{K}_{1m} - \boldsymbol{K}_{2m}\boldsymbol{D}_0^{-1}\boldsymbol{C}_0\right)\tau_m \mathrm{e}^{-\lambda\tau_m}\boldsymbol{v}_1\right] \\
&= \boldsymbol{u}_1^{\mathrm{T}}\boldsymbol{v}_1 + \boldsymbol{u}_2^{\mathrm{T}}\boldsymbol{v}_2 + \boldsymbol{u}_1^{\mathrm{T}}\sum_{i=1}^{m-1}\left(\boldsymbol{A}_i - \boldsymbol{B}_i\boldsymbol{D}_0^{-1}\boldsymbol{C}_0\right)\tau_i \mathrm{e}^{-\lambda\tau_i}\boldsymbol{v}_1 \\
&\quad + \boldsymbol{u}_2^{\mathrm{T}}(\lambda\boldsymbol{I}_{n_c} - \boldsymbol{A}_{cm})\tau_m\boldsymbol{v}_2 \\
&\quad + \boldsymbol{u}_1^{\mathrm{T}}\boldsymbol{E}_m\boldsymbol{D}_{cm}\left(\boldsymbol{K}_{1m} - \boldsymbol{K}_{2m}\boldsymbol{D}_0^{-1}\boldsymbol{C}_0\right)\tau_m \mathrm{e}^{-\lambda\tau_m}\boldsymbol{v}_1
\end{aligned}$$

leading to the conclusion that $N_2' = N$ and $D' = D$.

Therefore, the equation

$$\frac{\partial\lambda}{\partial\tau_{cm}} = \frac{\partial\lambda}{\partial\tau_{cm}'}$$

holds. In the special case of $\tau_{fm} = 0$, one has $\partial\lambda/\partial\tau_{cm}' = \partial\lambda/\partial\tau_m$. Besides, it can be known from Theorem 6.2 that $\partial\lambda/\partial\tau_{cm}' = \partial\lambda/\partial\tau_{fm}'$. Combining these equations completes the proof of Theorem 6.3. □

In this regard, let the feedback delay τ_{fm} is moved forward and merged with the control delay τ_{cm}, then (6.53) can be reduced to

$$\Delta\dot{\boldsymbol{x}}' = \tilde{\boldsymbol{A}}_0\Delta\boldsymbol{x}' + \sum_{i=1}^{m-1}\tilde{\boldsymbol{A}}_i\Delta\boldsymbol{x}_{di}' + \tilde{\boldsymbol{A}}_{dcm}\Delta\boldsymbol{x}'(t - \tau_{cm})$$

by utilizing Theorem 6.3, where

$$\begin{cases}
\boldsymbol{K}_0 = -\left(\boldsymbol{D}_0'\right)^{-1}\boldsymbol{C}_0' = \left[-\boldsymbol{D}_0^{-1}\boldsymbol{C}_0 \;\; \boldsymbol{0}\right] \\
\tilde{\boldsymbol{A}}_0 = \left(\boldsymbol{A}_0' + \boldsymbol{A}_{dfm}'\right) + \left(\boldsymbol{B}_0' + \boldsymbol{B}_{dfm}'\right)\boldsymbol{K}_0 = \begin{bmatrix} \boldsymbol{A}_0 - \boldsymbol{B}_0\boldsymbol{D}_0^{-1}\boldsymbol{C}_0 & \boldsymbol{0} \\ \boldsymbol{B}_{cm}\left(\boldsymbol{K}_{1m} - \boldsymbol{K}_{2m}\boldsymbol{D}_0^{-1}\boldsymbol{C}_0\right) & \boldsymbol{A}_{cm} \end{bmatrix} \\
\tilde{\boldsymbol{A}}_i = \boldsymbol{A}_i' + \boldsymbol{B}_i'\boldsymbol{K}_0 = \begin{bmatrix} \boldsymbol{A}_i - \boldsymbol{B}_i\boldsymbol{D}_0^{-1}\boldsymbol{C}_0 & \boldsymbol{0} \\ \boldsymbol{0} & \boldsymbol{0} \end{bmatrix}, \quad i = 1,\ 2,\ \ldots,\ m-1 \\
\tilde{\boldsymbol{A}}_{dcm} = \boldsymbol{A}_{dcm}' + \boldsymbol{B}_{dcm}'\boldsymbol{K}_0 = \begin{bmatrix} \boldsymbol{E}_m\boldsymbol{D}_{cm}(\boldsymbol{K}_{1m} - \boldsymbol{K}_{2m}\boldsymbol{D}_0^{-1}\boldsymbol{C}_0) & \boldsymbol{E}_m\boldsymbol{C}_{cm} \\ \boldsymbol{0} & \boldsymbol{0} \end{bmatrix}
\end{cases}.$$

Therefore, the corresponding characteristic polynomial can be expressed as

$$
\begin{aligned}
&|\Delta(\lambda)| \\
&= \left| \lambda \boldsymbol{I}_{n+n_c} - \tilde{\boldsymbol{A}}_0 - \sum_{i=1}^{m-1} \tilde{\boldsymbol{A}}_i \mathrm{e}^{-\lambda\tau_i} - \tilde{\boldsymbol{A}}_{\mathrm{dc}m} \mathrm{e}^{-\lambda\tau_{\mathrm{c}m}} \right| \\
&= \left[\begin{array}{c:c}
\begin{array}{c}\lambda \boldsymbol{I}_n - \left(\boldsymbol{A}_0 - \boldsymbol{B}_0 \boldsymbol{D}_0^{-1} \boldsymbol{C}_0\right) - \sum\limits_{i=1}^{m-1} \left(\boldsymbol{A}_i - \boldsymbol{B}_i \boldsymbol{D}_0^{-1} \boldsymbol{C}_0\right) \mathrm{e}^{-\lambda\tau_i} \\ -\boldsymbol{E}_m \boldsymbol{D}_{\mathrm{c}m} \left(\boldsymbol{K}_{1m} - \boldsymbol{K}_{2m} \boldsymbol{D}_0^{-1} \boldsymbol{C}_0\right) \mathrm{e}^{-\lambda\tau_{\mathrm{c}m}}\end{array} & -\boldsymbol{E}_m \boldsymbol{C}_{\mathrm{c}m} \mathrm{e}^{-\lambda\tau_{\mathrm{c}m}} \\ \hdashline
-\boldsymbol{B}_{\mathrm{c}m} \left(\boldsymbol{K}_{1m} - \boldsymbol{K}_{2m} \boldsymbol{D}_0^{-1} \boldsymbol{C}_0\right) & \lambda \boldsymbol{I}_{n_c} - \boldsymbol{A}_{\mathrm{c}m}
\end{array} \right] \\
&= \left| \lambda \boldsymbol{I}_{n_c} - \boldsymbol{A}_{\mathrm{c}m} \right| \times \\
&\left| \begin{array}{c} \lambda \boldsymbol{I}_n - \left(\boldsymbol{A}_0 - \boldsymbol{B}_0 \boldsymbol{D}_0^{-1} \boldsymbol{C}_0\right) - \sum\limits_{i=1}^{m-1} \left(\boldsymbol{A}_i - \boldsymbol{B}_i \boldsymbol{D}_0^{-1} \boldsymbol{C}_0\right) \mathrm{e}^{-\lambda\tau_i} \\
-\boldsymbol{E}_m \left(\boldsymbol{C}_{\mathrm{c}m} \left(\lambda \boldsymbol{I}_{n_c} - \boldsymbol{A}_{\mathrm{c}m}\right)^{-1} \boldsymbol{B}_{\mathrm{c}m} + \boldsymbol{D}_{\mathrm{c}m}\right) \left(\boldsymbol{K}_{1m} - \boldsymbol{K}_{2m} \boldsymbol{D}_0^{-1} \boldsymbol{C}_0\right) \mathrm{e}^{-\lambda\tau_{\mathrm{c}m}} \end{array} \right|.
\end{aligned}
\tag{6.80}
$$

It is known from (6.56) and (6.80) that the forms of the characteristic polynomial of the closed-loop power system in the three cases are the same, i.e., the existence of both $\tau_{\mathrm{f}m}$ and $\tau_{\mathrm{c}m}$ and the existence of only $\tau_{\mathrm{c}m}$, with the difference lying only in the delay constant in the last term of the second determinant.

6.3.3 An Example: Time-Delay Power Systems with the WADC Embedded

To facilitate understanding the aforementioned formulations of the index-1 Hessenberg DDAEs for time-delay power systems, two kinds of WADCs are taken here as examples to show how they are integrated into the open-loop power system.

6.3.3.1 Wide-Area PSS

To enhance the damping of an interarea oscillation mode, a wide-area PSS with the same lead-lag link as the conventional PSS is supplemented to the excitation system of the Generator i. The three typical wide-area feedback signals are: the relative power angle of Generator j with respect to Generator i, i.e., $\Delta\delta_{ji} = \Delta\delta_j - \Delta\delta_i$; the relative rotor speed of Generator j with respect to Generator i, i.e., $\Delta\omega_{ji} = \Delta\omega_j - \Delta\omega_i$; and the active power ΔP_{ij} on the tie line i-j.

Reformulating the wide-area PSS in the state-space form yields the coefficient matrices in (6.50) given by

$$\begin{cases} \boldsymbol{A}_{cm} = \begin{bmatrix} -\dfrac{1}{T_{\mathrm{w}}} & & & \\ -\dfrac{1}{T_{\mathrm{w}}} & -\dfrac{1}{T_5} & & \\ -\dfrac{\dfrac{T_1}{T_2 T_{\mathrm{w}}}}{} & -\dfrac{T_1 - T_5}{T_2 T_5} & -\dfrac{1}{T_2} & \\ -\dfrac{\dfrac{T_1 T_3}{T_2 T_4 T_{\mathrm{w}}}}{} & -\dfrac{T_3(T_1 - T_5)}{T_2 T_4 T_5} & -\dfrac{T_3 - T_2}{T_2 T_4} & -\dfrac{1}{T_4} \end{bmatrix}, \quad \boldsymbol{B}_{cm} = \begin{bmatrix} \dfrac{K_{\mathrm{S}}}{T_{\mathrm{w}}} \\ \dfrac{K_{\mathrm{S}}}{T_{\mathrm{w}}} \\ \dfrac{K_{\mathrm{S}} T_1}{T_2 T_{\mathrm{w}}} \\ \dfrac{K_{\mathrm{S}} T_1 T_3}{T_2 T_4 T_{\mathrm{w}}} \end{bmatrix}. \\ \boldsymbol{C}_{cm} = [0,\ 0,\ 0,\ 1],\ \ D_{cm} = 0 \end{cases}$$

It is assumed that the relative rotor speed $\Delta\omega_{ji}$, relative rotor speed $\Delta\delta_{ji}$ or the active power ΔP_{ij} is used as the feedback signal and its control output $U_{\mathrm{S}gi}$ (The subscript g means global signal, which is used to distinguish from the local PSS output variable $U_{\mathrm{S}i}$ of Generator i) is then added to the voltage reference V_{ref} of the exciter, as shown in Fig. 6.3. $U_{\mathrm{R}i}$ is the output variable of the integrated amplifying link of the excitation system of Generator i. U_{xi}, U_{yi}, U_{xj} and U_{yj} are the real part and imaginary part of the voltages at nodes i and j, respectively. In this case, one has

$$\boldsymbol{E}_m \boldsymbol{C}_{cm} = \begin{array}{c} U_{\mathrm{S}gi} \\ \downarrow \\ \begin{bmatrix} \vdots & \vdots & \vdots \\ \vdots & \dfrac{K_{\mathrm{A}}}{T_{\mathrm{A}}} & \vdots \\ \vdots & \vdots & \vdots \end{bmatrix} \end{array} \begin{array}{c} \\ \\ \leftarrow U_{\mathrm{R}i} \end{array}$$

$$\boldsymbol{K}_{1m} = [\ \cdots \ \underset{\delta_i}{\overset{}{-1}},\ \cdots\ \underset{\delta_j}{1},\ \cdots\ ] \text{ or } [\ \cdots\ \underset{\omega_i}{-1},\ \cdots\ \underset{\omega_j}{1},\ \cdots\ ],\ \ \boldsymbol{K}_{2m} = \boldsymbol{0}$$

(with $\delta_i \downarrow$, $\delta_j \downarrow$, $\omega_i \downarrow$, $\omega_j \downarrow$ marking the positions of the entries -1, 1, -1, 1 respectively)

or

$$\boldsymbol{K}_{1m} = \boldsymbol{0},\ \ \boldsymbol{K}_{2m} = \begin{bmatrix} \vdots \\ G_{ij}(2U_{xi} - U_{xj}) + B_{ij}U_{yj} \\ G_{ij}(2U_{yi} - U_{yj}) - B_{ij}U_{xj} \\ \vdots \\ -G_{ij}U_{xi} - B_{ij}U_{yi} \\ -G_{ij}U_{yi} + B_{ij}U_{xi} \\ \vdots \end{bmatrix}^{\mathrm{T}} \begin{array}{l} \\ \leftarrow U_{xi} \\ \leftarrow U_{yi} \\ \\ \leftarrow U_{xj} \\ \leftarrow U_{yj} \\ \\ \end{array}$$

where the arrow ($\rightarrow$) refers to the variable to which the row (column) of the matrix element corresponds, and the ellipsis ($\cdots$) means the zero element. K_A and T_A are the gain and time constant of the voltage regulator, respectively. G_{ij} and B_{ij} are the conductance and susceptance of the transmission line i-j, respectively.

6.3.3.2 Wide-Area Linear Quadratic Regulator

Suppose that a wide-area linear quadratic regulator (LQR) [27] is supplemented to the excitation system of the Generator i, whose wide-area feedback signal is $\boldsymbol{y}_{\mathrm{f}m} = [\Delta\delta_{ji},\ \Delta\omega_{ji},\ \Delta P_{ij}]^\mathrm{T}$. In this case, one has

$$
\boldsymbol{A}'_{\mathrm{dfc}m} = \boldsymbol{E}_m \boldsymbol{D}_{\mathrm{c}m} \boldsymbol{K}_{1m}
= \frac{K_\mathrm{A}}{T_\mathrm{A}}
\begin{array}{c}
\begin{array}{ccccccccc} & \delta_i & & \delta_j & & \omega_i & & \omega_j & \\ & \downarrow & & \downarrow & & \downarrow & & \downarrow & \end{array} \\
\left[\begin{array}{ccccccccc}
\cdots & \cdots & \cdots & \cdots & \cdots & \cdots & \cdots & \cdots & \cdots \\
\cdots & -k_1 & \cdots & k_1 & \cdots & -k_2 & \cdots & k_2 & \cdots \\
\cdots & \cdots & \cdots & \cdots & \cdots & \cdots & \cdots & \cdots & \cdots
\end{array}\right]
\end{array}
\begin{array}{c} \\ \\ \\ \leftarrow U_{\mathrm{R}i} \\ \\ \end{array}
$$

and

$$
\boldsymbol{B}'_{\mathrm{dfc}m} = \boldsymbol{E}_m \boldsymbol{D}_{\mathrm{c}m} \boldsymbol{K}_{2m} = \frac{k_3 K_\mathrm{A}}{T_\mathrm{A}}
\begin{array}{c}
\begin{array}{ccc} & U_{\mathrm{R}i} & \\ & \downarrow & \end{array} \\
\left[\begin{array}{ccc}
\vdots & \vdots & \vdots \\
\vdots & G_{ij}(2U_{xi} - U_{xj}) + B_{ij}U_{yj} & \vdots \\
\vdots & G_{ij}(2U_{yi} - U_{yj}) - B_{ij}U_{xj} & \vdots \\
\vdots & \vdots & \vdots \\
\vdots & -G_{ij}U_{xi} - B_{ij}U_{yi} & \vdots \\
\vdots & -G_{ij}U_{yi} + B_{ij}U_{xi} & \vdots \\
\vdots & \vdots & \vdots
\end{array}\right]^\mathrm{T}
\end{array}
\begin{array}{l} \\ \\ \\ \leftarrow U_{xi} \\ \leftarrow U_{yi} \\ \\ \leftarrow U_{xj} \\ \leftarrow U_{yj} \\ \\ \\ \end{array}
$$

where k_1, k_2 and k_3 are the optimal gains of the three wide-area signals, respectively.

6.3.4 Pseudo Delayed State Variables Introduced by Transforming DDAEs Into DDEs

This subsection focuses on pseudo delayed state variables of the time-delay power system. To be specific, when the wide-area feedback signal contains delayed algebraic variables, the transformation of DDAEs into DDEs undesirably introduces a large number of pseudo delayed state variables.

If the wide-area feedback signals contain only delayed state variables, such as the relative rotor angle or speed between two synchronous generators, then $\boldsymbol{B}'_{\mathrm{df}m}$ and $\boldsymbol{B}'_{\mathrm{dcf}m}$ are of all zeros. Thus, according to (6.54), one has $\tilde{\boldsymbol{A}}_{\mathrm{df}m} = \boldsymbol{A}'_{\mathrm{df}m}$ and $\tilde{\boldsymbol{A}}_{\mathrm{dcf}m} = \boldsymbol{A}'_{\mathrm{dcf}m}$, both of which have a small number of non-zero elements in the columns corresponding to δ_i, δ_j, ω_i and ω_j.

If the wide-area feedback signals involve delayed algebraic variables, e.g., the bus voltage and tie-line active power, then $\boldsymbol{K}_{2m}$, $\boldsymbol{B}'_{\mathrm{df}m}$ and $\boldsymbol{B}'_{\mathrm{dcf}m}$ have a small number of non-zero elements in the columns corresponding to U_{xi}, U_{yi}, U_{xj} and U_{yj}. At this point, the non-zero columns of $\tilde{\boldsymbol{A}}_{\mathrm{df}m}$ and $\tilde{\boldsymbol{A}}_{\mathrm{dcf}m}$ are determined by those of $\boldsymbol{C}_0$. According to (6.46), it further follows that the non-zero columns of $\tilde{\boldsymbol{A}}_{\mathrm{df}m}$ and $\tilde{\boldsymbol{A}}_{\mathrm{dcf}m}$ correspond to δ, $\boldsymbol{E}''_q$ (or $\boldsymbol{E}'_q$) and $\boldsymbol{E}''_d$ (or $\boldsymbol{E}'_d$) of all generators, which become pseudo delayed state variables.

6.4 Study Cases

All tests in this section are carried out using MATLAB on an Intel Core i5 4 $\times$ 3.4-GHz computer with 8GB of RAM. The tolerance for the IRA algorithm is 10^{-6} and the number of guard-vectors is $r + 3$, where r is the number of required eigenvalues. In the Newton's method, the maximum allowed number of iterations is 20 and the tolerance is specified to 10^{-6}.

6.4.1 Test Systems

The 2-area 4-machine test system (System I), 16-machine 68-bus test system (System II), Shandong provincial transmission system (System III) and the ultra-high-voltage (UHV) North China-Central China interconnected system (System IV) are employed in this section. Table 6.1 summarizes their basic data, including the wide-area feedback signals and numbers of delays, state and algebraic variables, delayed state and algebraic variables, and delayed augmented state variables, as well as the number of pseudo delayed state variables denoted as n'_2. If the wide-area feedback signal is the relative rotor speed or angle, the delayed state variables are associated with the output of the WADC and the rotor speeds or angles of the two related generators, which follows that $n_2 = 3$ and $l_2 = 0$. If the wide-area feedback signal is the active

Table 6.1 Basic data of test systems

System	Wide-area feedback signals	m	n	l	n_2	l_2	d_2	n'_2
I	Relative rotor speeds	2	56	22	3	0	3	0
II	Relative rotor speeds	2	200	484	6	0	6	0
III	Relative rotor speeds and angles	2	1128	4637	5	0	5	0
	Active power on the tie line	4	1136	4637	2	8	10	459
IV	Relative rotor speeds	4	80577	162718	5	0	5	0

power on the tie line, the delayed state variable corresponds to the output of the WADC and the delayed algebraic variables include the real and imaginary parts of voltages of the two related bus. At this point, $n_2 = 1$, $l_2 = 4$ and n'_2 is about three times the number of generators.

6.4.1.1 System I: The 2-Area 4-Machine Test System

As depicted in Fig. 6.8, the well-known 2-area 4-machine test system [24] is operating with Area 1 exporting 400 MW to Area 2. Each generator is equipped with a thyristor exciter with high transient gain and a PSS using the rotor speed as input.

To improve the damping of the inter-area oscillation mode with frequency of 0.5 Hz and damping ratio of 1.57%, a wide-area PSS is installed on Generator G_1, which uses the relative rotor speed between G_1 and G_3 as the wide-area feedback signal. The parameters of the controller include $K_S = 20$, $T_W = 5$s, $T_1 = T_3 = 0.0824$s, $T_2 = T_4 = 0.0303$s and $T_5 = 0.01$s. Time delays in feedback and output of the controller are assumed to be $\tau_f = 120$ ms and $\tau_c = 100$ ms, respectively.

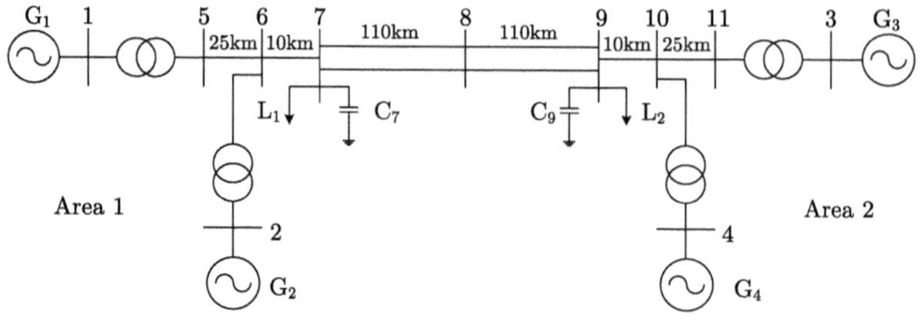

Fig. 6.8 One-line diagram of the 2-area 4-machine test system

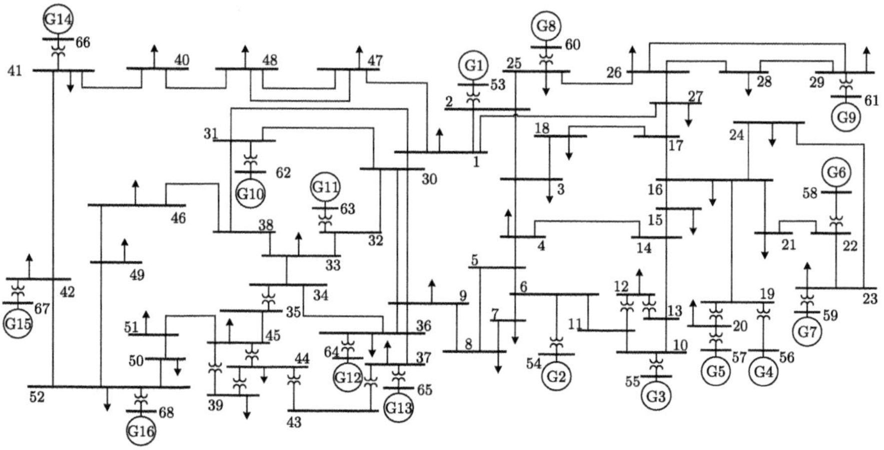

Fig. 6.9 One-line diagram of the 16-generator 68-bus test system

6.4.1.2 System II: The 16-Generator 68-Bus Test System

The data for the 16-generator 68-bus test system as shown in Fig. 6.9 can be found in [28]. Generators $G_1 \sim G_{12}$ are equipped with the IEEE DC1 exciter and PSS.

Eigen-analysis shows that the system has two inter-area oscillation modes with frequencies of 0.43 and 0.65 Hz, damping ratios of 0.08 and 1.21%, respectively. In the former mode, the dominating $G_1 \sim G_{13}$ oscillate against $G_{14} \sim G_{16}$, while the latter mode represents the oscillation between $G_1 \sim G_9$ and $G_{10} \sim G_{13}$. To enhance their damping, two wide-area PSSs are installed on G_2 and G_5, respectively, where the relative rotor speeds between G_2 and G_{15}, G_5 and G_{13} are used as wide-area feedback signals. The parameters of the two controllers are $K_{S,1} = 20$, $T_{W,1} = 10\text{s}$, $T_{1,1} = 0.411\text{s}$, $T_{2,1} = 0.479\text{s}$, $T_{3,1} = 1.0\text{s}$, $T_{4,1} = 0.155\text{s}$, $T_{5,1} = 0.01\text{s}$, $K_{S,2} = 20$, $T_{W,2} = 10\text{s}$, $T_{1,2} = 0.01\text{s}$, $T_{2,2} = 0.54\text{s}$, $T_{3,2} = 0.707\text{s}$, $T_{4,2} = 0.081\text{s}$ and $T_{5,2} = 0.01\text{s}$, respectively. The feedback and output time delays are $\tau_{f1} = 150\,\text{ms}$, $\tau_{c1} = 90\,\text{ms}$, $\tau_{f2} = 70\,\text{ms}$ and $\tau_{c2} = 40\,\text{ms}$, respectively.

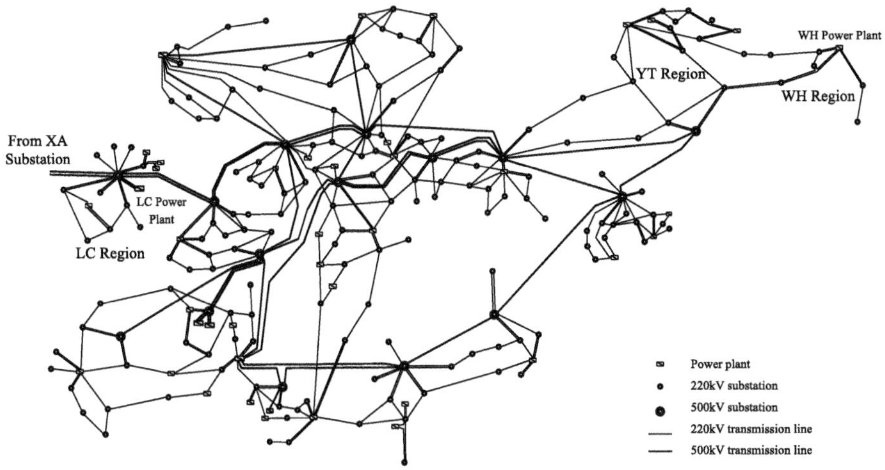

Fig. 6.10 Backbone of the Shandong provincial transmission system

6.4.1.3 System III: The Shandong Provincial Transmission System

The backbone of the Shandong provincial transmission system is shown in Fig. 6.10, which includes 516 buses, 936 transformer branches and transmission lines, 114 generators and 299 loads.

There exist two inter-area oscillation modes: in the one with frequency of 0.78 Hz and damping ratio of 6.49%, generators in the LC region oscillate against those in the eastern YT and WH regions; in the other with frequency of 0.97 Hz and damping ratio of 4.08%, generators in the LC region oscillate against those in the south-western HZ and JN regions. To improve the damping of the two modes of concern, two identical wide-area LQRs [27] are installed on units #l and #2 in LC power plant. Feedback signals are transmitted from unit #3 in WH plant to unit #1 (channel 1) and to unit #2 (channel 2) in the LC plant, consisting of the relative rotor speed and angle. The gains for the relative rotor speed and angle are 40 and 0.125, respectively. The integrated time delays of the two channels are $\tau_1 = 90$ ms and $\tau_2 = 100$ ms, respectively.

Wide-area feedback signals of the two controllers on units #l and #2 in LC power plant can also be selected as the active power on the 500 kV tie lines QZ-WF and JN-ZB, respectively. Let the wide-area feedback and control delays be $\tau_{f1} = 120$ ms, $\tau_{c1} = 100$ ms, $\tau_{f2} = 100$ ms and $\tau_{c2} = 70$ ms, respectively.

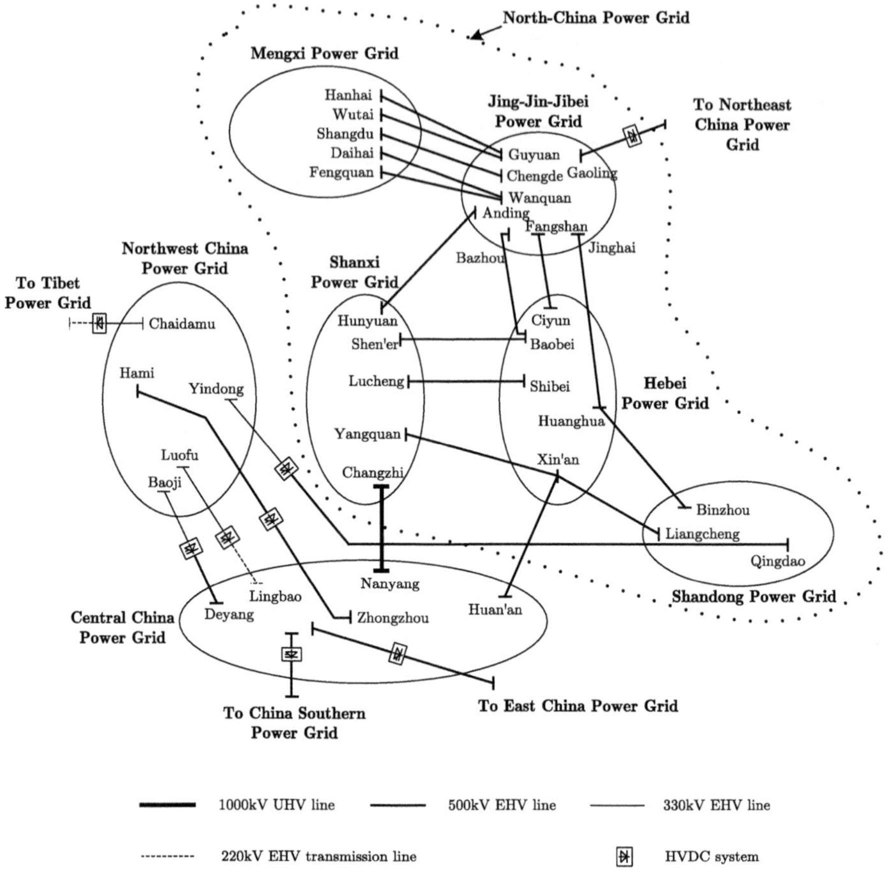

Fig. 6.11 Illustrative diagram of the UHV North China-Central China interconnected system

6.4.1.4 System IV: The UHV North China-Central China Interconnected System

The UHV North China-Central China interconnected power grid, as illustrated in Fig. 6.11, has 33028 buses, 2405 generators, 1991 induction motors and 16 HVDC transmission systems as well as 2287 exciters, 2117 governors and 1498 PSSs.

Eigen-analysis shows that the system has two inter-area oscillation modes with frequencies of 0.7727 and 0.4737 Hz, damping ratios of 2.12 and 2.99%, respectively. To enhance the damping of these modes, two wide-area PSSs are installed on unit #1 in Daihai power plant of Mengxi grid and unit #7 in Donghai power plant of Shandong grid. Feedback signals for the two controllers are the relative rotor speed between unit #1 in Daihai power plant and unit #1 in Gaoer power plant of Jing-Jin-Jibei grid and that between unit #7 in Donghai power plant and unit #2 in Hequ power plant of Shanxi grid, respectively. The parameters of the former are $K_{S,1} = 20$, $T_{W,1} = 10$s,

$T_{1,1} = 0.9979$s, $T_{2,1} = 0.5391$s, $T_{3,1} = 0.9152$s, $T_{4,1} = 0.5270$s and $T_{5,1} = 0.01$s, while those of the latter are $K_{S,2} = 19.9779$, $T_{W,2} = 10$s, $T_{1,2} = 0.5047$s, $T_{2,2} = 0.3882$s, $T_{3,2} = 0.6413$s, $T_{4,2} = 0.4366$s and $T_{5,2} = 0.01$s. The feedback and output delays of the two controllers are assumed to be $\tau_{f1} = 120$ ms, $\tau_{c1} = 100$ ms, $\tau_{f2} = 100$ ms and $\tau_{c2} = 80$ ms, respectively.

6.4.2 PIGD-PS Method

This subsection intensively validates the accuracy and efficiency of the PIGD-PS method as well as its applicability to large-scale time-delay power systems, and then investigates the impact of a large range of delays on small-signal stability of the system. In addition, the PIGD-PS method is further compared with the EIGD method presented in [29] and the DDE-based PIGD-PS method proposed in [30], to analyze the impact of the partial spectral discretization idea on its accuracy and the inefficiency of the DDE-based PIGD-PS method resulted from pseudo delayed state variables, respectively.

6.4.2.1 Accuracy Validation

The accuracy validation of the PIGD-PS method turns out to investigate the correctness of $\mathcal{A}_N$ in approximating $\mathcal{A}$, which can be achieved by comparing their eigenvalues. Specially, taking System I as an example, the dimension of $\mathcal{A}_N$ is $n + Nn_2 = 206$ for a large N of 50. Then, the QR algorithm is employed to compute all the eigenvalues of $\mathcal{A}_N$, namely $\hat{\lambda}$, which are essentially estimates to eigenvalues of $\mathcal{A}$ associated with the time-delay power system, i.e., λ. These estimates are further used as initial guesses for the Newton's method so that the exact values can be obtained, as shown in Fig. 6.12. In addition to a subset of λ, the sensitivities of these eigenvalues with respect to delays and the numbers of required Newton iterations are also given in Table 6.2.

It can be seen from Fig. 6.12 that, the eigenvalues $\hat{\lambda}$ of $\mathcal{A}_N$ distributed in the region of Re $\in [-46.82,\ 0]$ and Im $\in [0,\ 705.83]$ (i.e., the lower right of λ_{42}) and the neighborhood of $(-100,\ 0)$ coincide with the corresponding λ of $\mathcal{A}$. The accuracy of these estimates can be further quantified by the number of required Newton iterations. As can be seen in column 4 of Table 6.2, these estimates converge within 10 iterations, where the ones with higher imaginary parts and higher sensitivities need more iterations. Other $\hat{\lambda}$ that can not converge to the exact values are called spurious eigenvalues, which generally do not have any effect on the small-signal stability of the system.

The next step is to examine the relationship between the correctness of $\mathcal{A}_N$ in approximating $\mathcal{A}$ and the number of discrete points denoted as N. As depicted in Fig. 6.13, eigenvalues up to λ_{24} and λ_{36} can be accurately captured from $\mathcal{A}_N$, respec-

Table 6.2 Exact eigenvalues and their sensitivities to time delays as well as number of required newton iterations

i	λ_i	$\partial\lambda_i/\partial\tau$	Iter. No.
1	$0.08050 \pm j9.83237$	$4.2871 \mp j17.1867$	0
2	-0.05458	0	1
3	$-0.19161 \pm j0.02203$	0	2
4	$-0.19319 \pm j0.01298$	-0.0002	1
5	$-0.19342 \pm j0.02126$	$-0.0004 \mp j0.0001$	1
6	-0.20000	0	0
7	$-0.78888 \pm j4.02094$	$0.4615 \pm j1.5736$	1
8	$-1.93616 \pm j8.64020$	$0.0307 \pm j0.0526$	0
9	-2.66454	0.0008	0
10	-3.27535	-0.0407	0
11	-3.39866	-0.1866	0
12	$-5.86950 \pm j30.10290$	$-6.2871 \mp j121.8437$	0
13	$-7.58264 \pm j29.88100$	$53.2527 \pm j29.0406$	0
14	$-8.94727 \pm j1.81282$	$3.8360 \mp j35.5715$	0
15	$-12.18168 \pm j53.06187$	$100.9119 \mp j207.0969$	0
16	$-13.86612 \pm j15.52311$	$-2.5214 \mp j4.4732$	0
17	$-16.14307 \pm j3.35572$	$0.0034 \pm j0.0063$	0
18	$-17.38583 \pm j79.84739$	$132.9250 \mp j336.0173$	0
19	$-21.32808 \pm j107.59562$	$154.3991 \mp j467.5712$	0
20	$-24.43720 \pm j135.72620$	$170.2070 \mp j599.1163$	1
21	$-26.98770 \pm j164.03156$	$182.7046 \mp j730.3461$	0
22	$-29.14462 \pm j192.42718$	$193.0453 \mp j861.2860$	1
23	$-31.01116 \pm j220.87376$	$201.8691 \mp j991.9941$	1
24	$-32.65530 \pm j249.35106$	$209.5682 \mp j1122.5206$	1
25	-33.55651	0	0
26	$-34.12391 \pm j277.84778$	$216.4001 \mp j1252.9044$	2
27	$-35.45063 \pm j306.35727$	$222.5428 \mp j1383.1748$	2
28	$-36.66031 \pm j334.87538$	$228.1245 \mp j1513.3541$	2
29	$-37.77188 \pm j363.39946$	$233.2406 \mp j1643.4592$	2
30	-38.74571	-0.0003	1
31	$-38.80002 \pm j391.92775$	$237.9639 \mp j1773.5032$	2
32	$-39.75636 \pm j420.45905$	$242.3514 \mp j1903.4965$	2
33	$-40.65028 \pm j448.99251$	$246.4485 \mp j2033.4469$	2
34	$-41.48944 \pm j477.52755$	$250.2917 \mp j2163.3611$	2
35	$-42.28018 \pm j506.06374$	$253.9114 \mp j2293.2443$	2
36	$-43.02778 \pm j534.60076$	$257.3326 \mp j2423.1006$	2
37	$-43.73673 \pm j563.13840$	$260.5764 \mp j2552.9338$	2
38	$-44.41084 \pm j591.67647$	$263.6606 \mp j2682.7466$	2
39	$-45.05339 \pm j620.21486$	$266.6005 \mp j2812.5416$	3
40	$-45.66723 \pm j648.75345$	$269.4094 \mp j2942.3208$	3
41	$-46.25483 \pm j677.29218$	$272.0988 \mp j3072.0861$	3
42	$-46.81835 \pm j705.83099$	$274.6787 \mp j3201.8389$	4
43	$-47.35970 \pm j734.36985$	$277.1578 \mp j3331.5805$	4

(continued)

Table 6.2 (continued)

i	λ_i	$\partial\lambda_i/\partial\tau$	Iter. No.
44	$-47.88058 \pm j762.90870$	$279.5441 \mp j3461.3120$	5
45	$-48.38250 \pm j791.44755$	$281.8445 \mp j3591.0346$	6
46	$-48.86680 \pm j819.98635$	$284.0651 \mp j3720.7489$	5
47	-49.96897	0.0002	0
48	-50.73487	-0.0003	0
49	-51.43520	0	1
50	-100.00000	0	1
51	-112.70021	0	3

tively, when N is taken as 20 and 40. When using a very fine granularity of $N = 60$, only 10 more eigenvalues with imaginary part higher than λ_{36} can be exactly captured compare with the case $N = 40$. This is to say, $\mathcal{A}_N$ can approximate $\mathcal{A}$ with higher accuracy as N increases. In practice, a low granularity of $N = 20$ is sufficient already for small-signal stability analysis of the time-delay power system, due to that the larger N would not provide more useful information.

Last, Fig. 6.12 also gives the eigenvalues $\hat{\lambda}$ of the infinitesimal generator discretization matrix (of dimension $(N+1)n = 2856$) constructed by the EIGD method, to analyze the impact of the partial spectral discretization idea on the accuracy of the PIGD-PS method. It can be seen from the figure that the eigenvalues computed by

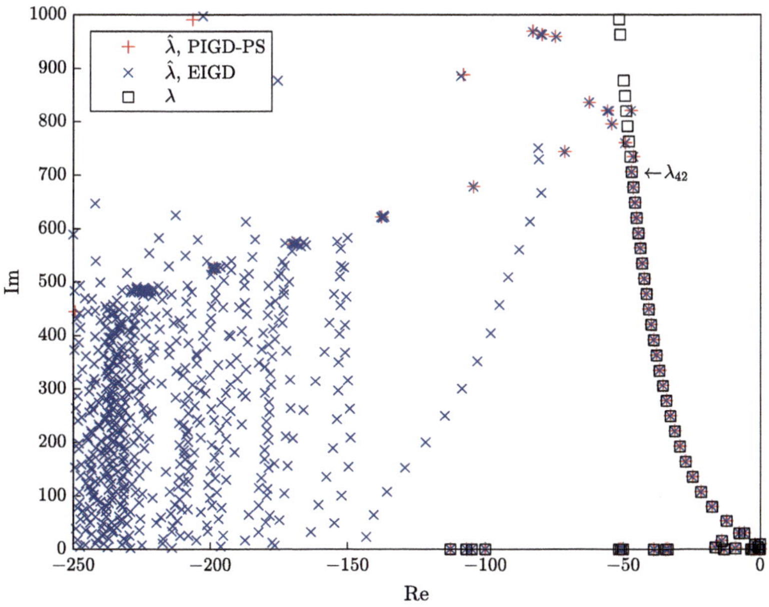

Fig. 6.12 Eigenvalues of $\mathcal{A}_N$ constructed by PIGD-PS and EIGD in case that $N = 50$ as well as the exact ones

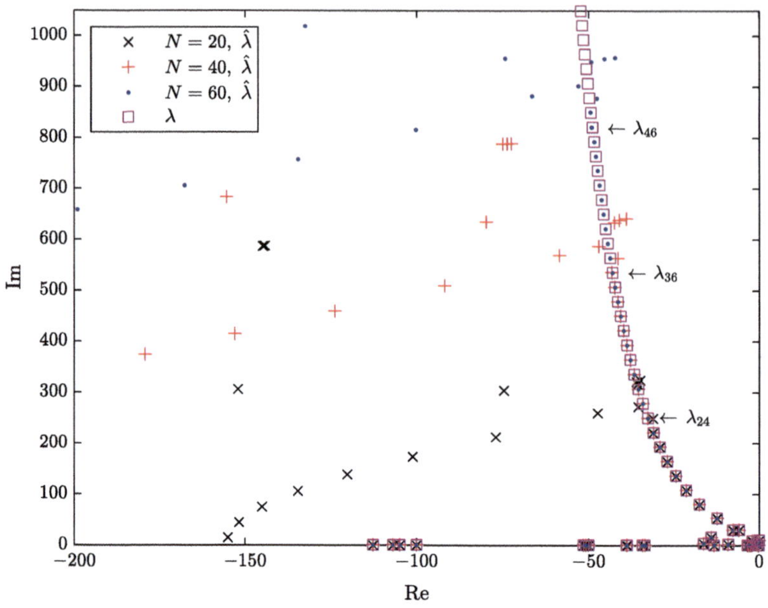

Fig. 6.13 Eigenvalues of $\mathcal{A}_N$ computed by PIGD-PS in cases $N = 20$, 40 and 60 as well as the exact ones

the PIGD-PS method are a subset of those obtained by the EIGD method. On the one hand, a portion of eigenvalues $\hat{\lambda}$ located on the right in the complex plane computed by the two methods basically overlap exactly. Taking $\lambda_{42} = -46.818 + \mathrm{j}705.831$ as an example, the moduli of the errors of the PIGD-PS and EIGD methods are 0.3965 and 0.2818, respectively. On the other hand, since the exact eigenvalues of tome-delayed systems converge to a finite chain of eigenvalues (see Sect. 2.2.2 and [31]), it can be claimed that the reduced eigenvalues of the PIGD-PS method compared with the EIGD method correspond to spurious eigenvalues located on the left of the complex plane and do not have any effect on the system stability.

6.4.2.2 Efficiency and Scalability Analysis

This section aims to analyze the ability of the PIGD-PS method to compute critical eigenvalues of large-scale time-delay power system after combining the shift-invert and Cayley transforms.

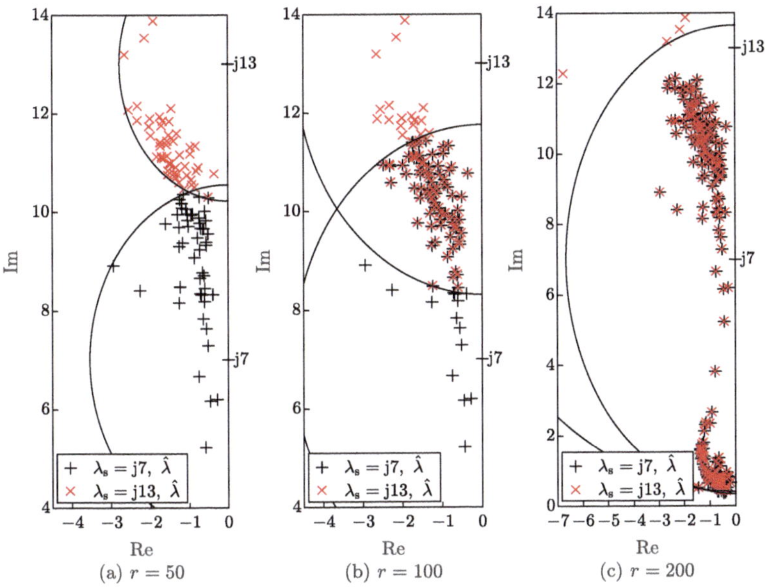

Fig. 6.14 Clusters of $r = 50, 100, 200$ eigenvalues of $\mathcal{A}_N$ around shift points $\lambda_s = \mathrm{j}7, \mathrm{j}13$ computed by PIGD-PS

The Combination with the Shift-Invert Transform

In order not to miss critical low frequency oscillation modes of System III with frequencies in the range of [0.1, 2.0] Hz (corresponding to eigenvalues with imaginary parts in the range of [0.628, 12.56] rad/s), two shift points $\lambda_s = \mathrm{j}7$ and j13 are selected. For each shift point, $r = 50, 100, 200$ eigenvalues $\hat{\lambda}$ of $\mathcal{A}_N$ constructed by the PIGD-PS method with $N = 25$ are calculated by applying the IRA algorithm, as plotted in Fig. 6.14. Table 6.3 summarizes the number of required IRA iterations and computational time.

As shown in Fig. 6.14a, the PIGD-PS method finds two clusters of eigenvalues centering about the two shift points with radii of 3.5501 and 2.7781, respectively. The two clusters are barely overlapped with only one eigenvalue in common. In Fig. 6.14b, two clusters of eigenvalues are located within circles centered at the two shifts with radii of 4.7569 and 4.6845, respectively, and overlapped with 83 eigenvalues in common. In this case, there are a total of 117 distinct eigenvalues, which means all 113 electromechanical oscillation modes of the system are captured. As shown in Fig. 6.14c, the PIGD-PS method finds two clusters of eigenvalues centering about the two shift points with radii of 6.6525 and 12.5892, respectively. There are 195 eigenvalues within the overlap of the two circles.

It can be observed from Table 6.3 that, on the one hand, the computational time of the PIGD-PS method increases with the number of required eigenvalues, since the dimension of the Krylov subspace generated by the IRA algorithm increases. On the

Table 6.3 Efficiency of PIGD-PS under different r and λ_s (N_{IRA}/(Time/s))

λ_s	r		
	50	100	200
j7	7/1.17	2/1.52	6/13.47
j13	4/0.70	2/1.28	12/24.03

Table 6.4 Efficiency comparison between PIGD-PS for System IV and traditional eigen-analysis for the delay-free system

Method	$r = 20$			$r = 50$		
	N_{IRA}	Time/s	Average Time/s	N_{IRA}	Time/s	Average Time/s
PIGD-PS	7	18.027	2.575	5	53.851	10.770
Delay-free system	6	18.549	3.092	5	52.330	10.466

other hand, there is no general pattern between the number of IRA iterations and the shift point or r. This is because the convergence rate of the IRA algorithm is related to the eigenvalue distribution, which can be expressed by the eigenvalue modulus ratio $\dfrac{|\lambda_{j+1} - \lambda_s|}{|\lambda_j - \lambda_s|}$ $(j < r)$ [32].

The efficiency and scalability of the PIGD-PS method can be further studied on System IV with 80577 state variables. Let $N = 20$, Table 6.4 summarizes the number of IRA iterations, computation time and average time for one IRA iteration required to solve for $r = 20,\ 50$ eigenvalues around shift points $\lambda_s = \mathrm{j}7$ by using the PIGD-PS method, as well as the computational efficiency in analyzing the corresponding delay-free system using the IRA algorithm.

As shown in the table, the average time for one IRA iteration of the PIGD-PS method is slightly higher than that of the traditional eigen-analysis for the delay-free system. This stems from that, the computational burden of one IRA iteration by the PIGD-PS method is determined by the MIVP operation related to Σ_N, which has the similar dimension and sparsity to the state matrix of the delay-free system. Moreover, the PIGD-PS method takes only 53.851s to compute 50 critical eigenvalues around the specified shift point of the very large-scale System IV, which is only 1.521s more than the traditional eigen-analysis for the delay-free system.

The Combination with the Cayley Transform

In order to ensure that the rightmost or least damped eigenvalues in the complex plane are preferentially obtained and not to be missed especially for large scale systems, the PIGD-PS method can also be combined with the Cayley transform and the coordinate rotation (hereinafter referred to as PIGD-PS-Cayley). Thereby, the rightmost or least-

Table 6.5 Efficiency of PIGD-PS-Cayley

Systems	Cases	r	s_1	s_2	N_{IRA}	Time/s
III	1	20	5	−5	77	13.75
	2		10	−10	63	13.60
	3		20	−20	87	18.23
	4	50	5	−5	40	18.08
	5		10	−10	33	16.27
	6		20	−20	34	18.65
	7	80	5	−5	26	26.65
	8		10	−10	21	22.89
	9		20	−20	23	25.99
	10	10	5	−5	94	433.76
	11		10	−10	161	588.66
	12		20	−20	3433	195.54
IV	13	20	5	−5	Failed	Failed
	14		10	−10	Failed	Failed
	15		20	−20	59	554.12

damped eigenvalue of $\mathcal{A}_N$ is transformed into the eigenvalues with maximum moduli of $(\mathcal{A}_N e^{-j\theta} - s_1 \boldsymbol{I})^{-1}(\mathcal{A}_N e^{-j\theta} - s_2 \boldsymbol{I})$, where s_1 and s_2 are parameters of the Cayley transform and θ is the rotation angle. Specifically, let $\theta = 8.63°$ and $s_1 = -s_2 = 5,\ 10,\ 20$, the IRA algorithm is used to compute the $r = 20,\ 50,\ 80$ eigenvalues of the transformed matrix associated with System III, followed by the Newton's method. The number of required IRA iterations and computation time for various cases are listed in Table 6.5, and Fig. 6.15 shows computed eigenvalues in case that $r = 20$ and $s_1 = -s_2 = 5$. Furthermore, in order to analyze the convergence characteristics of eigenvalues computed by the PIGD-PS-Cayley and PSOD-PS methods, the results of the latter with $Q = 5$, $M = N = 3$, $\alpha = 2$, $\theta = 8.63°$ and $r = 20,\ 50,\ 80$ are also given in Fig. 6.15 and Table 6.5.

It can be found from Fig. 6.15 that 20 eigenvalues calculated by the PIGD-PS-Cayley method with $s_1 = -s_2 = 5$ and $\theta = 8.63°$ are located on the right of the dashed curve $|\mu| = 1.0665$, where $|\mu|$ denotes the ratio of $|\lambda e^{-j\theta} - s_2|$ to $|\lambda e^{-j\theta} - s_1|$. In contrast, eigenvalues computed by the PSOD-PS method are located on the right of the line GH, which parallels to damping ratio line $\zeta = 15\%$. This phenomenon is consistent with the theoretical analysis in Sect. 3.4. As shown in Table 6.5, the PIGD-PS-Cayley method requires fewer IRA iterations when $s_1 = -s_2 = 10$.

Then, System IV is utilized to validate the computational reliability of the two methods in dealing with very large scale time-delay power systems. Here $\theta = 1.72°$ and the discretization-related parameters are unchanged. As shown in Table 6.5, when s_1 is equal to 5 and 10, the PIGD-PS-Cayley method is unable to compute $r = 20$ critical eigenvalues after 1000 IRA iterations and within 5000 s. Until a trial where s_1 is increased to 20, the method can capture 20 eigenvalues by taking 554.12 s.

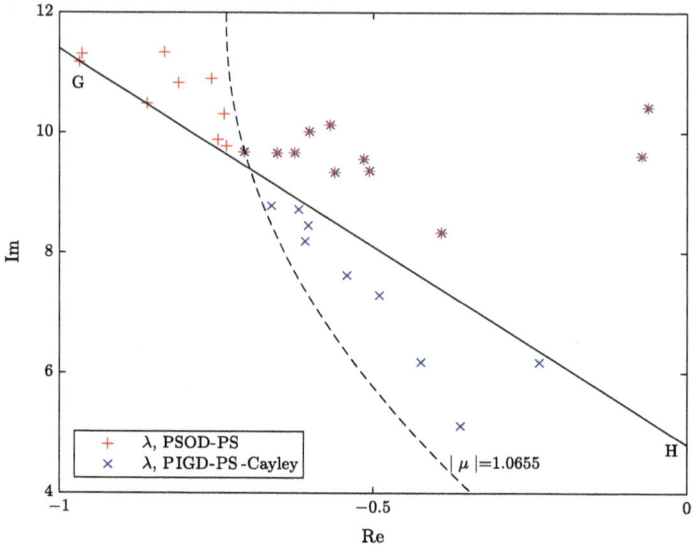

Fig. 6.15 20 eigenvalues computed by PIGD-PS-Cayley ($s_1 = -s_2 = 5$) and PSOD-PS

This reveals the weakness of Cayley transform, i.e., the convergence of the transformed eigenvalues largely depends on the shift points s_1 and s_2. Thus, one has to choose shift points by the trial-and-error method when utilizing PIGD-PS-Cayley to analyze large scale time delayed power systems. In contrast, the PSOD-PS method have the better computational reliability in computing a set of rightmost eigenvalues with damping ratios less than the specified threshold.

6.4.2.3 Inefficiency of the DDE-Based PIGD-PS Method

Under $N = 10,\ 20,\ 40$, $r = 20$ eigenvalues of System III around shift points $\lambda_s = \text{j}7,\ \text{j}13$ are required to be computed by the PIGD-PS method. Note that the wide-area feedback signals in the system here are chosen as delayed algebraic variables in order to facilitate the computational efficiency comparison with the PIGD-PS and DDE-based PIGD-PS methods. It should be also noted that when the wide-area feedback signals are delayed state variables, the two methods construct the essentially same infinitesimal generator discretization matrices and thus have the identical computationally efficient. In this regard, Table 6.6 gives the dimension of $\mathcal{A}_N$, the number of required IRA iterations and the CPU time consumption of the PIGD-PS and DDE-based PIGD-PS methods. For comparison purposes, the computational efficiency in analyzing the corresponding delay-free system using the IRA algorithm is also listed in the table. In addition, the metric "Speed Up" is defined as the speedup of the average time for one IRA iteration of the PIGD-PS method with respect to the DDE-based PIGD-PS method.

Table 6.6 Efficiency comparison between PIGD-PS and DDE-based PIGD-PS for the time-delay system and traditional eigen-analysis for the delay-free system

λ_s	N	Delay-free system			PIGD-PS				DDE-based PIGD-PS				Speed up
		Dim	N_{IRA}	Time/s	d_2	Dim	N_{IRA}	Time/s	n'_2	Dim	N_{IRA}	Time/s	
j7	10	1136	5	0.152	10	1236	5	0.179	459	5726	4	0.736	5.14
j13	10	1136	5	0.256	10	1236	5	0.259	459	5726	5	1.163	4.49
j7	20	1136	5	0.143	10	1336	5	0.186	459	10316	4	1.252	8.41
j13	20	1136	6	0.222	10	1336	6	0.266	459	10316	5	2.004	9.04

The following important conclusions can be drawn from Table 6.6.

1. The dimension of $\mathcal{A}_N$ constructed by the PIGD-PS method gets very close to that of the system state variables. When computing the same number of eigenvalues, the computational amount of the PIGD-PS method is comparable to that of the traditional eigenvalue analysis of the corresponding delay-free system.
2. The DDE-based PIGD-PS method is far less computationally efficient than the PIGD-PS method in analyzing System III where the wide-area feedback signals involve delayed algebraic variables. Specifically, the "Speed Up" in Table 6.6 can reach 9.04. This is due to the fact that the DDE-based PIGD-PS method suffers from pseudo delayed state variables, as shown in column 10 of Table 6.6. In this regard, the dimension of $\mathcal{A}_N$ constructed by the DDE-based PIGD-PS method is $n + Nn'_2$, which is much higher than that of the PIGD-PS method, i.e., $n + Nd_2$. At the same time, compared with the former, both of the discretization matrix's dimension and computational efficiency of the latter are negligibly affected by N. In other words, the PIGD-PS method is able to balance the contradiction between the accuracy and computational efficiency triggered by N.

6.4.2.4 Impact of a Large Range of Delays on System Stability

As the time delay is in fact non-deterministic [19, 20], the impact of a large range of delays on the small-signal stability of the system is investigated here by conducting 1000 Monte Carlo runs. To be specific, let τ_1 and τ_2 associated with the two wide-area LQRs of System III be modeled as the random variable with mean of 250 ms, covariance of 70 ms, and correlation coefficient of 0.9, as show in Fig. 6.16. In each Monte Carlo run, the eigenvalue with the smallest damping ratio computed by the PSOD-PS method is depicted in Fig. 6.17.

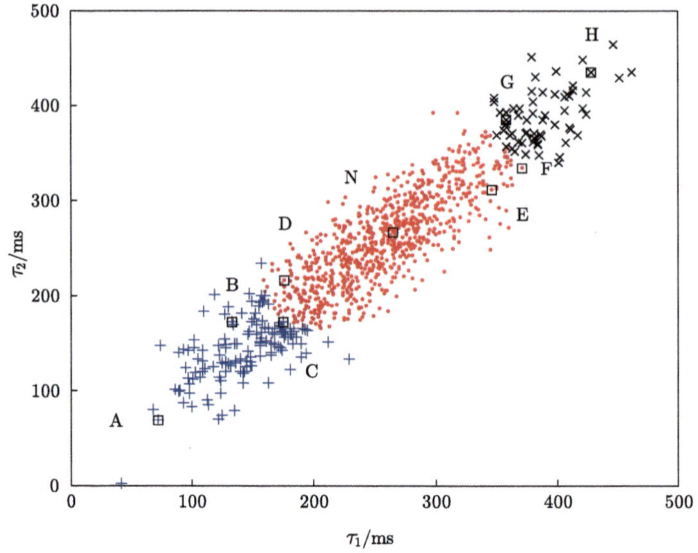

Fig. 6.16 Random time delays τ_1 and τ_2, where points A~H and N are randomly selected for illustration. In the lower-left (red) and upper-right (magenta) zones, the system is stable, while in the central zone (blue), the system is unstable

According to the trajectory of the critical eigenvalue shown in Fig. 6.17, the delay space shown in Fig. 6.16 can be divided into three zones. In the lower-left zone, shown in red, the system stability deteriorates when τ_1 and τ_2 increase. At the point $\mathrm{C} = (\tau_1 = 175$ ms, $\tau_2 = 172$ ms), the rightmost eigenvalue, i.e., $\mathrm{C}' = \mathrm{j}10.5514$, is on the imaginary axis. In the central zone in blue, the system loses its stability. However, when the two delays continue to increase and exceed the boundary of the central zone, the system regains stability. As an illustration, the rightmost eigenvalue is once again on the imaginary axis, namely $\mathrm{F}' = \mathrm{j}9.7254$, at the point $\mathrm{F} = (\tau_1 = 371$ ms, $\tau_2 = 334$ ms). The rationale behind it is that the exponential terms in the characteristic equation of the time-delay system can be represented by trigonometric functions, resulting in that the impact of delays on the small-signal stability of the system is inherently periodic.

In addition, sensitivities of eigenvalues A′~H′ to τ_1 (represented by green arrows) and τ_2 (represented by black arrows) are also shown in Fig. 6.17. For better illustration, their magnitudes are normalized by 10. It is clear that eigenvalue sensitivities coincide with the change of the eigenvalue trajectory.

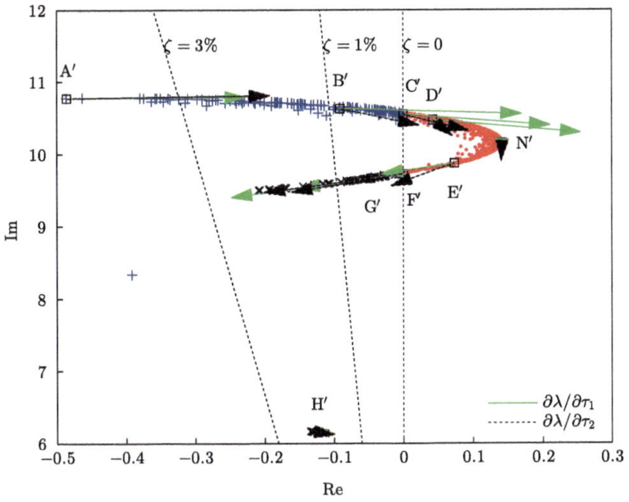

Fig. 6.17 Trajectory of the critical eigenvalue and the corresponding sensitivities under different τ_1 and τ_2. Green arrows denote eigenvalue sensitivities to τ_1, and black for τ_2. Three dotted lines are composed of points with constant damping ratios 0, 1 and 3%, respectively

6.4.3 PSOD-PS Method

This subsection aims at investigating the accuracy and efficiency of the PSOD-PS method as well as its applicability to large-scale time-delay power systems.

6.4.3.1 Accuracy Verification

The accuracy of the PSOD-PS method, interpreted as the correctness of $\boldsymbol{T}_{M,N}$ in approximating $\mathcal{T}(h)$, can be studied by comparing the their eigenvalues.

Taking System II as an example, given $M = N = 3$ and $h = 0.0153$s, $Q = \lceil \tau_{\max}/h \rceil = 10$ and the dimension of $\boldsymbol{T}_{M,N}$ is $n_1 + (QM + 1)d_2 = 380$. Here the well-known QR algorithm is adopted to compute all the eigenvalues of $\boldsymbol{T}_{M,N}$, i.e., $\hat{\mu}$. The corresponding estimates $\hat{\lambda}$ of the system in the complex plane can be obtained according to the spectral mapping relation $\hat{\lambda} = \ln \hat{\mu}/h$, a part of which can converge to the accurate ones, namely λ, by the Newton's method. Subsequently, the exact values μ of $\hat{\mu}$ is obtained according to $\mu = \mathrm{e}^{\lambda h}$. Figures 6.18 and 6.19 provides the estimates $\hat{\mu}$ and the exact values μ, and the estimates $\hat{\lambda}$ and the exact values λ, respectively.

Firstly, as shown in Fig. 6.18, the estimates $\hat{\lambda}$ on the lower right side of λ_{115} are nearly identical to the corresponding λ with the error less than the convergence accuracy of 10^{-6}. These λ correspond to μ of $\mathcal{T}(h)$ with moduli greater than $|\mu_{115}| = 0.5196$, as shown in Fig. 6.19.

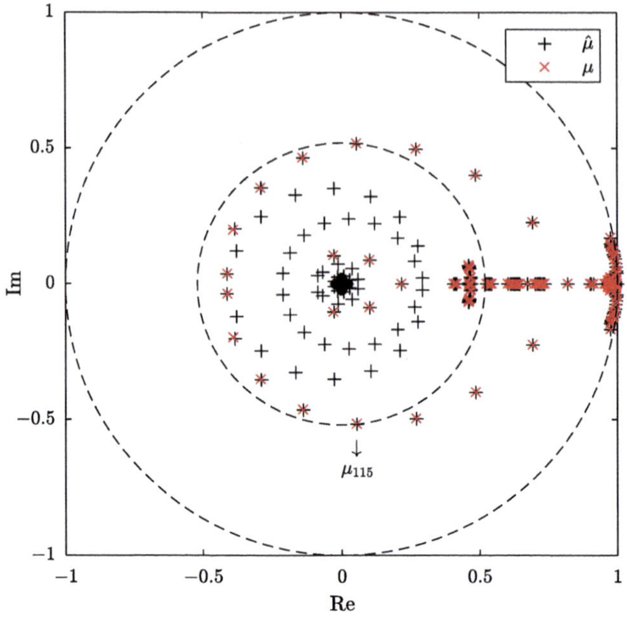

Fig. 6.18 Eigenvalues $\hat{\mu}$ of $\boldsymbol{T}_{M,N}$ constructed by PSOD-PS in case $M = N = 3$ and $h = 0.0153$s as well as the exact ones μ

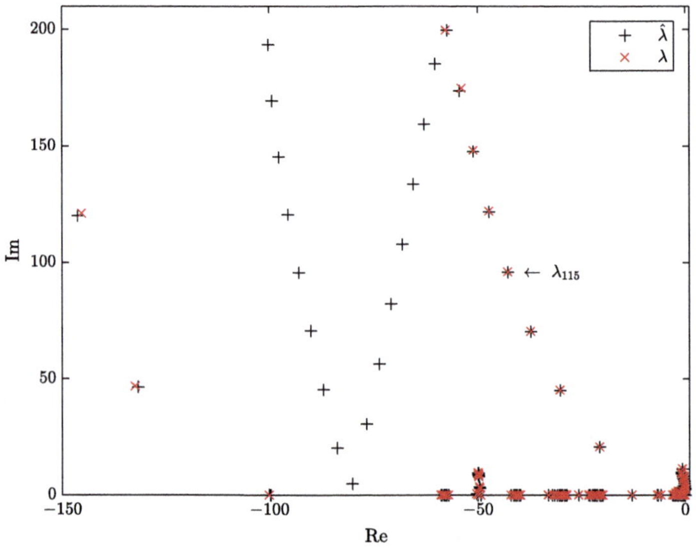

Fig. 6.19 Estimates $\hat{\lambda}$ of the system as well as the exact ones λ

Secondly, it can be found from Fig. 6.19 that, except for estimates $\hat{\lambda}$ in the region of Re $\in [-57.44,\ 0]$ and Im $\in [0,\ 199.60]$, three eigenvalue estimates $\hat{\lambda}$ with Re $\in$

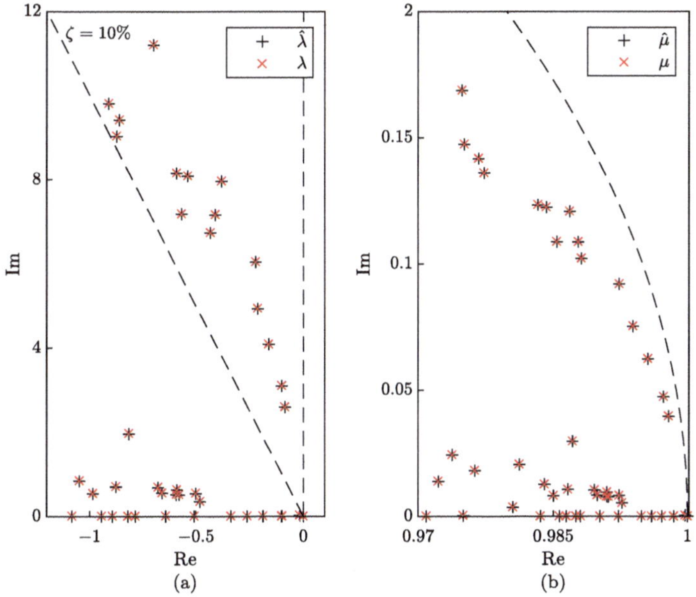

Fig. 6.20 Zoomed-in views of Figs. 6.18 and 6.19

[−150, 100] can also converge to the corresponding exact values. That is to say, the eigenvalues λ with large moduli in of System II lie asymptotically along two exponential curves, which varies from the one exponential curve of System I as shown in Figs. 6.12 and 6.13. The cause of this phenomenon has been elaborated in Sect. 2.2.2.

Next, since the most critical eigenvalue of $\mathcal{T}(h)$, i.e., $\mu = 0.9997 + \text{j}0.0003$, has a modulus less than 1, the corresponding λ is located on the left-half plane and the system is determined to be stable. This is also clearly illustrated in Fig. 6.20, i.e., the zoomed-in views of Figs. 6.19 and 6.18.

In addition, with regard to the effects of the partial spectral discretization idea on the accuracy of the PSOD-PS method, the conclusions are directly given here without the specific results. Compared with the PSOD-PS method, the SOD-PS method finds more μ near the origin of the z-plane. According to the spectral mapping relation, these μ correspond to λ of the system located on the left-half of the complex plane, which do not provide any useful information related to the small-signal stability of the system. That is, the partial spectral discretization idea does not involve any simplification and does not cause a loss of accuracy of the PSOD-PS method.

6.4.3.2 Effectiveness of Preconditioning

The effectiveness of the rotation-and-multiplication preconditioning on the PSOD-PS method is validated by comparing the number of IRA iterations and computational time to compute $r = 15$ key eigenvalues of System II, as listed in Cases 1–6 of Table 6.7.

In Case 1, no preconditioning is applied. In Cases 2–6, different preconditioning techniques, including the eigenvalue multiplication, coordinate rotation and rotation-and-multiplication, are applied, where $\alpha = 1,\ 2$ and $\theta = 5.74°,\ 8.63°$. It should be pointed out that, eigenvalues obtained in Cases 1 and 2 are the rightmost ones with the largest real parts due to the rotation angle $\theta = 0°$, while eigenvalues obtained in Cases 3–6 are critical electromechanical oscillation modes of the system with damping ratios less than 10 or 15%. In addition, for comparison purposes, M in Cases 2, 5 and 6 is double as those in Cases 1, 3 and 4, so that the dimension of the preconditioned $\boldsymbol{T}_{M,N}$ in each case is equal to 380.

By comparing and analyzing the performances of the PSOD-PS method in Cases 1–6, following conclusions can be drawn. First, the coordinate rotation preconditioning is indispensable for the PSOD-PS method. Specifically, it is difficult for IRA to terminate without this operation (see Cases 1, 3 and 4). Second, it can be found by comparing Cases 3 and 5 with 4 and 6 that N_{IRA} and computational time are largely reduced when additional eigenvalue multiplication technique is adopted. Third, Cases 5 and 6 with rotation-and-multiplication are the most efficient, followed by Cases 3 and 4 using the coordinate rotation. In summary, the rotation-and-multiplication preconditioning is preferable for efficient eigen-analysis of large-scale time-delay power systems.

6.4.3.3 Performance in Cases of Large Delays

The performance of the PSOD-PS method is further investigated in the cases of large time delays, In Cases 7–10 of Table 6.7, the dimensions of the preconditioned $\boldsymbol{T}_{M,N}$ are identical to those in Cases 1–6 by adjusting h.

In Cases 7 and 8, the feedback delays in the two wide-area channels are assumed to be $\tau_{\text{f1}} = 0.57$s and $\tau_{\text{f2}} = 0.73$s, respectively, while the control delays τ_{c1} and τ_{c2} remain. $r = 15$ estimates $\hat{\lambda}$ of System II obtained by the IRA algorithm converge to the exact values after less than 10 Newton iterations, as shown in Fig. 6.21a. The largest modulus of obtained μ is $|\mu|_{\max} = 1.0016$ and the corresponding $\lambda = 0.0199 + \text{j}4.1202$, the system is therefore unstable.

In Cases 9 and 10, extremely large time delays $\tau_{\text{f1}} = 1.14$s and $\tau_{\text{f2}} = 1.34$s are assumed. Note that this may not happen in practical applications because the delays are far beyond the confidence levels of various communication links [21]. As seen from Fig. 6.21b, 15 eigenvalue estimates $\hat{\lambda}$ obtained by the IRA algorithm converge to their exact ones λ after less than 16 Newton iterations. At this regard, the largest modulus of obtained μ equals to $|\mu|_{\max} = 0.9996$ and the system mode with the

Table 6.7 Efficiency of PSOD-PS

Cases	τ_{f1}/s	τ_{f2}/s	M	N	h/s	α	θ/°	λ_s	Dim	PSOD-PS		
										N_{IRA}	Time/s	$\Delta\lambda_{max}$
1	0.15	0.07	3	3	0.0153	1	0	–	380	1078	9.85	7.5835×10^{-6}
2	0.15	0.07	6	3	0.0153	2	0	–	380	706	6.03	−0.0015 + j0.0003
3	0.15	0.07	3	3	0.0153	1	5.74	–	380	22	0.70	−0.0078 − j0.0602
4	0.15	0.07	3	3	0.0153	1	8.63	–	380	22	0.48	−0.0140 − j0.0860
5	0.15	0.07	6	3	0.0153	2	5.74	–	380	14	0.36	−0.0078 − j0.0603
6	0.15	0.07	6	3	0.0153	2	8.63	–	380	11	0.27	−0.0140 − j0.0861
7	0.57	0.73	6	3	0.08	2	5.74	–	380	5	0.13	0.1219 + j0.0637
8	0.57	0.73	6	3	0.08	2	8.63	–	380	4	0.12	−0.1514 + j0.0399
9	1.14	1.34	6	3	0.14	2	5.74	–	380	4	0.13	−0.0585 − j0.3739
10	1.14	1.34	6	3	0.14	2	8.63	–	380	3	0.10	−0.1358 − j0.3552

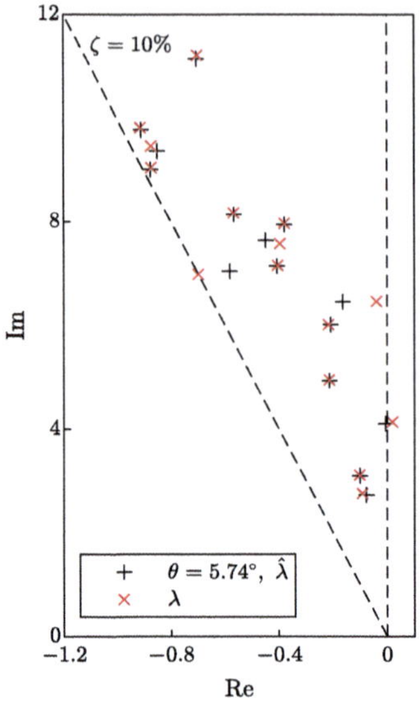

(a) $\tau_{\text{f1}} = 0.57$ s, $\tau_{\text{f2}} = 0.74$ s

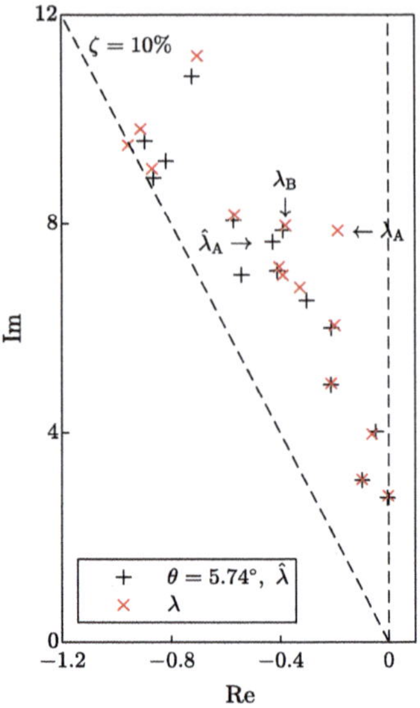

(b) $\tau_{\text{f1}} = 1.14$ s, $\tau_{\text{f2}} = 1.34$ s

Fig. 6.21 $r = 15$ eigenvalues $\hat{\lambda}$ of the system with large time delays obtained by PSOD-PS

least damping is $\lambda = -0.0005 + \text{j}2.7874$, which implies that the system regains the small-signal stability. It can also be noticed from Fig. 6.21b that, although several estimates $\hat{\lambda}$ are far from the corresponding exact values, the Newton correction still converges. This is because the convergence of Newton's method depends not only on the eigenvalue estimates but also on the eigenvector estimates. Taking $\hat{\lambda}_{\text{A}}$ in the figure as an example, although it is closer to λ_{B}, $\hat{\lambda}_{\text{A}}$ is still able to converge to the more distant λ_{A} rather than the closer λ_{B} since the error between the eigenvector estimate $\hat{\boldsymbol{v}}_{\text{A}}$ and the exact value $\boldsymbol{v}_{\text{A}}$ is very small.

6.4.3.4 Effect of Delay Approximations on Accuracy of PSOD-PS

In this subsection, the effect of delay approximations (3.54) in the coordinate rotation preconditioning on the PSOD-PS method is studied by analyzing the resulted maximum error $\Delta\lambda_{\max} = \max\limits_{i=1-15} \left\{\hat{\lambda}_i - \lambda_i\right\}$, as listed in column 13 of Table 6.7.

In Cases 1 and 2, the coordinate is not rotated and the magnitudes of $|\Delta\lambda_{\max}|$ are in the orders of 10^{-6} and 10^{-3}, respectively. In Cases 3–6, $|\Delta\lambda_{\max}|$ is almost determined by the rotation angle θ, and are dominated by their imaginary parts. Specially, when $\theta = 5.74°$ and $8.63°$, $\Delta\lambda_{\max}$ are about $-0.0078 - \text{j}0.0602$ and $-0.0140 - \text{j}0.0860$, respectively. In contrast to Cases 3–6, the major terms of $|\Delta\lambda_{\max}|$ in Cases 7–10 with large time delays become their real parts. This is because, besides the increased

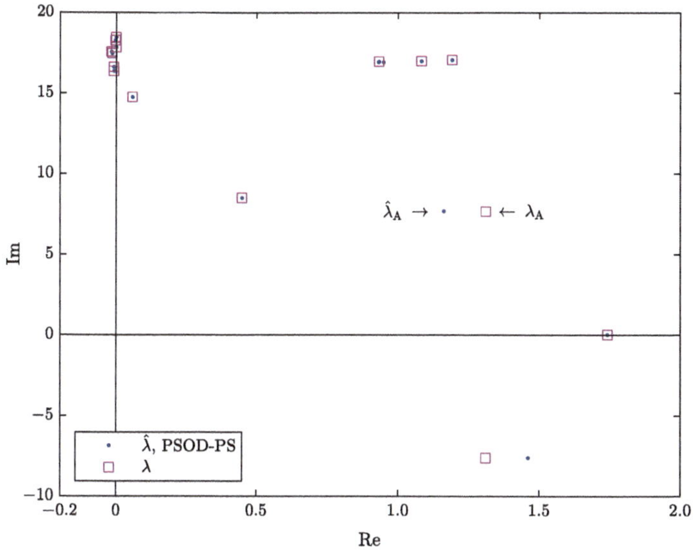

Fig. 6.22 $r = 10$ eigenvalues with the smallest damping ratios obtained by PSOD-PS

time delays, $\Delta\lambda_{max}$ are also largely affected by the eigenvalue sensitivity with respect to time delays.

In a word, the delay approximations in the coordinate rotation preconditioning do result in bias in estimating the eigenvalues of time-delay power system. The bias increases with the magnitudes of time delays and rotation angle and is largely affected by eigenvalue sensitivities with respect to time delays.

6.4.3.5 Scalability Analysis to Large-Scale Systems

This section investigates the performance of the PSOD-PS method in fast and reliably determining the stability of the large-scale System IV with 80577 state variables by computing few critical eigenvalues. Given $Q = 10$, $M = N = 3$, $h = 0.006$s, $\alpha = 2$ and $\theta = 2.87°$ (i.e., $\zeta = 5\%$), the dimension of the preconditioned $\boldsymbol{T}_{M,N}$ is 80757, and the PSOD-PS method requires 134 IRA iterations and 1168.82 s to compute $r = 10$ eigenvalues of the system with the smallest damping ratios.

As shown in Fig. 6.22, the obtained modes are dominated by hydraulic generators which are with small rated capacities in Mountain areas of Central China and connected to the main grid through long distance transmission lines. It can be seen from the figure that the estimates obtained by the PSOD-PS method coincide with the exact ones except $\hat{\lambda}_A$, whose error is due to its large sensitivity to delays.

References

1. Alur, R. (2015). *Principles of cyber-physical systems*. Boston: MIT Press.
2. Roy, S., & Das, S. K. (2020). *Principles of cyber-physical systems: An interdisciplinary approach*. Cambridge: Cambridge University Press.
3. Yu, X., & Xue, Y. (2016). Smart grids: A cyber-physical systems perspective. *Proceedings of the IEEE, 104*(5), 1058–1070.
4. Niculescu, S.-I. (2001). *Delay effects on stability: A robust control approach*. Heidelberg: Springer.
5. Gu, K., Kharitonov, V. L., & Chen, J. (2003). *Stability of time-delay systems*. Boston: Springer.
6. Phadke, A. G., & Thorp, J. S. (2008). *Synchronized phasor measurements and their applications*. New York: Springer.
7. Lu, C., Shi, B., Wu, X., et al. (2015). Advancing China's smart grid: Phasor measurement units in a wide-area management system. *IEEE Power and Energy Magazine, 13*(5), 60–71.
8. The Department of Energy's Office of Electricity. Synchrophasor technologies and their deployment in the recovery act smart grid programs. (2013).
9. Ali, I., Aftab, M. A., & Hussain, S. M. S. (2016). Performance comparison of IEC 61850-90-5 and IEEE C37.118.2 based wide area PMU communication networks. *Journal of Modern Power Systems and Clean Energy, 4*(3), 487–495.
10. Karlsson, D., Hemmingsson, M., & Lindahl, S. (2004). Wide area system monitoring and control-terminology, phenomena, and solution implementation strategies. *IEEE Power and Energy Magazine, 2*(5), 68–76.
11. Terzija, V., Valverde, G., Cai, D., et al. (2011). Wide area monitoring, protection and control of future electric power networks. *Proceedings of the IEEE, 99*(1), 80–93.
12. Pierre, B. J., Wilches-Bernal, F., Schoenwald, D. A., et al. (2019). Design of the Pacific DC intertie wide area damping controller. *IEEE Transactions on Power Systems, 34*(5), 3594–3604.
13. Aboul-Ela, M. E., Sallam, A. A., McCalley, J. D., et al. (1996). Damping controller design for power system oscillations using global signals. *IEEE Transactions on Power Systems, 11*(2), 767–773.
14. Chow, J. H., Sanchez-Gasca, J. J., Ren, H. X., et al. (2000). Power system damping controller design-using multiple input signals. *IEEE Control Systems Magazine, 20*(4), 82–90.
15. Kamwa, I., Grondin, R., & Hebert, Y. (2001). Wide-area measurement based stabilizing control of large power systems—A decentralized/hierarchical approach. *IEEE Transactions on Power Systems, 16*(1), 136–153.
16. Dotta, D., Silva, A. S. E., & Decker, I. C. (2009). Wide-area measurements-based two-level control design considering signal transmission delay. *IEEE Transactions on Power Systems, 24*(1), 208–216.
17. Wilches-Bernal, F., Pierre, B. J., Elliott, R. T., et al. (2017). Time delay definitions and characterization in the pacific DC intertie wide area damping controller. In *IEEE Power & Energy Society General Meeting* (pp. 1–5). Chicago.
18. Cheng, L., Chen, G., Gao, W., et al. (2014). Adaptive time delay compensator (ATDC) design for wide-area power system stabilizer. *IEEE Transactions on Smart Grid, 5*(6), 2957–2966.
19. Zhang, F., Sun, Y., Cheng, L., et al. (2015). Measurement and modeling of delays in wide-area closed-loop control systems. *IEEE Transactions on Power Systems, 30*(5), 2426–2433.
20. Stahlhut, J. W., Browne, T. J., Heydt, G. T., et al. (2008). Latency viewed as a stochastic process and its impact on wide area power system control signals. *IEEE Transactions on Power Systems, 23*(1), 84–91.
21. Naduvathuparambil, B., Valenti, M. C., & Feliachi, A. (2002). Communication delays in wide area measurement systems. In *Proceedings of the 34th Southeastern Symposium on System Theory* (pp. 118–122). Huntsville.
22. Myrd, P., & Sternfeld, S. (2022). Smart grid information sharing webcast: Synchrophasor communications infrastructure [EB/OL]. June 01, 2025. https://smart.grid.epri.com/doc/Synchrophasor-Communications-Infrastructure.pdf.

23. Sipahi, R., Niculescu, S.-I., Abdallah, C. T., et al. (2011). Stability and stabilization of systems with time delay. *IEEE Control Systems Magazine, 31*(1), 38–65.
24. Kundur, P. (1994). *Power system stability and control*. New York: McGraw-Hill.
25. Martins, N. (1986). Efficient eigenvalue and frequency response methods applied to power system small-signal stability studies. *IEEE Transactions on Power Systems, 1*(1), 217–224.
26. Zhang, F. (2005). *The Schur complement and its applications*. Boston, MA, USA: Springer.
27. Preece, R., Milanovic, J. V., Almutairi, A. M., et al. (2013). Damping of inter-area oscillations in mixed AC/DC networks using WAMS based supplementary controller. *IEEE Transactions on Power Systems, 28*(2), 1160–1169.
28. Rogers, G. (2000). *Power system oscillations*. New York: Kluwer Academic Publishers.
29. Ye, H., Liu, Y., & Zhang, P. (2016). Efficient eigen-analysis for large delayed cyber-physical power system using explicit infinitesimal generator discretization. *IEEE Transactions on Power Systems, 31*(3), 2361–2370.
30. Mou, Q., Ye, H., & Liu, Y. (2020). Enabling highly efficient eigen-analysis of large delayed cyber-physical power systems by partial spectral discretization. *IEEE Transactions on Power Systems, 35*(2), 1499–1508.
31. Vyhlidal, T., & Zitek, P. (2009). Mapping based algorithm for large-scale computation of quasi-polynomial zeros. *IEEE Transactions on Automatic Control, 54*(1), 171–177.
32. Stewart, G. W. (2001). A Krylov-Schur algorithm for large eigenproblems. *SIAM Journal on Matrix Analysis & Applications, 23*(3), 601–614.

GPSR Compliance

The European Union's (EU) General Product Safety Regulation (GPSR) is a set of rules that requires consumer products to be safe and our obligations to ensure this.

If you have any concerns about our products, you can contact us on ProductSafety@springernature.com

In case Publisher is established outside the EU, the EU authorized representative is:

Springer Nature Customer Service Center GmbH
Europaplatz 3
69115 Heidelberg, Germany

Batch number: 10291550

Printed by Printforce, the Netherlands